The Death of Ramón González

The Death
The Modern

by Angus Wright

Flagman setting the pilot's course in the Culiacán Valley tomato fields. The spray is a mixture of acutely toxic organophosphate insecticides and carcinogenic fungicides. The flagman is sprayed repeatedly throughout the day and wears none of the protective equipment and clothing specified by law and the manufacturer's recommendations.

# of Ramón González
Agricultural Dilemma

University of Texas Press, Austin

First paperback printing, 1992

Requests for permission to reproduce material from this work
should be sent to Permissions, University of Texas Press, Box
7819, Austin, Texas 78713-7819.

∞ The paper used in this publication meets the minimum
requirements of American National Standard for Information
Sciences—Permanence of Paper for Printed Library Materials,
ANSI Z39.48-1984.

Library of Congress Cataloging-in-Publication Data

Wright, Angus Lindsay.
    The death of Ramón González : the modern agricultural
dilemma / by Angus Wright. — 1st ed.
        p.   cm.
    Includes bibliographical references.
    ISBN 0-292-71566-8 pbk.
    1. Agricultural laborers—Health and hygiene—Mexico.
2. Pesticides—Health aspects.   3. Pesticides—Environmental
aspects.   I. Title.
RC965.A5W75   1990
363.17'92'0972—dc20   90-12039
                                                              CIP

To Howard and Thelma,
to Joseph and Jessica,
and to the Mixtec people

Mixtec man mixes acutely toxic insecticides and carcinogenic fungicides from concentrates. The canal serves as drinking water supply for farm worker camps.

Young men from Durango work in paraquat spraying crew in the Culiacán Valley with none of the mandated protections.

Contents

Acknowledgments   ix

Introduction: Mexico and the Pesticide Crusade   xi

1. The Death of Ramón González   1

2. The Road to Culiacán   10

3. Doctors and Bureaucrats   51

4. Going Home   87

5. King Eight Deer, the Plagues, and the Devastation
   of the Land   122

6. Technology and Conflict   140

7. Consumers, Workers, Growers, and Experts   188

8. Theory and Consequences   222

9. The Modern Agricultural Dilemma   244

10. Points North   286

References   303

Index   325

Young girl working in San Quintín tomato fields as they are being sprayed with parathion and fungicides.

Farm worker camp in San Quintín, Baja California. Photo by Michael Kearney.

Acknowledgments

Research and writing of this book extended over a ten-year period and involved work in various regions of Mexico and the United States. I can only hope to acknowledge here those who made the most important contributions to the work, necessarily leaving out dozens of people who facilitated the study in various ways. Thanks are due all of those named in the text who consented to interviews. Some people could cooperate with the work only on terms of anonymity—a special thanks here to all those people, anonymous and otherwise, who took risks to assist me. I also owe a very special debt of gratitude to all my friends and family who, in every instance, gave their support and encouragement even when it meant sacrifice or inconvenience to them.

The Council for the International Exchange of Scholars provided a Fulbright Senior Research Award for 1983–84, essential to the study. The staff of the Fulbright office in Mexico City went beyond the call of duty to be helpful on more than one occasion. California State University, Sacramento, granted a difference-in-pay leave for the same period, in addition to partial release-time from teaching in 1987–88. Beginning with a generous invitation to a conference in 1980 in Culiacán, and extending through 1985, the Universidad Autónoma de Sinaloa and members of its staff extended many forms of hospitality and guidance in assisting the study. Dr. Lilia Albert and other staff members at the former Instituto Nacional de Investigaciones sobre Recursos Bióticos in Xalapa, Veracruz, lent time, energy, and insights toward the research. Victor Clark Alfaro contributed generously of his time and knowledge in Tijuana. Dr. Javier Trujillo Arriaga was a young government agronomist when I first met him and interviewed him for this study—he is now a professor at the Graduate School of the Mexican Agricultural University in Montecillo. He helped with the study in innumerable ways, from 1983 until his critical comments on the final text. Monica Moore, executive director of the North American Office of the Pesticide Action Network, her staff, and board members have also been especially helpful.

In Culiacán, João Carniero e Lima and family, Ricardo Mimiaga Padilla, Florencio Posadas, and the staff of energetic agronomists associated with the state university's Servicios Sociales provided important guidance and assistance, as did Benito García of the independent farm worker's union and Héctor Miguel González of the national vegetable growers' association. Salvador Calderón and his staff at SEDUE made a place for me to work in their crowded offices as well as helping in many other ways. Elsewhere in Mexico, Arturo and Lili Lomelí, David Barkin, Iván Restrepo, Monique Mitastein, and Enrique Astorga Lira were particularly helpful.

This work is a synthesis drawn from many disciplines and would not have been possible without assistance from many teachers and friends who helped me shape my ideas. Some of those who have made the strongest contributions are Charles Stansifer, Richard Graham, the late Charles Gibson, Richard Sheridan, Jercy Romanowski, Miguel Altieri, Rayna Rapp, Wes Jackson, Charles Washburn, Valerie Anderson, Harold Kerster, Susan McGowan, Ephraim Kahn, Bruce Jennings, Ralph Lightstone, William Kistner, Tomás Martínez Saldaña, Carol Nagengast, Sean Swezey, Doug Murray, Valerie Wheeler, Richard Wiles, Howard Wright, Bruce Wright, and Mary Mackey. Professors Steven Mumme and Billie Dewaldt served as reviewers for the University of Texas Press and offered many helpful suggestions for the final text. Teresa May, executive editor at the UT Press, was a model of critical insight and enthusiastic support in the last years of the project—I owe her a special thanks.

Dr. Michael Kearney of the Department of Anthropology of the University of California, Riverside, assisted me with amazing generosity and relentless energy. In many ways, this book is his, considering his multiple contributions to it, including his critical comments. His dedication to his profession and to the Mixtec community also served as a constant inspiration.

The Mixtec people, from San Jerónimo Progreso in Oaxaca, to Culiacán, Tijuana, Riverside, and Fresno, made this book possible by sharing their experiences and something of their lives with me. The family of the man in this book known as Ramón González was open and helpful, as was Quirino Torres and the entire town council of San Jerónimo Progreso. I can only hope that the book itself will serve as a fitting tribute to the dignity and courage of the Mixtec people—dignity and courage that survives against the terrible challenges presented to the Mixtecs in both Mexico and the United States.

Introduction: Mexico and the Pesticide Crusade

*I*n 1962, Rachel Carson's *Silent Spring* galvanized the American public into an awareness of the dangers of synthetic chemical pesticides. Her book was so sweeping in its indictment of the way society was undermining the health of nature that she is often given credit for starting the modern wave of environmental awareness that began in the 1960s. More specifically, *Silent Spring* led to a decade of intense political infighting and scientific research resulting in the virtual banning in most industrialized nations of many persistent pesticides such as DDT. As a result of the limitations on the use of persistent pesticides, the peregrine falcon, the California brown pelican, and other species began to make an encouraging comeback. The revival of many of the declining bird populations constituted a heartening sign that humankind could recognize its errors and make amends to nature.

However, the growth of the pesticide industry continued, doubling and redoubling output since *Silent Spring*. Seeing their markets for the persistent pesticides drying up in the more prosperous nations, chemical companies shifted their sales to the Third World, selling there what they could no longer sell in the United States and Europe. There was some justification for this, particularly in the dependence of tropical and subtropical nations on cheap pesticides for the control of malaria and other insect-borne diseases. But the social and ecological conditions of the countries of the South tended to magnify rapidly the ill effects of dangerous pesticides. Since the most lucrative market for pesticides in the Third World was for pesticides to be applied to crops exported to richer nations, consumers in the United States and Europe began to fear that what they had prohibited at home was returning to them on the fruits and vegetables they imported from abroad. David Weir and Mark Shapiro's *Circle of Poison* (1981) documented the basis for such fears and aroused the public to a new level of awareness.

More important, the banning of some pesticides did not mean that those still allowed for use were safe or benign. Unfortunately, in many instances, pesticides that presented serious long-term environmental dangers and the possibility of cancer and other long-latency diseases were traded in for pesticides that broke down fairly rapidly in the environment under most conditions but that were far more dangerous to people, plants, and wildlife that came into immediate contact with them. This was especially tragic in the tropics, where pest problems can be particularly threatening, and where exceptionally large numbers of rural residents and farm workers came into direct contact with the chemicals due to the almost total lack of safety precautions or effective regulation. David Bull, in his little-known but excellent book *A Growing Problem: Pesticides and the Third World Poor* (1982), documented the tragedy, including the irresponsible behavior of chemical companies and governments. More specialized literature has subsequently pointed out the virtual total lack of meaningful regulation of the international pesticide trade and the dire consequences for people and environment of nations with little effective regulatory capacity of their own (Bull 1982; Boardman 1986; Wiles 1983).

Recently, dramatic events have forced an even more intense awareness of the dangers of pesticides. On December 3, 1984, a poisonous cloud of gas began to penetrate the houses of tens of thousands of mostly very poor people in the outskirts of Bhopal, India. Isocyanate gas used by Union Carbide Corporation in its local plant to manufacture the widely used pesticide carbaryl, or Sevin, left thousands dead (according to the Indian government, 2,500 dead; according to more carefully done surveys, about 8,000) and 300,000 sick and injured. Meanwhile, Robert Metcalf, the inventor of carbaryl, has been among the many pest management specialists to call for a radical turn away from reliance on chemicals to protect our crops (Kurzman 1987; Ottawa Declaration 1986).

In California, a new state law that requires monitoring of water wells to detect pollutants led to the revelation that about a fifth of the wells in the state's Central Valley were contaminated with DBCP. DBCP had been used heavily throughout the world to kill nematodes, a kind of worm in the soil. In 1977, men working in the Occidental chemical plant that manufactured the chemical found that they were sterile, and further investigation showed that DBCP was the cause. DBCP has been shown to be carcinogenic at very low levels of exposure as well as mutagenic. The state of California, and after a two-year delay the U.S. Environmental Protection Agency, banned the use of the chemical, but it is very stable in soils and will continue to travel through the ground to contaminate water sources for years, if not decades. Researchers also revealed that Dow and Shell had known of the capacity of the chemical to cause reproductive problems for nineteen years before the workers found they were sterile. The scientist who had conducted the original tests was careful to make the fine distinction that his research did not show that DBCP caused sterility, merely that it caused the testicles of mice fed the substance to atrophy. The use in the late 1970s and early 1980s of DBCP in banana plantations in Costa Rica apparently caused sterilization in banana workers (Gips 1987: 147–156; Thrupp 1988; Weir 1989).

In October 1986, the Sandoz Company warehouse in Basel burned, sending 66,000 pounds of pesticides into the Rhine River while bringing attention to the fact that the nearby Ciba-Geigy plant, for reasons that were unclear, was releasing large quantities of highly toxic substances into the Rhine at the same time. At least half a million fish died and towns were forced to bring water in by rail or truck until the toxic chemicals partially cleared from the river. One important consequence of the incidents in Switzerland was that the growing myth that pesticides were only a problem when handled by ill-trained people in the Third World, an explanation frequently used to explain Bhopal, was definitively put to rest (Weir 1987; Postel, in Brown 1988).

We now have a very impressive literature and body of scientific data that call into serious question our global dependence on pesticides. Every year chemical companies manufacture about five billion pounds of substances chosen for their ability to kill some form of life and intended to be cast rather freely into the air, soil, and water of our planet. For all our concern with toxic waste dumps and irresponsible disposal of hazardous materials, corporations, governments, and international agencies encourage the deliberate dispersal of this fantastic quantity of biocides into the environment. Since *Silent Spring*, dozens of popular and technical books and thousands of articles have detailed convincing ways to grow crops and control disease with sharply reduced use of pesticides. Rachel Carson herself concluded from her survey of nonchemical means of pest control that while pesticide use was sometimes justified, "We allow the chemical death rain to fall as though there were no alternative, whereas in fact there are many, and our ingenuity could soon discover many more if given opportunity" (1962: 12). The range of nonchemical alternatives available, in spite of the near monopoly on pest control research exercised by the chemical industry, is far greater than it was a quarter of a century ago when Carson came to that conclusion. (Recent summaries are National Academy of Sciences 1989; Dover 1985; Gips 1987; Hansen 1986; classic works include Flint and van den Bosch 1981.) But the pesticide industry continues to grow more rapidly than the rest of the global economy, and international agencies operate on the assumption that it will continue to grow (Boardman 1986; UNFAO 1985).

When the growth of a technology seems immune to the consequences of disastrous experience and the invention of inexpensive and effective alternatives, we have to begin to ask why. The eminent pest management specialist, the late Robert van den Bosch, concluded that the explanation was a cynical conspiracy on the part of chemical companies, able to corrupt or intimidate government officials and scientists. Others have elaborated at length on modified versions of this theory (van den Bosch 1978). Economists have shown that much of the growth of the industry, especially in the Third World, is due to large government subsidies of pesticide use, in nine Third World nations surveyed amounting to an average of 44 percent of the retail price (Repetto 1985). Clearly, there is an enormous institutional commitment to pesticide use that has a momentum of its own.

I do not reject van den Bosch's conspiracy theory—in its broad outlines I think it can be taken for granted. Any industry with a profitable product to sell will do all it can to protect and expand the market and will not hesitate to attempt to dominate scientific research and governmental actions related to the product. But to dwell on this alone is to leave very important questions unanswered. There are many ways to skin a cat, and many potential technologies that can earn a profit. Many of the alternatives to pesticides create opportunities for making profits. Why do particular technologies seem to prevail over others? I believe that for very large conspiracies to work over large geographical areas and for decades at a time, the conspiracy must be transformed into something else—a belief system, an ideology, a world view, a concept of proper professional behavior, even a crusade. Too many people are uncomfortable with playing the role of conspirator. There must be a way for Lady Macbeth to wipe out or obscure her damned spot. There must be something positive and uplifting to believe in. Such is the case with the promotion of pesticides.

A group of critics of the pesticide industry went to visit the Ciba-Geigy plant near Basel in 1985. Ciba-Geigy executives tried to put the minds of the critics to rest, and one way they did so was to show the visitors the organic gardens the executives themselves tended in their yards and to show that the executives' children attended the strongly environmentally minded Waldorf schools founded by Rudolf Steiner. But, one had to understand, agriculture was not yet ready to make the transition, especially in the Third World, where the population explosion and dire poverty created an unusually strong need for the production miracles that only pesticide use could bring.

The roots for such beliefs lie in the Mexican soil. There, largely North American scientists pioneered a program of agricultural modernization that was to be called the Green Revolution. But more than new seed varieties and chemical-dependent technologies came out of this self-styled missionary effort. Far more important was the refinement in Mexico of a political strategy bolstered by an ideology of technological development. The ideology had been taking form among agricultural researchers in the United States and then was adapted for use in Mexico and other poor countries. Much of the latter half of this book is concerned with an understanding of how and why this came to pass and, most of all, with its results. By understanding what has happened in Mexico, it is easier to identify the obstacles to the transition from pesticide dependence to a healthier, more diverse agriculture.

Persistence of dangerous technologies may also be based on severe social imbalances that allow hazardous technologies to prevail because the pattern of rewards and penalties associated with the technology is rooted in the distribution of power in the society. The Mexican case is an excellent illustration, and most of the first half of the book is dedicated to showing who suffers and who gains from things as they are. Because radical imbalances of power are partly maintained, and sometimes eroded, by culture and cultural differences among people in a society, I have tried to give my reader a feeling for the cultural context of pesticide use in Mexico. The farm workers who are poisoned have very different belief systems from the ones held by the growers, chemical salesmen, investors, government officials, and researchers who perpetuate the abusive use of pesticides. These differences are at times obstacles to positive change, and they may also serve as powerful tools of change.

In particular, I came to believe that the amazing capacity of many of Mexico's poorest people for sustained effort in a single direction and for solidarity among themselves under trying circumstances could together be a significant force toward a healthier society and a more sustainable agriculture. But those people will need allies. I was also impressed that among Mexico's professional classes there is a substantial number of insightful and dedicated people who want to see change and who are prepared to seek out alliances to help create it.

A more important point I want to make to readers of this book in the United States is that a realistic picture of people and their dilemmas in countries like Mexico is essential to the project of humanizing the policies of our own institutions as they affect other countries. For that reason, too, I have put into my narrative a good deal of material that is about the character of the people themselves, diverse as the people of Mexico are. It would be difficult and perhaps dangerous to think about how they might resolve the political questions and technological choices before them without having some feel for what the people themselves are like.

A great deal of the pioneering work on seed varieties, chemical use, and machine use that occurred in Mexico constitutes a kind of model that has been used in dozens of other nations. The model partly came from the United States, but as we shall see, it was reshaped in Mexico and returned to the United States as one of the key influences on the modernization of agricultural technology. For economy of time and space, I have not drawn out all the implications of this story for the United States. But there are many excellent studies of the crisis of American agriculture that highlight such crucial problems as high capital costs; environmental deterioration of farmland through erosion, salinization, compaction, and chemical overload; pesticide and chemical fertilizer pollution of lakes, streams, and groundwater; unhealthy working conditions for farm workers, farmers, and farm families; dependence on a severely narrowed and destabilizing genetic base in major crops; dependence on nonrenewable mineral and energy resources; the destruction of rural communities; and the increasingly concentrated control of the nation's food supply (see, for example, Strange 1988; Berry 1978; Barney 1980; Jackson et al. 1984; Center for Rural Affairs 1984; Moses 1988; Wasserstrom and Wiles 1985; USDA 1980; Sheridan 1981).

In many ways, the story of Mexican agricultural development reveals the roots of problems in the United States. In addition, by seeing the way institutions and groups have worked in Mexico, we can get a fresh perspective on the politics of our own farm economy and rural communities.

The cultures of Mexico are ancient, and so is agriculture. Mexico is one of the great world centers of the domestication of human food crops, most notably maize, or corn. For ten thousand years, apparently, Mexicans have been working on the design of agricultural technologies. In the rush to modernize Mexico, much was lost or ignored about what Mexicans had to teach us about agriculture. Much of what we know as modern agriculture developed out of Mexican crops and techniques, either through forming deliberate strategies to eliminate and replace Mexican ways considered backward or through using Mexican seed varieties or cultivation techniques as the resource for the design of modern methods. By digging briefly into the Mexican past, we understand more clearly what has been done in global agriculture and what could be done for the better.

The changes in Mexico associated with rising chemical dependence in agriculture are changes that must be understood by those north of the border. The exodus from the Mexican countryside and the explosive growth of the entire population, combined with a formerly dynamic but currently stagnating economy, are all intimately connected with the project of agricultural modernization. These things also mean two vital things to those who live in the United States. There will be a rapidly growing flow of Mexicans into the United States, and there will be insistent demands for radical political changes in Mexico of a kind that the government of the United States is likely to see as extremely threatening to the security of the United States. Based on past performance, the U.S. government will very possibly act precipitously and unwisely in the face of such change. In any case, the combination of political change in Mexico and a migration of more Mexicans into the United States will present transcendently important and complicated challenges to the government and people of the United States. Surprisingly enough, the story of how pesticides came to be so badly abused in Mexico is a good place to begin in coming to an understanding of the nature of these challenges.

Because it is diverse, because it is ancient, because it is filled with conflict, and because for all its problems it yet flies on the amazingly strong and durable wings of constant hope and good will so characteristic of many Mexicans, Mexico is simply and eternally fascinating. I hope that beyond all the sometimes too-earnest purpose behind the writing of this work, the reader will share with me the excitement of coming to know the great beauty of the Mexican land and people.

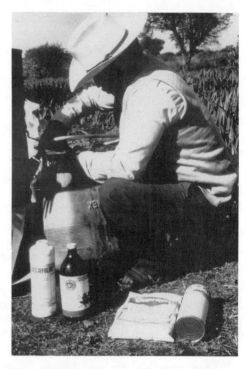

A newborn Mixtec baby in the Colonia Obrera Mixtec settlement. Photo by Michael Kearney.

Man attempts to repair leaking spray tank that has been soaking his back with parathion, acutely toxic as a skin-contact poison. His crew is spraying commercial flower fields in the Puebla Valley.

The church in San Jerónimo. The clock, newly installed, was a gift to the village from a group of San Jerónimo men who wanted to commemorate a successful year as migrant workers.

Like the eroded fields of the Mixteca, the devastating deforestation of upland Chiapas causes economic losses that lead to new emigration out of the region.

A Zapotec man who has planted his field using the traditional mix of corn, beans, and squash. He reports few pest problems.

Young Mixtec man in crew spraying parathion and fungicides in San Quintín tomato fields.

The Death of Ramón González

Tijuana

San Quintín

BAJA CALIFORNIA

Sea of Cortés

SONORA

SINALOA

Culiacán Valley • Culiacán

LA LAGUNA

MEXICO

EL BAJÍO

Tula

Chapingo

Mexico City • Cholula

Puebla

Tehuacán

Atencingo

Huajuapan de León

Oaxaca

MICHOACÁN

GUERRERO

MORELOS

LA MIXTECA

Isthmus of Tehuantepec

Isthmus of Tehuantepec

San Cristóbal de las Casas

CHIAPAS

SOCONUSCO

Huajuapan de León

Tonalá

Silacayoápan

San Jerónimo Progreso

Yanhuitlán

Nochixtlán

LA MIXTECA

Oaxaca

STATE OF OAXACA

1
The Death of Ramón González

I first heard of the death of Ramón González from Michael
Kearney, an anthropologist friend who had spent several years studying the Mixtec people of southern Mexico. My own interest in the
Mixtecs had developed out of my research on pesticide use in Mexico. I was doing most of the fieldwork portion of my study in the
Culiacán Valley, located six hundred miles nearly straight south of
Nogales, Arizona. Although Culiacán is about thirteen hundred
miles north of the Mixtec heartland in the southern state of Oaxaca,
the Mixtecs and the closely related Zapotecs made up the majority
of migrant farm workers in the Culiacán Valley. Doctors in Culiacán
had told me that most pesticide poisoning victims were Mixtecs or
Zapotecs (Pérez 1984; Velázquez 1984).

I explained to Kearney that I had found out a great deal about how
pesticides were used and abused in Mexico and in Culiacán in particular. I had interviewed Mixtecs and other farm workers who had
suffered various degrees of poisoning and had put this material together with information from doctors, nurses, social workers, farm
labor organizers, foremen, and agronomists. This allowed me to con-
struct a picture of how poisoning incidents occurred and how people
reacted to them. I wanted to know more, however, about the circumstances surrounding fatal cases. I had obtained anonymous de-
scriptions of fatal cases from doctors, but since they could not give
out names I could not interview others who might have known
about the cases or been affected by them. I asked Michael if he knew
of any fatal pesticide poisonings among the people he worked with.

"Yes," he said, "I think so, but it's not certain that the death was
caused by pesticides. I knew a young man named Ramón González
who died in Culiacán in early 1981, and his father and some of his
friends believe that he died of pesticide exposure. A doctor who
worked on Ramón told them he thought it was pesticide poisoning.
Other people in the Mixtec community think he died of a karate

blow, so it's not certain. But it might be worth looking into. I can give you the names of various people who were there who could tell you more."

From his notes, Dr. Kearney could tell me the following about Ramón González, whose name is changed here. Ramón was the son of Guadalupe Pérez y González. She had become pregnant during the months when she had gone to work in the sugarcane fields of the state of Veracruz in 1961. She gave birth to her first child in San Jerónimo Progreso, her home village in the southern Mexican state of Oaxaca. Ramón grew up speaking the language of the village, Mixteco, and slowly he learned Spanish as he traveled with his father as a migrant agricultural worker. His father, Irineo, liked to play the tuba as part of the village band, and he also played the guitar. Ramón learned to play the guitar from his father, and while in San Jerónimo they would sometimes spend their evenings with friends playing and singing. Ramón liked to play with his nine younger brothers and sisters and his parents frequently entrusted him with their care. He loved to play soccer and joined a game whenever he got the chance. Kearney knew him as a hardworking, pleasant young man who got along well with everyone.

As a teenager, Ramón and his father hitchhiked, rode buses and trains, and walked great distances in search of migrant farm labor jobs. On one occasion, they went with a group of friends from San Jerónimo to Riverside, California, where they picked oranges and lemons. That season they stayed for a time in Tijuana, where many people from San Jerónimo had built homes for themselves in Colonia Obrera. In February 1981, Ramón, his father, and an uncle were working in the Culiacán Valley in the state of Sinaloa, fourteen hundred miles north of their home in Oaxaca. They were picking tomatoes that would be sent by truck to supermarkets in the United States. One day in early February, Ramón returned from the fields feeling ill. He took a bath in one of the irrigation canals, and became very sick. Two days later, on February 9, at the age of twenty, Ramón González died. While some say that he may have died of a karate blow while practicing with a friend, it is difficult to fit this in with his returning ill from the fields and becoming very sick after bathing in the canal (Kearney 1983, 1984b; Kearney and Nagengast 1988; G. P. González 1984, 1988; C. González 1988).

Most of the people who had been working with Ramón in Sinaloa's Irrigation District Number Ten believe that he died from the chemicals sprayed on the tomatoes to kill insects. Many farm workers they knew had at one time or another become sick while working in contact with the pesticides. When workers had gone to the federal government clinic, the doctors had sometimes told them they were suffering from pesticide poisoning. In some cases, they had been given an injection of atropine, an antidote to many insecticide toxins, and were told to rest a few days before returning to work. Some had suffered much more serious illness, and some had been irritated by skin rashes, headaches, mental confusion, and respiratory problems that they had come to associate with working around pesticides. Ramón's friends and father thought Ramón's complaints and behavior were typical of pesticide poisoning, and one of the doctors who attended Ramón told his father that Ramón was suffering from pesticide poisoning. Based on that, what they knew of Ramón's illness and death, and their own experience and knowledge, it seemed clear to these friends that Ramón had died of pesticide poisoning (Kearney 1983, 1984b; Kearney and Nagengast 1988).

There are strong reasons to suspect that pesticides were the cause of Ramón's death. The World Health Organization of the United Nations uses very conservative assumptions in its estimates of pesticide deaths, relying on extrapolations from data more than a decade old and based on superficial surveys in which only relatively well-regulated countries replied (UNWHO 1973). Based on these figures, David Bull of the United Nations Environmental Liaison Centre estimated in 1982 that more than 13,000 people a year die of accidental pesticide poisoning, most of them farm workers. About 750,000, using the same assumptions, are thought to suffer from acute pesticide poisoning short of death. A UN report in 1985 put the number of poisonings at 1 million, with 20,000 estimated fatalities (Levine 1985). The widely varying figures reflect the multiple difficulties of proper diagnosis, inadequate and noncomparable reporting systems, and a variety of biases at work in the field, in the clinic, and among those who gather statistics. There is general agreement that pesticide poisoning constitutes a serious public health problem, especially among farm workers (Bull 1982: 37–38; Goldenman and Rengam 1986). About half the documented deaths occur in the Third World, although it is the highly industrialized countries that account for more than two-thirds of all pesticide use. But in more prosperous temperate countries, pesticide use is generally regulated more tightly, and ecological conditions do not encourage the kind of lavish use of large quantities, high concentrations, a variety of mixtures, and frequent applications that those who use pesticides under tropical conditions often favor. In addition, the enormous growth in the use of the most acutely toxic pesticides has occurred during a

time when a large share of agriculture traditionally requiring intensive use of migrant labor crews has moved to Third World countries
where labor is cheaper and more easily recruited and controlled.

The statistics on the numbers of pesticide injuries and deaths,
however, are obviously unreliable. There are many good reasons to
believe that pesticide deaths and injuries in all countries, and especially Third World nations, are many times higher than those cited
by the World Health Organization, as will be discussed in Chapter 3.

The prime chemical suspects in the death of Ramón González are
a group of highly active chemicals that are very effective in killing
pests but that break down rapidly in the environment under most
conditions. Among these are the organophosphate insecticides, developed in the early twentieth century as nerve gases to be used in
warfare. The German government did additional research on organophosphates for possible use in World War II, and close chemical relatives of the insecticides are stockpiled by modern nations for possible use in war. The organophosphates have various advantages for
use in agriculture, but many of them are extraordinarily toxic to human beings and animals. They have come to replace other pesticides
that are less acutely toxic but which present a variety of other difficulties that have led to their frequent replacement by organophosphates. Many chemicals of a class called carbamates (aldicarb, for
example) are as acutely toxic as the most toxic organophosphates
and have also come to replace the persistents. This transition is discussed in detail later (Hayes 1982: 182–183, 339, 260–264; Duoll,
Klaasen, and Amdur 1980: 385–386; Sax 1979: 373).

Another of the prime suspects in Ramón's death would be the
highly active, acutely toxic, nonpersistent herbicide paraquat. Paraquat is widely used around the world to kill weeds and to defoliate
cotton for mechanical picking. It has also been used in Mexico,
the United States, Colombia, and other countries to kill marijuana
plants found by drug control agents. Paraquat in the quantities to
which farm workers may easily be exposed is capable of killing
people in a matter of hours or days after contact. It can also cause
various illnesses that may not show up for several years, including
the proliferation of tissue in the lungs leading to suffocation and, in
severe cases, death (Duoll, Klaasen, and Amdur 1980: 390–392).

The use of agricultural pesticides remains a burning public policy
issue in the United States, where their use is relatively well regulated. Contrary to popular belief, there are now more migrant farm
workers in the United States than twenty years ago, in spite of
mechanization of many field tasks. In some predominantly agricultural counties of California, there has been a 60 percent growth in

the numbers of both permanent and migrant farm workers from 1965 to 1985. High value specialty fruit and vegetable crops make up a more important share of agricultural production than previously, and these crops require more workers and heavy pesticide applications (Palerm 1987). The rates of pesticide poisonings of farm workers have been steadily climbing in California, where the best statistics are available, and in the United States as a whole (Barnett 1988; Wasserstrom and Wiles 1985). More than 20 percent of California's groundwater wells are seriously contaminated with pesticides, as are those in many other regions of the nation (USEPA 1987). On August 21, 1988, Cesar Chavez, leader of California's United Farm Workers Union, completed a month-long hunger fast to publicize his organization's boycott against table grapes aimed at limiting the continued use of highly toxic pesticides that endanger farm worker lives. The attendance by various national political figures and celebrities at the mass where he ended his fast signals that the long-simmering debate in the United States over pesticide safety will come to a boil (*Sacramento Bee*, August 22, 1988, p. 1).

Such chemicals as the organophosphates and paraquat, which we will take a more detailed look at later, represent a constant danger to farm workers, especially in regions like Mexico's Culiacán Valley. The pesticides are especially dangerous to those on the bottom of the labor market, people like Ramón González who are desperate for work, have traveled great distances to find it, and cannot afford to turn away from a job or to complain about the conditions once they have started work. In Culiacán, the majority of these people, and they make up the majority of the migrant labor force, are Mixtec and Zapotec Indians like Ramón. They have traveled well over a thousand miles to find work in Culiacán planting, cultivating, and harvesting the tomatoes, cucumbers, summer squash, bell peppers, chili peppers, eggplants, and other fresh vegetables that people in the United States eat in the winter months when there are few places such vegetables can be grown in the United States. People like the Mixtecs and the Zapotecs provide the bulk of the human labor that makes possible what has become known as "the global supermarket."

The Culiacán Valley is a superb example of the producing regions that feed the appetites of those who shop in the global supermarket. Many of those who make agricultural policy in countries all around the world look to the Culiacán Valley as a model of agricultural development. Culiacán proves that relatively poor countries can achieve agricultural modernization in a relatively short period of time, attracting large amounts of foreign and domestic investment and producing high profits and impressive export earnings in hard currencies like the dollar. The Culiacán Valley and a few other smaller regions in the state of Sinaloa provide about a third of everything Mexico earns in export agriculture. It is a linchpin of Mexican development (Alanis et al. 1982; Mares 1987).

People all over the world look to Culiacán not only because of its success as a region but also because to many, Mexican agricultural development in general seems like a model worthy of imitation. Agricultural modernization in Culiacán is part of an ambitious global project to transform agriculture. In one sense, this project began with the enclosures of communal land in England, from the fourteenth to the early nineteenth centuries, associated with the change away from agriculture for peasant and local village self-subsistence toward a commercial agriculture suited to industrialization and urban growth. Names such as that of the farmer-innovator "Turnip" Townsend and phrases such as "scientific agriculture" began to be part of a clearly identifiable trend in western Europe during the eighteenth century. However, as agriculture came to be more commercial and more machine-dependent in the United States and Europe, by the twentieth century there were those who worried that tropical and semitropical countries lagged too far behind in this transformation of rural life and agricultural production. Driven by a variety of motives that we will examine in later chapters, they began to look for specific ways to modernize agriculture in nonindustrialized nations and improve it in rich nations.

In 1941, the Rockefeller Foundation and the Mexican government launched an agricultural research effort that came to be known as the Green Revolution. Officers of the foundation and the researchers they employed hoped that the program would succeed well enough in Mexico that it would subsequently serve as a model for agricultural development all over the world. For better or worse, their hopes were realized. Many of the techniques and much of the Green Revolution approach to agricultural innovation spread not only to other poor countries but to the richest and most developed nations as well (Wright 1985; Jennings 1988; Karim 1986).

Plant breeding formed the basis of the Green Revolution project. By selecting plants and testing them out in the great diversity of conditions prevailing in Mexico, agricultural researchers developed a new generation of crop seeds that were generally capable of producing higher yields on a given piece of land than more traditional seed varieties. In most cases, however, the new seed varieties could not produce more unless they were combined with lots of water, large quantities of fertilizers, and chemical pesticides. The new agricultural development model depended on irrigation projects to make water supplies more dependable and on cash expenditures for agrochemicals. These requirements brought with them the necessity of an infusion of private and public financing on a grand scale. Profound and lasting social and economic changes followed.

One of the most obvious effects was that traditional farmers with little control over capital were pushed aside by those who could command capital. Water and chemical requirements led to vast environmental changes. The Green Revolution became one of the most important global agents of social and environmental change in the decades after the end of World War II.

Considered one of the great success stories of the Green Revolution, Culiacán serves certainly to tell us much of what is best about the changes introduced by the worldwide project of agricultural modernization. Certainly, few regions of the world are as favored by nature as the Culiacán Valley, with its easily dammed and channeled mountain rivers running out of the majestic Sierra Madre Occidental, its rich alluvial soils, and its semitropical climate. At the end of the valley are coastal lagoons teeming with crabs, shrimp, and fish life, and a large share of the waterfowl of the North American western flyway winter in and around the valley.

Few regions in relatively poor countries have been so favored by government policies providing irrigation projects, rail lines, highways, communication, and social services. Culiacán has highway and rail connections across relatively flat country to Nogales, Arizona, six hundred miles north, and thus to the lucrative markets of the United States. Private investors were quick to appreciate the potential of such natural advantages and governmental largess. It is little wonder that the valley has come to be considered a model for others. However, for regions without all the advantages of Culiacán, the road to success is narrower and more perilous. For that reason, any failures in Culiacán must be pondered carefully by others who would imitate its successes.

The death of Ramón González in Culiacán raises some serious issues. Are pesticide-caused illnesses and deaths among farm workers a frequent occurrence where chemical-dependent crops are grown? Are they inevitable? Do the chemicals that sometimes injure farm workers cause significant damage to animals and plants? Do the chemicals seriously contaminate the soil, air, and water of the area, and what are the consequences of pesticide contamination? Are the damages caused by pesticides limited to immediate and identifiable injury to people and the environment, or are there less obvious problems that might be more serious in the long run? Can the problems involved in pesticide use be reasonably tolerated or resolved, or are the difficulties so severe that they make production dependent on agrotoxins inherently unstable and unreliable?

In the course of this investigation it became obvious that pesticide use is intimately tied up with the whole project of agricultural modernization. As we shall see, many of the crops in which the worst pesticide abuses occur are not those that were developed in the Green Revolution research effort begun in Mexico in the 1940s. But the Green Revolution established an approach and a set of concepts about modernization that have come to dominate agricultural policy, so much so that the term Green Revolution is often used somewhat inaccurately simply to mean agricultural modernization in nonindustrialized countries or, a little more precisely, the promotion of new seed varieties strongly dependent on synthetic agrochemicals. In any case, the approach that developed out of the particular character of the Green Revolution effort in Mexico formed a pattern for agricultural modernization globally in the last half of the twentieth century. The pattern nearly always included strong reliance on synthetic chemical pesticides. Norman Borlaug, who won the Nobel Prize for his Green Revolution plant-breeding work in Mexico, puts it clearly, "We can either use pesticides and fertilizers at our disposal or starve" (Boardman 1986: 5). The investigation of the death of Ramón González is inevitably an inquiry into the reasons modern agriculture has become so pesticide-dependent.

Pesticides are only one way of dealing with pests. Ancient cultures as well as modern science have developed many other ways to protect crops from rodents, birds, insects, spiders, mites, fungus, and weeds. Each of these ways, like pesticides, commits those who farm and those who depend on farming to things other than the technique itself. Because pest management means intervention into agricultural systems that are complex and tied up with human society, every technique of pest management has some implication for other aspects of the agricultural system and for society. To question chemical pesticides as the overwhelmingly dominant means of protecting crops is to raise an amazing variety of other questions about the way we do things.

For example, pesticides have made it possible, for a time at least, to implant intensive agricultural production very rapidly and profitably into regions previously forbidding to ambitious commercial farmers. By combining the power of pesticides with new irrigation projects, governments and investors created vast wealth in new regions while older, more traditional regions suffered from a corresponding lack of attention. Ramón González was born in one of these neglected traditional regions and died in a frontier of profitable agribusiness opened up by government-financed irrigation works and the lavish use of new pesticides. This aspect of his life and death in itself is symptomatic of the changes that are occurring in Mexican life as a result of agricultural modernization in general and pesticides in particular.

Ramón González's family are Mixtecs, a people whose homeland is centered in the northern part of the state of Oaxaca and whose culture runs back at least one thousand years. In Oaxaca, they have lived by taking most of what they needed from the local environment and trading for the rest in local, regional, and even international markets. They now find it impossible to carry on the tradition on these terms and must earn much of the living for the family through migrant farm work. This puts them in regions like the Culiacán Valley, far from the support of their village communities and desperate for income. Why is it that people like Ramón can no longer make a living from their own land and must work instead where they own nothing and control nothing and where their only apparent future is to move on to work in yet some other alien and unfriendly land? The investigation into the death of Ramón González is not meant to point to individuals who may have been responsible in a narrow sense for his death, nor is it meant to simply identify the cause of his death. This is an investigation, in the broadest sense, into the circumstances of his death for the sake of learning what such circumstances might mean to his family and friends, his community, and his nation. And because the kind of agricultural development that set the terms of his life and death is much admired and very widely imitated around the world, it is an investigation into those technologies and policies that shape the lives of those of us still living.

2
The Road to Culiacán

The Frontiers of Agribusiness

In 1540, Spanish explorer Francisco Vásquez de Coronado set out from Culiacán to find the Seven Cities of Cíbola. Coronado found little of interest to the Spanish conquistadores of his day, and Culiacán remained a kind of border town for centuries, representing the last large outpost of settled Spanish culture on the Pacific Gulf Coast of New Spain (Nakayama 1983: 91). Four hundred years after Coronado's expedition, in 1940, the town of Culiacán still had no more than 16,000 inhabitants. It was a sleepy little state capital, forty-five miles from the coast, nestled into the foothills of the Sierra Madre Occidental. Much of the land of the valley that ran down to the gulf, as well as most of the foothills, was covered by a dense subtropical thorn forest so teeming with wildlife that it was a favorite resort of wild game hunters from the United States.

Forty-five years after that, in 1985, Culiacán boasted a population of half a million. Virtually all the valley land and most of the foothills had been cleared for agriculture. Culiacán had become less the northern outpost of Spanish or Mexican culture than the southern outpost of the cultural and economic forces of the United States, which tend to exercise a special influence over northern Mexico. The governor of the state in 1985 had achieved his prominence through ownership of the John Deere agricultural implement dealership. Some of his friends were wealthy from renting the services of their fleets of tractor-trailer rigs to growers shipping vegetables north to the United States. Local businessmen and farm owners gathered in the mornings at "Chic's," a brightly lit, formica-topped, orange-and-green-vinyl-upholstered chain restaurant, for breakfast. There they discussed local political problems, news from the vegetable brokers in Arizona and Los Angeles, or stock market and currency quotations from New York, London, and Tokyo. Prosperous housewives jogged along the riverside boulevard in their fashionable designer-named jogging suits. Middle-class youths gathered in the coffeehouse of the state theater and arts complex to discuss their homework, their music and dance lessons, and the political enthusiasms of their professors at the enormous state university located in the expanding suburbs at the edge of town. Factory workers labored around the clock at the Perkins diesel engine manufacturing plant and at dozens of shops and warehouses devoted to serving the booming agribusiness of the Culiacán Valley.

The multimillion-dollar annual harvest of the Culiacán Valley includes so many vegetables for export to the United States that Culiacán alone accounts for about a third to a half of the tomatoes, cucumbers, bell peppers, summer squash, zucchini, eggplants, and chili peppers sold in the United States between December and May. Valley farmers also grow safflower for oil, sugarcane, alfalfa, wheat, soybeans, and corn. The Culiacán Valley and its continued productivity are always a central concern of investment bankers and government economists a thousand miles south in Mexico City (Mares 1987; Alanis et al. 1982: 7–28; Rama and Vigorito 1979: 275, 295, 301; Cook and Amon 1988: 23–29).

There is a violent but prosperous underground economy in Culiacán as well. Hidden fields on hillsides and in small valleys are planted to marijuana and the opium poppy. Culiacán is often said to be the most important Western hemisphere source of heroin in the lucrative markets of the United States, and it is an important center for marijuana sales as well. In an otherwise quiet suburb called Tierra Blanca, the nights are frequently punctuated by shots from pistols and small machine guns as drug merchants and buyers quarrel over their business. This was true even in the early 1980s, when Culiacán authorities proudly announced that they had driven most of the trade out of town. In April 1989, when Mexican authorities arrested drug kingpin Felix Gallardo, they took the precaution of first arresting all eighty members of the Culiacán city police force, the police chief, and the state chief of police—they quickly released the ordinary policemen but detained the two chiefs for a longer time (Walker 1981; *Sacramento Bee*, April 10, 1989, p. A8).

Culiacán has been strongly connected to the economy of the United States for a long time. Relatively small-scale irrigation projects built in the nineteenth century provided water for sugar, cotton, and fresh vegetable production. Seventy-five percent of the irrigated land was owned by U.S. firms before the Mexican Revolution of 1910. The land laws of the 1917 Constitutions made such ownership illegal, but U.S. companies are still strongly present in Sinaloa.

Some land is owned by *prestanombres*, name loaners, who provide a front for foreign owners. Most of the financing for the vegetable farms comes from U.S. firms—according to the head of the National Union of Vegetable Growers, about 90 percent (Hamilton 1982: 45; González 1987; Mares 1987 provides a detailed history of land, labor, and water development in modern Sinaloa).

Battles for control of the land have been prolonged and bitter. The corruption of the Mexican land reform system makes it possible for large landholders to operate under the cover of legal forms intended for cooperative owners or small private landowners. Small landholders and peasant groups and communities have fought back, and landholders have retaliated through their own private security forces, *la guardia blanca*, with the help of the state and municipal police force at times. Through the 1960s and 1970s these conflicts were often violent, with land invasions by peasants, kidnappings, and murders. But the real control the large landholders have is their integration with the private and public credit institutions, their control over the market, and their ability to corrupt or intimidate agrarian reform officials. Through the exercise of these various forms of control, they have largely excluded true small holders and cooperative farms from the valley in violation of the spirit and often the letter of the agrarian reform legislation. The most frequent legal facade for these operations is to rent land from the legal small holders or cooperatives, but the fact and conditions of rental are completely determined by the large landholder, who uses corruption and the threat of force to obtain rental agreements from people who know from experience that if they do not cooperate, they cannot succeed as farmers in the face of the opposition of the large operators. The small holders and cooperative members "end up as wage labor peons working on their land" (Millan Echeagaray, in Cecena Cervantes, Burgueno Lomeli, and Millan Echeagaray 1974: 86–103; see also Mares 1987).

The growth and wealth of the Culiacán Valley are built on the foundations of the Mexican federal irrigation projects. The first unit of the projects flowing out of new dams in the Sierra Madre opened in 1948. At the time of the death of Ramón González in 1981, the irrigation projects supplied water reliably to 600,000 acres of land, 100,000 of them devoted to intensive vegetable production. In the mid-1980s, the Ministry of Agriculture expected to add more than 200,000 acres to the projects (Secretaría de Agricultura y Recursos Hidráulicos 1977; Morelos González 1984).

Ramón González and his father, Irineo, saw little of the flashy wealth generated by all this growth in Culiacán. About 140,000 migrant farm workers came in those years (the figure in 1987 was approaching 250,000) to plant, cultivate, and harvest the vegetables to be exported north. Ramón lived in the squalid camps, or *campamentos*, as they are called locally, with the other migrants. The *campamentos* are open-sided sheds, overcrowded and unhealthy places that even the growers agree are a disgrace. Local doctors declare that the conditions of the camps and the low wages of the workers make the *campamentos* inevitable nurseries of intestinal parasites, respiratory disease, infections of every kind, malnutrition, and pesticide poisoning. The growers protest that it is not their social obligation to provide housing for workers, that the government should do so, and that they should be thanked for providing anything at all (Posadas 1982; González 1987).

The best of the camps are built of corrugated sheet metal on steel frames, sold to growers by firms who sell them as poultry sheds. Each shed, or "gallery" as they are known locally, has a center wall of sheet metal and two long rows of cubicles, one row running along on each side of the center. The cubicles back to the center wall and have a wall on each side but are open on the fourth side. Farm workers often build a makeshift fourth wall from packing crates or blankets. The older camps are constructed on a similar plan, but the frame is of wood and the walls are made of corrugated, semirigid black tar paper.

From six to twelve people usually live in each nine-by-twelve-foot cubicle. The cubicles have dirt floors, sometimes with small drainage streams, often contaminated with human waste and agricultural chemicals, running through them during and after rain storms. There is no ventilation except through whatever space is left clear on the open fourth wall. Families often cook inside the cubicle.

The first time I entered one of these cubicles was in June 1980. The vegetable harvest was nearly complete for the year, and only a few migrant workers remained in the *campamentos*. The father of the family had severe laryngitis and spoke to our party of local anthropologists and visiting professors and social workers with difficulty. He also had a severe skin rash covering much of his face and running down his neck to where it disappeared under his jacket. His wife and three children were lying on the floor in the cubicle, each of them suffering from some combination of cold and flulike symptoms. The only furniture in the cubicle were metal pesticide containers used as stools, and another identical pesticide container was being used for drinking water storage. The labels on these containers

said "Toxaphene," a chlorinated camphor insecticide derived from a
pine tar; it is a pesticide that was determined to be carcinogenic and
otherwise highly hazardous and was severely restricted in the United
States in the early 1980s (USEPA Toxaphene data sheet). A spray gun
containing a DDVP compound (a moderately hazardous insecticide)
lay next to the improvised cooking stove. I asked what the spray gun
was for. "Flies. The flies are unbearable in here. We have to use this
spray gun all the time," he said as he picked up the gun and gave it
a few pumps toward the back of the cubicle, as if to demonstrate
its use.

I asked if he was aware that the chemical in the gun and the
chemical residues in the containers in the room could be dangerous
to himself and his family. "So they say," he croaked through his rasp-
ing throat. "So they say. But sir, we are poor people. Very poor people.
We must do what we can. The flies, no sir, you cannot live with
these flies. And we have nothing to buy pots and pans with. We pick
up the containers left in the fields because they are free. This is it.
This is our life, as you see it here." He said this as forcefully as he
could, and then bowed his head.

That day we saw many variations on such conditions. One of our
guides was a farm labor organizer, a Mixtec Indian like the other
workers we interviewed. His name was Benito, and he was greeted in
a friendly way by everyone living in the *campamentos.*

We visited a *campamento* where electric light poles had been set
up and bare wires hung down from the top of them. Benito explained, "You notice that in this camp there are some improvements—the latrines are built over excavated pits instead of going directly into the irrigation canals. There is a system to bring water
from a well instead of directly from the canal. It isn't working right
now, but perhaps we will be able to use it later. These improvements
were won as the result of a strike in 1978. These were some of our
demands that the growers agreed to, and some have kept some of
their promises." He looked at the ground and laughed. "But their
promises are sometimes strange. You see the bare wires hanging
down from the poles. Those are there because we demanded outdoor
electric lighting in the camps, like street lights. So, the growers,
some of them, have put in these poles and electric lines. But they say
they promised 'electrification' and that we have it. Meaning, the
lights themselves are our problem, since they have provided the
electricity. We don't know yet what we will do about this" (Posadas
1982; Mares 1987).

Benito explained that during the strike he had been picked up
by the police as a strike organizer. They had first threatened to

drown him in a canal, holding him under for long periods. Then they threw him in jail for two weeks, and nearly every day someone came in to beat him with a two-by-four. "They were mean, too," he added. "They would aim for the spots that were already sore from the earlier days' beatings. This was very painful." Benito explained that his limp and other health problems dated from that time two years earlier.

Early in the day, someone had asked if he was fearful of visiting the fields. "Not when I am with a number of other people. Or, if I can stay in sight of a number of workers at all times. But look behind you." About fifty yards behind us was a pickup truck with the logo of a local farm. Two men rode in it. We were to see them following us all through the several hours of our visit to the *campamentos*.

"That is a foreman and his *pistolero*," Benito explained. "They will be with us, close behind, all day. If I were to be isolated, out of sight in the middle of a field somewhere, I would not be long for this world, I think." He laughed heartily as he finished this sentence.

My next visits to the *campamentos* of Culiacán were over a period of eight months in the 1983–84 agricultural season, during the peak of the season when the camps were teeming with tens of thousands of workers. This time, I traveled alone, working under a U.S. Fulbright grant for international scholarly exchange.

There were some improvements in the *campamentos*. There were fewer tar paper buildings and more of the sheet metal poultry sheds in their place. A few camps had functioning well water systems, and more of them had pit latrines away from the canals. Some even had streetlamps attached to the overhead poles. (I was told by Florencio Posadas, a sociologist at the local state university, that brief work stoppages at some of the camps had achieved this result from the growers.)

In spite of these improvements, in many important ways the conditions of the camps were more hazardous to the farm workers. This was especially true because of a major change that had been made in the kind of pesticides most frequently used by the growers. Pesticides that presented one set of hazards had been replaced by other pesticides that brought equal or greater danger of a different kind. The implications of this transformation cannot be understood without a brief description of the characteristics of some of the most widely used pesticides.

One way of classifying chemical pesticides is according to the time it takes them to break down into other compounds once they are released into the environment. There are the *persistent* pesticides that have half-lives—the amount of time it takes for half the material to break down into something else—measured in months, years, or decades. Among these persistents are most of the chlorinated hydrocarbons such as DDT, BHC, Toxaphene, aldrin, dieldrin, endrin, chlordane, lindane, and many others. It was the persistents that were most widely used in Culiacán until the late 1970s and the early 1980s (Ussher 1984).

There is a second class of pesticides that are *nonpersistent*, with half-lives of hours, days, or weeks. Among these are the organophosphates such as parathion, methamidophos, guthion, malathion, and carbamates such as aldicarb. The herbicide paraquat is another nonpersistent. There are also nonpersistent chlorinated hydrocarbon insecticides such as endosulfan. The nonpersistent pesticides began to replace the persistents in Culiacán beginning in the late 1970s. By 1983, the nonpersistents had become dominant (direct observation 1980–1987; Ussher 1984; Aviles González 1984; Sánchez 1984; Vargas 1984; González 1987; Wright 1986; USEPA 1983; USFDA n.d.).

The nonpersistent chemicals are, in general and with important exceptions, much more acutely toxic than the persistents. It is the nonpersistent pesticides such as parathion and paraquat that are most likely to cause both moderate and severe poisonings within hours after contact. The persistents, such as DDT, are not likely to cause such immediate symptoms from the quantities normally encountered in field work. The persistents are more worrisome for their possible long-term effects such as cancer and liver disease and, most notably, for their tendency to cause reproductive failures in animal species, including birds and mammals (Duoll, Klaasen, and Amdur 1980; Hayes 1982).

Although the persistent DDT has only 2 to 5 percent of the acute toxicity of the nonpersistent parathion, DDT is the chemical that has been much more widely banned or restricted. The reason is that its persistence means that once released into the environment, it will stay in the environment for decades, and it will accumulate in some animals much more than others. In such animals, notoriously the peregrine falcon, pelicans, eagles, seals, and other carnivorous predators, the accumulation of the persistent pesticide may lead to reproductive failures and a range of chronic disease syndromes. A nonpersistent such as parathion, on the other hand, will be a much more severe problem at the time and in the place of its release, but its rapid decay presumably assures that its overall effect on the environment will be less lasting, less complex, and more easily managed. It is for this reason that environmentalists in particular have worked hardest to limit the use of the persistent pesticides, even though the nonpersistents are typically much more acutely toxic to humans and most animals (Marco, Hollingworth, and Durham 1987; Ragsdale and Kuhr 1987).

Parathion and other highly active nonpersistent pesticides are also especially dangerous because they are readily absorbed through the skin and retain most of their potency when taken into the human body in this way. Comparison of the most commonly used persistent insecticide, DDT, to parathion, the most commonly used acutely toxic nonpersistent, reveals that parathion poses a much greater immediate threat to mixers, applicators, field workers, and rural residents than DDT. Parathion is almost certainly the insecticide responsible for more acute poisonings than any other. Hayes, a standard authority in the field of pesticide toxicology, writes, "In the USA, parathion has been the main . . . cause of cropworker poisoning" (1982: 382–383). Parathion is thought to account for 80 percent of pesticide poisoning cases in Central America, where it has also been held responsible for massive semiannual fish kills in El Salvador (Wolterding 1981). In contrast, acute poisoning incidents are very rare for DDT, aside from suicide attempts and accidental ingestion of large quantities. The transition from persistents like DDT to nonpersistents like parathion means that the long-term ramifications of pesticide use are not so great, but the immediate consequences for people and wildlife are likely to be much more severe so long as the pesticides remain in use.

Between my first visit to the fields of Culiacán in 1980 and my later visits in 1983 and later, a major shift occurred in pesticide use patterns, from heavy use of the persistents to the substitution of the nonpersistents. This shift occurred, as we will discuss in more detail below, for two main reasons. First, pests were becoming strongly resistant to the persistents because the persistents had been used more commonly than the nonpersistents. The persistents are typically cheaper than the nonpersistents, which accounts for the fact that they were the group used most in the early decades of synthetic pesticide use. This in turn explains why pests in most regions of the world adapted to the persistents first before they adapted to the nonpersistents. With time, though, growers were forced to shift to the generally more expensive nonpersistent pesticides as resistance grew in target pest populations.

The second reason for shifting to the nonpersistents has to do with the political and public relations problems of Culiacán growers in relation to their markets in the United States. As the persistents, such as DDT and the "drins" (aldrin, dieldrin, endrin), were more and more heavily restricted in the United States and other highly developed countries due to concerns over their environmental effects and their possible role in long latency diseases such as cancer, journalists and other researchers noted that these chemicals were still being commonly used in Mexico and other Third World countries. Chemical companies were marketing the persistents even more aggressively in the Third World in order to recover investments lost in richer countries (Ussher 1984; González 1984; Weir and Shapiro 1981; Bull 1982).

Investigators also began to notice that fruits, vegetables, and meats imported into the United States from Mexico and other Third World countries carried significant residues of the persistents. In a muckraking book called *The Circle of Poison* (1981), journalists Weir and Shapiro called public attention to the fact that consumers in the United States were eating pesticides that had been manufactured in the United States and Europe, exported to places like Mexico because the chemicals were no longer allowed in the country of manufacture, and imported back into the United States and Europe as residues on food. This image of a circle of poison captured the imagination of many consumers and the attention of national and state legislators. Residue inspection programs at the border were somewhat improved, some consumers turned away from imports, and the Culiacán growers became seriously concerned about the loss of markets in the United States.

Mexican growers talked to Mexican public officials, who in turn talked to their counterparts in the United States. Agencies in the two countries signed technical agreements for the exchange of information, and technical advisers from the U.S. agencies offered their services to Mexico under the terms of the agreements. Upon consultation, the judgment that many growers had made on their own took hold—if the growers would switch even more rapidly than they already were switching (as a response to resistance in pests) to the nonpersistent pesticides, they would run fewer risks of losing out to competing growers in other countries and in Florida. With the combined incentive of responding to pest resistance and concerns of consumers in the United States, the Culiacán growers made a decisive turn to the use of the nonpersistent, very acutely toxic pesticides (Ussher 1984; USEPA 1983; USFDA n.d.).

The logic of this shift was straightforward. Since the nonpersistents break down more rapidly, it would be easier to deliver vegetables to distant markets with low residue levels. Pesticides applied on crops would be largely decomposed by the time they reached the consumer. And since the nonpersistents were still largely permitted in the United States, any residue problems that did exist would not be different in kind from the residues encountered on domestic produce.

The trend toward the very acutely toxic nonpersistents in Culiacán had immediate and tragic consequences for farm workers. Most of the nonpersistent pesticides, as mentioned, are very acutely toxic. International, industry-supported standards require that anyone using such chemicals as parathion, methamidophos, guthion, phosdrin, or aldicarb should be thoroughly protected against contact with the poison through inhalation, ingestion, or skin absorption. In most cases, this means that people working with the chemical or working in fields recently sprayed with the substance should wear rubber shoes, a rubber apron or rubber coveralls, a hat, preferably of rubber or vinyl, and a mask or respirator (University of California Division of Agricultural Sciences 1980).

In casual observation of hundreds of pesticide spray applications and systematic observation of fifty-two operations in the agricultural season of 1983–84, I never observed workers using any of this protective gear. In observation of ten spray operations in 1987, I observed one instance in which mixers using very acutely toxic materials were equipped with masks that they put on for a few minutes from time to time, and the pilot whose plane was spraying the mixture wore a respirator. In the other nine instances, foremen had made no provision for protective measures or gear. Even in the 1987 case of the mixers and pilot who had some protection, the pilot sprayed his highly toxic brew in a field only a few hundred yards from a *campamento* housing hundreds of people, and an old man and a young girl carrying a baby walked along the border of the field as it was being sprayed, with the material visibly drifting over them.

Although lack of proper gear is routine, it is certainly not the only abuse of safe pesticide use practices one observes in the fields of Culiacán. Growers have a choice of three different methods for pesticide application. They may order spraying by light aircraft, by tractor-drawn rigs, or by backpack spray rigs carried by individual workers, usually working in crews.

Aerial applicators operate in almost complete disregard for internationally recognized safety practices. In three instances, I observed aircraft spraying directly over crews of twenty to thirty workers— the sprays consisted of organophosphate insecticides mixed with copper or manganese-based fungicides. A television crew from CBS in Los Angeles filmed such an incident during the time I was doing fieldwork in Culiacán. A social worker employed by the growers' association told me, "We try to tell them how to protect themselves, to wear the proper gear, and to exhale when the plane passes over them in the field" (Duran and Mitchell 1984; Velázquez 1984).

A common sight in Culiacán is a man standing at the edge of a field with an upraised flag, signalling to an approaching spray plane. The *bandalillero*, or flagman, is assigned the task of marking the boundary of the last pass taken by the pilot in spraying the field. Without such a moving marker, the pilot cannot determine where to fly without gaps and overlaps in the spray pattern. The flagman is sprayed dozens, even hundreds of times a day. Most flagmen run at the last minute to stay out of the thickest of the pesticide fog, but they nonetheless suffer heavy exposure. By universally recognized standards of safe practice, the *bandalillero* should be wearing a mask or respirator approved for use with the type of pesticides being used, and he should be well covered from head to foot in protective, impermeable gear. In observation of twenty-three *bandalilleros* at work in the Culiacán Valley, I never observed the use of any protective clothing other than a light cotton bandana worn across the face.

One of the most difficult safety and economic problems of using aerial rather than ground spray techniques is the control of what is called "drift," the pesticides carried outside of the target field. Under normal circumstances in developed countries, 20 to 80 percent of sprays applied land outside the targeted field and may injure non-target crops twenty miles downwind. Under ideal conditions, using advanced techniques, drift can be limited to 10 percent of the product applied hitting areas outside the targeted field, and under especially adverse conditions, investigators have shown that 95 percent can miss the field. If pilots are careless or their equipment is not properly maintained and calibrated, the rates can go well above the average of 40 or 50 percent drift beyond the field boundaries. In Culiacán, there are virtually no controls on pilots to encourage them to minimize drift. One former pesticide pilot in Culiacán told me, "These men do not even know how to calibrate their equipment in most instances, and you can be sure that it is very rare for anyone to do it." Calibration insures the proper rate of release of the spray and controls the size of the droplet, thus allowing adjustment to atmospheric conditions. If it is not done, the farmer will be paying for sprays that do relatively little damage to pests and that insure high rates of contamination of surrounding areas (Guenzi and Beard 1974: 108; Henderson 1968; Akesson, Yates, and Christensen 1978; Carneiro 1983, 1984).

Atmospheric conditions are an important consideration in Culiacán. As in many valleys that run down to the sea, the typical day is calm in the morning hours, but by afternoon, strong convection currents caused as the land heats faster than the surface of the sea create strong winds. In Culiacán, it is common for planes to continue their spray operations until two or three in the afternoon, when winds are often reaching twenty miles an hour or higher. A pilot said to me, "Of course, under such conditions, we are wasting the grower's money as well as causing pollution, but there is a lot of hurry and there is not much supervision. The services of the pilots are very much in demand, and they are going to log all the hours they can."

Since the *campamentos* are in almost all cases surrounded on two, three, or four sides by fields running right up to the living areas, the lack of drift control means that everyone living in the camps is exposed to a variety of acutely toxic substances on a more or less regular basis, even in the case of children too young to work in the fields.

In February 1984, the town council of Montelargo complained that fumigation pilots were daily washing out the spray tanks of their planes in the local canal that served as the drinking water supply for the town. They lodged complaints with federal authorities, who took no action until the townspeople took their case to the government-controlled newspaper in Culiacán, which published a front-page story with a photograph of a pilot washing his tanks out in the canal. In this unusually well-documented case, the federal authorities withdrew the license of the pilot but would not give out any information on how long the suspension would last (*El Sol de Sinaloa*, February 15, 1984; meeting of Sanidad Vegetal and SEDUE officials, February 21, 1984).

Pilots are legally required to operate from runways or roadways specifically approved for their use. In the Culiacán Valley, however, pilots often use private lanes, levee top roads, and even public roads to land and take off without such approval, or they use approved sites in ways that are strikingly dangerous. For example, on February 9, 1984, the pilot of a fumigation plane was landing and taking off from the dead end of a levee top road near the little town of Moroleón in the Culiacán Valley. A Mixtec man mixed chemicals into a metal barrel, using a small gasoline motor to pump water from the large irrigation canal between the levees to mix with the commercial pesticide powders and liquids. A hose connected to the barrel allowed the man to pump the mixture into the spray tanks of the small plane when it landed periodically for reloading.

The mixture loaded into the plane for spraying across nearby vegetable fields contained pesticides with the commercial names

Tamaron, Bionex, Manex, and Cupravit. This mixture represents various acute and long-term hazards to humans and animals exposed to it. The legally approved labels on the containers of the first two chemicals warn that anyone working with the chemicals should be wearing protective clothing, including neoprene gloves, rubber boots, a respirator or mask approved for use with toxic powders and organic vapors, and rubberized overalls. These chemicals are restricted in the United States to handling by people specifically certified for such tasks. On this day in Moroleón, the illiterate Mixtec who prepared the mixture wore open plastic thongs, polyester pants, an open cotton shirt, and a baseball cap. He had none of the required protective clothing and equipment and no safety training. He told me that he had no idea what the chemicals were he was working with—he invited me to read the labels for myself, because "I cannot read the labels and the boss doesn't tell me because I don't know anything about it."

Tamaron is the trade name of Bayer, a West German firm with manufacturing facilities in Mexico, for methamidophos. Methamidophos is an organophosphate insecticide of high acute toxicity. It is about equal to parathion in acute toxicity to humans and experimental animals, and while both parathion and methamidophos are unusually toxic on skin contact, experiments have shown methamidophos to be an especially hazardous skin contact poison because human skin absorbs it at an unusually high rate. In the United States, methamidophos has been responsible for serious poisoning incidents in humans, poultry, sheep, cattle, and various kinds of wildlife.

Bionex is the trade name used by Planters Products in the Philippines for an organophosphate insecticide, azinphos-ethyl. The acute toxicity is high, although only about a fifth that of methamidophos or parathion, and it is also a dangerous skin contact poison. All the precautions that apply to methamidophos should also be used with the somewhat less hazardous azinphos-ethyl (Hayes 1982; Duoll, Klaasen, and Amdur 1980; USEPA data sheets on individual pesticides).

The two fungicides, Manex and Cupravit, are low in acute toxicity, and agronomists in Culiacán as in the United States often refer to them as "harmless." They are not harmless. Manex, manganese ethylenebisdithiocarbamate, has breakdown products that are suspected of causing cancer, mutations, and damage to fetuses in the womb. Manex itself, if exposure is repeated over a period of months or years, can cause a form of poisoning that is very difficult to diagnose because the symptoms resemble a number of other diseases that affect the central nervous system. The victim may appear to be "punch drunk" due to nervous system damage involving permanent and severe disabilities (Sax 1979: 786; Ritchie 1985). There is substantial disagreement among toxicologists about the effects of the copper-based fungicide Cupravit, but the substance is known to cause both skin and eye irritation. Estimates of its acute toxicity vary widely but fall within the general range of DDT (Ritchie 1985; National Academy of Sciences 1987: 127–130; Hayes 1982: 4–6, 607–608). Separately, each of these chemicals deserves cautious handling, and the two insecticides require very specialized precautions and gear, but as a mixture of all four, there is very little that science can tell us about possible combined or synergistic effects.

Aside from the lack of proper protective clothing and gear and the ignorance of possible dangers on the part of the worker, there were many other violations of safe practice in this instance. The concentrated pesticide powders were scattered about on the ground from normal spillage as the mixer worked rapidly to prepare for each landing of the plane. When the plane landed for a load, it was positioned in such a way that when the pilot started the motor, the wind from the propeller blew the powders up into the face of the mixer and into the canal. The pump drawing water from the canal had no automatic shutoff valve to insure that if the motor or pump failed, the mixture from the barrel would not be siphoned into the canal. The canal serves as the water supply for camps housing thousands of migrant workers. Empty but unrinsed pesticide buckets had been thrown into the drainage ditch on the dry side of the levee, while others had been thrown into a wire-enclosed open trailer. Tire marks in the drying mud made it clear that the trailer had not been moved for at least three days.

I asked the pilot if he could describe anything about the special dangers of working with these particular pesticides. He was quite friendly and told me that he thought it was a good idea that someone was looking into the way pesticides were being used in the valley. When I asked if he could say anything about the particular hazards of the chemicals he was spraying, he replied that he could not. "Those things are the business of my boss," he said with a smile and a wave as he climbed into the cockpit to unload yet another tankful.

During the ten-month agricultural season from late August into early June, such scenes as this one from Moroleón are common. They are especially frequent from mid-January until early in May, the heaviest spraying season. The multiple violations of ordinary safety precautions, as in this example, are a daily routine during this season. While not every aerial spray operation contains every violation mentioned here, I was never able to find an example of an aerial spray application in which there was a serious attempt to follow safety procedures. Every operation involved serious safety failings.

Some growers run their own airplanes and control the application procedures completely while others contract for spraying by independent companies. A pilots' association occasionally protests the conditions under which pilots must work, but the pay is high by Mexican standards and safety-inspired protests have never been persistent. As in the United States, it is very difficult for pilots to obtain any kind of life insurance, and a dead or injured pilot who committed errors leading to a crash is likely to be seen as negligent and not as a victim of pesticide poisoning. This is the case even though studies under the much less hazardous working conditions in the United States show that serious accidents and near misses involving pesticide pilots are very often the result of intoxication from pesticide exposure (Hayes 1982: 357).

It is difficult to say whether aerial application of pesticides involves more or less hazard than the other commonly used application technique in the Culiacán Valley—use of backpack spray rigs carried by workers walking through the fields.

The use of backpack spray technique is attractive to the growers for several reasons. Under some conditions, it is cheaper than aerial spraying, and it does not rely on the availability of a plane and pilot. People using backpack spray rigs can achieve a fuller penetration of the foliage of growing plants than an overhead aerial application, assuring that the spray is more likely to contact the target pest. (Aerial spraying by helicopters is competitive with hand spraying in this sense because the wind created by the propeller blade of the helicopter drives the spray down and outward very effectively, giving an excellent penetration. Helicopters are seldom seen in the Culiacán Valley, however, probably because of the higher cost and the scarcity of pilots.) Backpack spraying allows herbicides to be directed at the weeds growing between rows while minimizing the threat of damaging contact with the vegetable plants. Backpack sprayers can be used in areas too small or too irregular for efficient aerial spraying. While sprayers often receive a small bonus over the wages of other field workers, labor costs can usually be kept at three to five dollars per day per worker. In sum, the use of hand spraying with backpack rigs is usually cheaper and almost always more flexible and more discriminating than aerial sprays (Ussher 1984; Vargas 1984).

Backpack applications are discriminating enough that spraying is only done to the edge of a field and at no more than a few feet off the ground, unlike the aerial applicators, who often spray somewhat beyond the field boundary and in any case at an altitude that insures a relatively high rate of drift into adjoining areas. For this reason, backpack operations offer a degree of safety to farm worker *campamentos*, small towns, and rural residents compared to aerial applications. But this advantage is offset by many hazards peculiar to the backpack technique.

Description of particular operations highlights many of the problems routinely encountered in the fields of the Culiacán Valley. On a bright, chilly morning in early February 1984, a crew of eleven young men worked spraying paraquat in a field across from Campo Patricio. Long rows of tomato plants had reached well above waist height, the plants supported by wires and posts. Small, green tomatoes weighed heavily on the vines. The crew walked down the rows with backpack tanks connected to nozzles equipped with plastic hoods to direct the spray only at the ground between the rows, minimizing contact with the tomato plants, which themselves can be damaged or killed by paraquat application to their leaves. The hooded nozzles were also effectively directing the spray at the feet of the men as they walked along. Some wore traditional open leather sandals called huaraches, others wore plastic thongs, and others wore cloth athletic shoes, visibly drenched by the liquid paraquat spray. Upon exhausting their tanks, the men walked to the edge of the field, where they refilled the tanks from a fifty-five-gallon drum used to mix the paraquat concentrate with water. As the men poured and mixed the solutions, they splashed the liquids on themselves and on anyone standing near. They agitated the mixture in the barrel with a big, crooked stick, but on occasion impatiently dipped in a hand to achieve a faster mixing effect.

The men were between the ages of fifteen and twenty-two. They told me that they lived in the state of Durango, across the crest of the Sierra Madre to the east of Sinaloa. They traveled as a crew specializing in pesticide application work—they had been working together for three months and some hoped to work their way north to end the season in July and August near Fresno, California before returning home to Durango.

"Have any of you experienced any health problems during the time you have been working at this?" I asked, as the men assembled around the barrel to talk to me.

"What kind of problems?" one man asked in return.

"Anything at all, or anything that you thought might have been caused by the pesticide sprays you work with."

"No, never!" one man answered, so sharply that he visibly startled some of the crew members. He turned out to be the foreman of the crew. The other crew members looked around quickly at each other and bowed their heads.

"No one, none at all?" I asked.

"Not a thing, nobody," the foreman said forcefully.

We all stood silently for a time, one of the men tracing patterns in the ground with a pointed toe.

A man who had not yet spoken broke the silence. "Yes, there have been those with problems," he said quietly.

"Like who?" the foreman challenged the man aggressively.

"There have been several, as you know," the other replied in a soft voice.

"Name one! Who do you mean?"

"Well, there was Alfonso. He had to drop out of the crew. He got sick from the sprays and couldn't work anymore, had to go home. He's still sick, I heard." The other men nodded affirmatively.

"And there have been others," the man added. There was a quiet, shy chorus of "yes, that's so," "that's the truth," "yes, others," and simply "yes, yes" from the other men.

The foreman gathered his composure. "Sure, but those were just the weak ones!" he concluded heartily.

I asked if they knew any of the specific dangers of the chemical they were working with. Three of the men looked to the label on the can, trying to puzzle out for me the information on the label. They were unable to read many of the words and gave up trying after some considerable stumbling apparently embarrassing to the men. The foreman made no effort to read the can label, saying simply that only the agronomist in charge could tell me that, and he was back at the farm headquarters. The men started walking back into the field, resuming their spraying.

Paraquat, the chemical herbicide they were using that day, is an extraordinarily dangerous pesticide. It is famous among rural people around the world for its high acute toxicity, so much so that it is now said to be the leading choice of people committing or attempting to commit suicide in the Third World—a poison readily available in significant quantities without prescription or questions from a pharmacist. Paraquat used in the fields can cause a variety of illnesses ranging from skin rashes to severe illness to death upon hours or days. Skin contact as well as ingestion or inhalation can cause these effects, as with the organophosphate insecticides. But paraquat has some specially dangerous characteristics of its own. With frequent exposure over a period of two or three years, exposure at a level that might never cause noticeable immediate effects, paraquat can cause the proliferation of lung tissue, reducing the ability of the lungs to absorb oxygen. The emphysemalike symptoms can lead to significant disability or to death. Paraquat is also a probable carcinogen, and almost certainly disrupts reproductive functions and causes birth defects (Duoll, Klaasen, and Amdur 1980: 390–392; USEPA data sheet).

The label on the can of paraquat concentrate that these men were mixing into the water in the barrel that morning near Campo Patricio clearly stated the industry-approved precautions for paraquat use. Anyone exposed to the chemical should be wearing rubber boots, rubber gloves, a hat, rubber apron or coveralls, and a mask or respirator. None of the men wore any of this protective gear.

In addition, while paraquat breaks down rapidly, it is very toxic to fish and a danger in human water supplies, and the can's label cautioned against allowing it to contaminate water. A small irrigation ditch running along the field was receiving direct spray from the men's equipment as they neared the end of each row.

In most developed countries, paraquat applicators must be trained and licensed for the work. These men did not know the name of the chemical they were working with and were unable to read the complex information on the can's label, in spite of apparently good-willed efforts to do so when I asked about the chemical they were using.

In another case, only about two miles from the first, on another beautiful day in February, a larger crew of nearly two dozen men sprayed a mixture of fungicides and acutely toxic insecticides on a field of bell pepper plants. In this case, the mixture had been made up earlier and was carried to the field in a small tank truck. The workers filled their backpack spray rigs from hoses coming out of the truck's tank. A good deal of the mixture spilled on the hands and clothes of the workers as they refilled their small tanks, and puddles formed on the ground.

An agronomist supervised the operation. I asked him what chemicals they were using, and he told me they were applying a mixture of five chemicals. Cupravit and Maneb, copper- and manganese-based fungicides, were mixed with three nonpersistent insecticides. The insecticides were malathion, an organophosphate of low acute toxicity, parathion, an organophosphate of high acute toxicity, and endosulfan. Endosulfan is an organochlorate that is very acutely toxic and nonpersistent, in contrast to most other organochlorates that are lower in acute toxicity but highly persistent. Experiments with lab animals have shown endosulfan to double in toxicity with moderate protein deprivation—a condition nearly universal among farm workers. This mixture would require all the precautions mentioned in the cases above and poses significant acute and long-term hazards to those exposed to it. None of the workers wore any protective clothing beyond a cloth hat and a thin cotton handkerchief over the nose and mouth (Hayes 1982: 250–252, 379–385, 333).

The pepper plants had grown to about shoulder height to most of the men. They walked quickly up and down the rows, holding five-nozzle tubes at the height of the top of the plants. Frequently, one would pass on one side of a row walking south while another man passed on the other side of the row walking north. When this happened, each received a direct dose of the mixture in the face from the other's spray.

I asked the agronomist what insects he was spraying against. He named eight pests; that is, all the major pests of the bell pepper in the Culiacán region. I asked what population levels of these insects he had found in the field, and he replied that he had not done any sampling, that this was "purely preventive."

Universal standards of good pest management require that chemical sprays be used only when it has been determined by systematic sampling that potentially damaging population levels of given pest species actually exist in the field. This prevents wasteful spraying and unnecessary hazards from spray operations, and it greatly reduces the speed at which pest populations will develop resistance to chemicals. Preventive spraying, or "scheduled spraying" as it is often called, means that the pests are undergoing a continuous process of evolutionary adaptation to the chemicals used rather than suffering high mortality from less frequent selective spraying. The use of preventive or scheduled spraying is routine in the Culiacán Valley, as in this instance in bell peppers (Flint 1981: 31–50, 121–180).

While talking to the agronomist, an inspector from the federal Mexican plant protection agency (Sanidad Vegetal) charged with enforcing pesticide regulations stopped his truck by the side of the field and approached us. After introductions, he asked the agronomist what they were using, and the agronomist gave him the same list of chemicals he had given me. Incredibly, the inspector said, "Nothing harmful, then." The agronomist replied, "Right, nothing harmful."

The inspector got back into his truck beside his passenger, a uniformed federal soldier carrying a carbine, and left. Upon passing by the same field about two hours later, I noticed the inspector and the soldier walking up and down the rows, examining the area under the pepper plants meticulously. I was later told by another agronomist with experience in the valley that this was typical of inspections carried out to determine whether marijuana plants were being grown, concealed by vegetable plant rows grown to a height great enough to hide other plants. It is notable that in daily inspections of pesticide spray operations in the Culiacán Valley amounting to a total of forty days scattered through the agricultural season, this is the only time I saw a Sanidad Vegetal inspector in the field. The exception was the time the local head of Sanidad Vegetal arranged a tour of the valley for me conducted by one of his subordinates—that day was the only day during the height of the spraying season when I failed to observe any pesticide application operations.

The examples of backpack operations given here are typical—in no case did I ever observe any protective gear or clothing being used, and other abuses of normal precautions were routine. I never observed a case where even a casual effort had been made to follow accepted safety practices.

The third method of pesticide application, tractor-drawn spray rigs, is used primarily early in the season before the height of the rows makes it difficult to pass through with machinery. At least three farms owned "spider" rigs, with extendable legs able to carry the spray tank and equipment above the level of high rows. In 1983–84, I only saw these in operation on three occasions and was unable to make systematic observations due to the difficulty of approaching the men involved as they passed hurriedly up and down the rows. It may be supposed that this technique offers potential safety advantages over aerial and backpack operations, since drift can be controlled more effectively than it can from an airplane and fewer men are exposed to the chemicals because only one or two workers are needed as opposed to large crews. The other dangers of contamination of those who mix the material, of those who ride the tractor, and of those exposed to pesticides as they work in the fields after spraying is completed or through contamination of air and water would remain otherwise roughly the same. A health worker in the San Quintin area of Baja California, insisting on anonymity, showed me slides he had taken of spider rigs spraying what he believed to be parathion directly over farm workers, although I have never observed this particular abuse. In the 1983–84 season, tractor rigs played a small role in pesticide application, but by February 1987, growers boasted that pesticide safety in the fields had been improved by the introduction of more tractor operations. Although no numbers were available to confirm this, field observation did show that tractor applications were more common than they had been three years earlier.

Dr. Luis Morales, director of federal medical services for the state of Sinaloa (accounting for most public health services and virtually all medical care available to the poor), points out that it may be beside the point to focus on the safety violations of pesticide application procedures. The real danger, says Dr. Morales, comes from the ubiquity of pesticides in the air, soil, and water through most of the year. He believes that aerial application poses the worst human health hazard, because of the inevitable drift of pesticides into farm worker camps and rural towns. "It is the more or less continual exposure of everyone who lives in rural areas of Sinaloa to a very wide variety of agricultural chemicals that poses the real danger to public health, and it is a danger that is almost impossible to quantify or understand in detail. There are so many different chemicals, each with its own effects, some of which we may understand and others of which we know little or nothing. And the chemicals interact with the circumstances of poverty, the lack of sanitation, and the extraordinary range of diseases in a subtropical environment—we do not ever hope to have a comprehensive picture of the role pesticides play in human health given the complexity of the problems of epidemiological analysis in such an environment" (Morales 1987).

One public health doctor who works under Dr. Morales's supervision expressed her opinion that in many instances "farm workers are poisoned by hunger." This young woman, a specialist in occupational health and safety, has been involved in a study not yet available to the public of the unusually high rate of leukemia among children in the farm labor camps. While she believes that it is possible that the leukemia is linked to the pesticides used or to solvents used in pesticide mixtures (benzene, in particular, a known carcinogen), she sees the problem of pesticide poisoning as significant primarily when it is understood as part of the larger context of the farm workers' lives. Dr. Castro is an attractive woman in her thirties, energetic and forceful, with large glasses and the fashionable dress of a woman from Mexico City. She walks fast and talks fast and speaks to us in her tiny office overflowing with medical journals and reports.

"They die of hunger, because they have so little to eat that they often go to the vegetable fields to have something to put in their stomachs. Often the vegetables are heavily contaminated, and they die from eating them. They are not paid enough to eat regularly, you see."

Dr. Castro puts her hand to her head and breathes deeply for a moment, then leans forward across her desk. "Look, the problem of pesticide poisoning per se, as far as we are able to see it and diagnose it properly, which is another question, is not particularly notable among all the diseases from which the migrants suffer. It does not, if one examines the figures of the clinics, call attention to itself. The question, as far as pesticides is concerned, is more a matter of figuring how these chemicals relate to all the other disease syndromes and, ultimately, to the living conditions of the migrants as a whole." She throws up her hands and gazes heavenward for a moment. "Really, you can imagine, this is a very difficult question" (Castro 1987).

Dr. Morales, a calm and cautious man, meticulously dressed under the monogrammed white lab coat he wears in his administrative offices, is convinced that the relationship between pesticides and other diseases is quite strong. "Obviously, for instance, in the case of the dermatitis we encounter so frequently in the camps. But especially, the respiratory diseases. We are completely convinced that many of the pesticides make people much more susceptible to streptococcus infections. One that is very common in the camps is rheumatic fever, which, you probably realize, leaves many of its victims with permanently weakened hearts. But who can put a number on it, one that would be meaningful in terms of talking about the dangers of this or that pesticide or in terms of pesticides as a whole? The problem is too complex, epidemiologically."

The most common form of acute pesticide poisoning in the fields of Culiacán, as in most agricultural regions of the world in the last decade, is caused by exposure to organophosphate insecticides and carbamates, whose action is similar to organophosphates. Organophosphates, like most insecticides, are central nervous system poisons, but the organophosphates have their own characteristic way of acting upon the nervous system. Specifically, they reduce the level of acetylcholinesterases in the body and in the synapses of the nervous system. Acetylcholinesterase (ACHE) is a word that describes a family of chemical enzymes that break down acetylcholines, chemicals formed within synapses to translate electrical impulses received in the synapse into a chemical form. The acetylcholines, chemical messages, formed at one end of the synapse in response to electrical stimulus, travel the microscopic distance across the synapse, where they in turn stimulate other electrical discharges that are sent to the brain. The enzymes, the ACHEs, then clear away the acetylcholine message chemical to make way for new message chemicals to be formulated and transmit new impulses across the gap within the synapse. If the ACHEs are not present in sufficient quantities to clear the old message chemicals, the messages traveling across the gap become increasingly incoherent. In mild cases, this causes the nervous system to stimulate various excitation effects in the body—profuse sweating, severe cramps, headaches, vomiting, narrowed pupils—and in severe cases the body begins to shut down, as evidenced by coma and/or death. Organophosphate insecticides reduce ACHE levels in insects, fish, birds, reptiles, amphibians, and mammals, including human beings.

ACHE level reduction is the primary way organophosphates weaken and kill their intended insect victims and their unintended human victims. Their reliability, at appropriate doses, in killing human beings has made the more acutely toxic organophosphates attractive to the military of various countries, including Nazi Germany and the modern superpowers, all of which have at times stocked them for possible use in chemical warfare. There is, in other words, no doubt about their toxic effects on humans. What makes life difficult for the doctors and the poisoning victims they treat is that as a central nervous system poison, organophosphates, as well as many other pesticides, manifest themselves in symptoms that vary from flulike syndromes to respiratory problems to mental disturbances.

Doctors do have available tests to determine ACHE levels, one set that tests red blood cell levels and the other that samples plasma ACHE levels. Some of the standard testing methods are much more reliable than others, but tests with lower reliability continue to be widely used because of cost or convenience factors. It is usually on the basis of a combination of standard organophosphate poisoning symptoms and tests revealing a severely reduced level of ACHEs that doctors base a diagnosis of pesticide poisoning. Unfortunately, many victims incorrectly assume that they are suffering from some other problem, or they are reluctant, for a variety of other reasons, to seek medical help. Many conscientious doctors misdiagnose pesticide poisonings, and there are good reasons to believe that less conscientious doctors frequently see pesticide poisonings through the spectacles of preconceptions about farm workers or through a loyalty to or pressures from growers who do not want to admit to their own liability in poisoning cases. A California study done in the early 1970s showed that it was probable that only about 1 percent of the acute pesticide poisoning cases that occur among farm workers are diagnosed as such (Duoll, Klaasen, and Amdur 1980: 365–375; Metcalf 1982; Kahn 1976; Barnett 1988).

Among other difficulties of diagnosis is the fact that ACHE levels vary from one individual to the next and over time within a single person. Without a series of tests previous to the poisoning incident to establish the normal basal level in a given person, it is very difficult to prove that an apparently low level of ACHE is due to pesticides. This is one reason few growers or chemical companies have ever been held legally responsible for pesticide poisoning.

Further complicating the matter is the fact that ACHE levels can be reduced by malnutrition, illness, physical fatigue, or mental stress. Such conditions are virtually universal among farm workers during the agricultural season.

Fighting Pests

Ingeniero Luis Sánchez is in charge of the farm and packing shed that sends Ritz brand tomatoes to the U.S. market. The title "Ingeniero" refers to his degree in agronomic engineering. Licensed agronomists like Luis Sánchez work as foremen, managers, and consultants to the farms of the Culiacán Valley. Most of these men have a specialty—in Sánchez's case it is plant pathology. Some growers have earned their own four-year agronomy degrees as Sánchez did, and among both owners and hired agronomists the most common specialties are ones like plant pathology and entomology, directed toward pest management.

Sitting around the kitchen table in the front office of the farm and packing shed, drinking coffee out of big white mugs, with the chatter of a two-way radio communicating with foremen in the field in the background, Sánchez tells me something about the importance of pesticides in farm management in the Culiacán Valley.

"We spend from 15 to 35 percent of our total production costs on pesticides and pesticide applications, depending on the conditions from year to year. This year, our costs for pesticides and application will come to something like $1,800 per hectare (more than $700 per acre), so you can see that we have a powerful incentive not to overuse pesticides. But we have some very special problems here in Culiacán that require heavy pesticide use."

Sánchez excuses himself for a moment to go to the radio to talk to a foreman who has called in. "In that case, ease off on the irrigation for the day and we'll take another look at it later in the week," he advises the man.

Returning to our conversation, Sánchez goes on. "You see, we very seldom have a freeze. That means we have no killback of the pest organisms in the winter as you do in the north. We have a ten-month season that begins with planting in August and doesn't end until the last harvest in late May or early June. We stagger plantings for continual harvest, and each farmer has the legal obligation to burn old plants as they go out of production, in order to deny harboring places to pests. The government is supposed to ensure that this happens, but a lot of times guys don't do it and the inspectors don't catch it or say nothing about it. This irritates a lot of us who have to pay the price and who try to do our part for our neighbors' sake. But if the inspectors don't do their job, how can we expect everybody to stay on top of it out of their own good will alone?

"Another thing: this is a tropical environment for all practical purposes. We cut these fields out of dense scrub forest. We are planting what amounts to a monocrop. What that means is clear enough. We have an ideal environment for a wide variety of pests. A lot of those pests were here when we started. And with tens of thousands of acres in vegetables, we have created an even better environment for the pests than what was here before. It has to be a continual battle. That's what makes our job what it is."

I ask, "How many sprayings do you do in that ten-month season?"

Ing. Sánchez takes a long drink of coffee and looks at me intently. Then he stares out the window. "It varies," he says.

"An average, a rough idea," I suggest.

"Hard to say, depends on conditions."

"What will it be this year?"

He sighs heavily. "For us here this year, about thirty-five applications, more or less. Well, it ranges from twenty-five to fifty applications per year. Depending. In the fall, we usually have about one spraying a week, and then as it gets hotter in the spring and we get the buildup of pests over the season, we have to go to two times a week for a few months."

Later, we go for a visit to the packing plant. A truck driver dumps tomatoes into an agitated bath of water laced with chlorine. As the tomatoes are brought out of the bath, they are dried with blasts of warm air. They travel through a series of Rube Goldberg belts and devices along which young women sort them and pack them in boxes. Sánchez explains that an average of 70 percent of the tomatoes picked in the valley will be rejected for export because they don't meet the cosmetic standards enforced by the U.S. Food and Drug Administration. The tomatoes must not show any blemishes or irregularities in shape. When they are packed, they are packed very precisely by size and they are all light green in color, looking like perfect, machine-made items out of a factory, not like something that grows with all the irregularities of nature or ordinary agriculture. The cosmetic standards are written under pressure from U.S. growers trying to limit competition and hold prices up by restricting supply.

Ing. Sánchez proudly shows me the manifests that show that he is getting twice the average acceptance rate—60 rather than 30 percent. That will increase his boss's profits enormously, because what does not go north must be sold in the national Mexican market at prices less than a third what they will fetch at the U.S. border, less than the costs of production. The riper culls will go to local livestock growers, where they are worth as much as they would earn being sold elsewhere in Mexico. I ask if one of the methods for improving the acceptance rate is the use of higher amounts of pesticides to avoid the blemishes and irregularities that pests can introduce. Sánchez says that yes, this is certainly the case.

His answer is consistent with a well-known study done in California by the late University of California, Berkeley, professor, Robert van den Bosch, who concluded that fully 25 percent of the pesticides used in California horticulture were used to ensure the high cosmetic appearance of fruits and vegetables, having no effect on yields or nutritional quality. With the rigorous enforcement of cosmetic standards at the U.S. border by the FDA and with the special conditions favoring pests in the Culiacán Valley, it is reasonable to suppose that the corresponding figure for the Culiacán Valley might be higher than 25 percent. (For a recent survey of such research, see Feenstra 1988; van den Bosch 1978: 197; California Department of Food and Agriculture 1978: chap. 4.)

Other agronomists working in the valley agree that the cosmetic standards are an important incentive for high rates of pesticide use. Ing. Mayra Aviles González, an entomologist who works for the federal agricultural experiment station in the Culiacán Valley, says that it is "the anxiety of the growers about losses due to pests and pest-induced cosmetic damage that has a great deal to do with irrationally high rates of pesticide use."

When I asked her about the numbers of sprayings per season, she replied, "That's supposed to be a secret around here."

"Other agronomists have told me it's in the range of twenty-five to fifty applications per ten-month season. Would it be wrong to quote them to that effect?"

Ing. González grinned. "That would not be wrong."

"How much of that could be avoided?"

"Well, that depends on a lot of assumptions. But we have shown in experimental plots here that we can get the same production and cosmetic results with about half the amount of spraying that is typical. Our conclusions, based on numerous experiments that all show more or less the same thing, is that about 50 percent of the prevailing average pesticide application rate is simply irrational."

"But with such high costs, why wouldn't the growers latch onto those results and reduce their spraying accordingly?"

"Hard to say. But I think it's anxiety. And a lack of trust of what we tell them, and what their own technical advisors tell them. They look at it this way: with cosmetic damage, with crop loss, they may lose some indeterminate amount of their crop. In spite of the expense, they would just rather be safe than sorry."

She added later on in the conversation that she hoped that with more experience using technical advisors, the growers would begin to trust them more and there would be a corresponding reduction in spray rates. "We already see this happening to some extent."

I asked Ing. González if there was any attempt at the experiment station to move to a greater use of biological controls; for example, the introduction of wasps that parasitize pest insects or of bacteria that prey on tomato worms and other caterpillar pests.

"Yes, there has been a little of that. There's some of it going on right now. But its use is very limited here. We see no chance of shifting out of heavy chemical dependence."

"Why?"

"Look, we know about the kind of integrated pest management programs that have been worked out in California and elsewhere for some of the vegetable crops we grow here. And we would love to experiment more with a whole range of biological controls." She paused for a moment and looked off into space. "That's what my specialty was in school, after all. I'd love to do something with it. But here's the truth of the matter: the spraying rates in this valley are so high that we have to consider that for all practical purposes the only organisms that can survive in the field environment are the vegetable plants and their pests. Try to introduce organisms that parasitize or prey on the pests, and they don't survive well enough to be practical."

"But wouldn't it be possible to move through a transition period where you would slowly reduce pesticide application rates and gradually introduce biological controls?" I asked.

She smiled a rueful smile. "No, it wouldn't."

"Why not?"

"Because to do that you need coordination. To have coordination you need authority. We have no authority. You can't move through such a transition in one field or in one farm when the pesticides are so heavily used in surrounding fields—the spraying of one farmer negates the efforts of the one who is trying to get off the treadmill. And we have no authority, no way to move the valley farmers as a group in a systematic way. So, we will continue to rely on chemicals."

Ing. González had already showed me some disturbing statistics she and her colleagues had developed in their studies. They showed that insects adapted rapidly to new pesticides here in the Culiacán

Valley as they have everywhere where pesticides have been used. She told me that the Culiacán Valley had to abandon any significant cotton plantings in 1965 because pests of cotton had become so resistant to the pesticides available. She spoke of the problem of secondary pest outbreaks—organisms that had not been known as serious pests became very damaging to crops after pests were introduced that eliminated the organisms that had preyed on them. Predators of pests nearly always succumbed to pesticides earlier and at faster rates than pests themselves, because the predators had to eat seven to ten units of body weight in pests to gain one unit of body weight for themselves—being at a higher level in the food web or chain meant that pesticides concentrated in predators much more rapidly than in pests. Ing. González and I discussed the statistics developed at the University of California at Riverside that showed that more than 450 insects in the world had developed significant pesticide resistance and that many species of weeds, fungus, spiders, and other pests also showed high levels of resistance. A National Academy of Sciences scientific study concluded that pest populations resistant to one or more pesticides develop resistance to other pesticides much more rapidly and that pests can retain inherited resistance over long periods of time. "Hence primary reliance on chemical control strategies over the long run will depend on a steady stream of new compounds with different modes of action that can also meet regulatory requirements and economic expectations—an unlikely prospect in many pest-control markets" (National Academy of Sciences 1986).

"So," I asked, "this seems like a losing battle. If you cannot make a transition to biological controls, and if you are getting ever higher rates of resistance and secondary pest outbreaks, doesn't it seem as though this kind of agriculture is doomed in Culiacán?"

"No, I don't think so. There will always be a company coming up with a new chemical, something for which there is no resistance. I have faith in science. That's what it's all about, isn't it?" she asked (Aviles González 1984).

Not all the agronomists in the valley are so optimistic. One agronomist, a plant pathologist, repeated to me the story of cotton production falling to pest resistance. He said he thought it was obvious that by the end of the century there would be no more large-scale commercial vegetable production in the Culiacán region due to pest resistance. This agronomist was the sales manager of a pesticide sales firm. I asked him how he felt about selling pesticides when he thought that it was a self-defeating proposition to continue to rely on them.

"Look," he said, "you see my son waiting out there for me to take him to his baseball practice, right? Well, that's it. I've got a family to support, don't you? Isn't that what it's all about?" (Fernández Flores 1984).

An Environment Drenched in Pesticides

Agriculture is not the only activity in and around the Culiacán Valley. The Culiacán River and the irrigation canals of the valley lead to a series of lagoons on the Sea of Cortés, or Gulf of California, as it is generally known in the United States. These lagoons have been known for centuries for their rich bird and fish life. About 50 percent of the Mexican West Coast shrimp fishery depends on the lagoons of the state of Sinaloa. Fishermen go out from towns like Altata and Laguna de Chiricahueto in big fiberglass launches for fish and shrimp. Official complaints by fishermen's cooperatives assert that agricultural drainage is killing the fishery, especially for shrimp. Officials at the fisheries school in Mazatlán believe that this may well be an important factor in the decline of the shrimp fishery but say that the government is unwilling to allow studies that might determine whether this is the case. The increase in the number of Mexican, Japanese, and U.S. fishing boats, ever larger and more mechanized, in the Sea of Cortés certainly provides an alternative explanation for the decline of the fishery. Another factor is the change introduced by the withdrawals of the main river running into the sea, the Colorado, for agriculture and urban uses in the United States and the Mexicali Valley—the water reaching Mexico is of poor quality, and in most years the U.S. withdrawals have been so great that the river no longer reaches the sea at all, depriving the sea of nutrients once brought down by the Colorado. Irrigation projects in the Mexican states of Sonora and Sinaloa that reduce the amount of water and natural nutrients running into the sea while delivering agricultural fertilizers, pesticides, and industrial wastes also affect the system. Nonetheless, shrimp fishermen in the towns of southern Sinaloa believe that the annual timing of the fish and shrimp kills they observe suggests pesticides as a major problem. Fishermen told me that they had been used to bringing back tons of shrimp when in many cases now they bring in only a few pounds. This issue, which was becoming heated in 1984, had become even more so by 1988, at which time fishing associations had hired biologists to begin investigations of pesticide contamination of the shrimp fishery (*El Sol de Sinaloa*, February 7, 1984; interviews, Altata, Laguna, 1984; Calderón Vega 1984; López Ballieau 1989).

The state of Sinaloa is a major overwintering spot for waterfowl that spend their summers in the United States and Canada. The organization of duck hunters, Ducks Unlimited, which has a Mexican chapter and active Mexican members, complains that pesticide use in the valley is killing off a significant share of the ducks that winter in the valley, with a resultant loss of revenue to the state from the hunters who have come for decades to hunt wildlife in Sinaloa (*El Sol de Sinaloa*, February 21, 1981). A study by a Mexican graduate student at the University of California, Davis, found residues of some persistent pesticides in ducks of the Culiacán Valley, but unfortunately his study was designed to test for the kind of persistent pesticides that were no longer commonly used in the valley during the time he took his samples. His study emphasizes the importance of a much closer coordination between biological researchers attempting to assess the environmental consequences of pesticide use and people engaged in agricultural research who are more aware of agricultural practices (Morales 1987).

Local health officials worry that the recent epidemic of malaria and other insect-borne diseases in the state of Sinaloa is closely related to the use of agricultural pesticides, creating a more rapid adaptation of mosquitoes to pesticides. Health officials rely heavily on pesticides for control of malaria. While they have found signs of pesticide poisoning among their personnel who work with pesticides, especially malathion, an organophosphate of low acute toxicity, they remain convinced that the overall benefits of continued spraying far outweigh the costs for the present. They are frankly stumped by the question of what they will do when insects have attained higher levels of resistance, as has occurred in southern Mexico (Morales 1987).

There is no way to know what the overall effects of pesticide use in the Culiacán Valley may be. In the best-studied regions of the world, such as the San Joaquin Valley of California, the complexity of the interactions between people and their environment makes it very difficult to know what pesticides may be doing to plants, animals, and human beings. Cancer clusters have shown up in heavily sprayed areas of the San Joaquin, but ongoing studies remain inconclusive. While such classic examples of pesticide abuse as the melancholy tale of mistake upon mistake in Clear Lake, California show that pesticide use can be far-reaching and drastic in its environmental consequences, each pesticide is different and each use of each pesticide has its own consequences. The same can be said for the economic and public health disasters produced by widespread pesticide use on monocrops in the Sudan, Peru, Central America, Malaysia, Indonesia, and Egypt—the particular ecological and health results vary greatly with local conditions. Migratory waterfowl and migratory farm workers are difficult to study, and only a relatively few people show strong concern for either. Pesticide studies have always been highly politicized because of the high economic stakes involved to agribusiness and the sense of outrage among those who consider that they or the environment are suffering from the callousness or greed of others. All of these factors are especially strong in the Culiacán Valley—thousands of tons of acutely toxic pesticides are used in the valley each year and yet there are very few studies of the consequences and nothing like a study that could approach an overall assessment of the consequences. As in other agribusiness regions of the world, a massive experiment is being carried out on human beings and wild nature. Aside from the ethical questions that immediately arise from such an experiment, there is no one systematically collecting data to evaluate the results of this dubious experiment.

What we do know about the hazards of pesticides, however, tells us that the daily conditions of pesticide use in the Culiacán Valley can be expected to cause serious human health and environmental problems. All relevant professional associations and the chemical industry itself hold the conditions of field use that occur routinely in Culiacán to be very dangerous. While growers and government officials deny that such abuse occurs at all, no one argues that it would be acceptable.

Pesticides in Other Regions

Unfortunately, the reckless and abusive use of pesticides occurs in many regions and under a variety of circumstances in Mexico. Conditions very similar to those in the Culiacán Valley prevail in the vegetable fields of San Quintín in Baja California. In San Quintín, growers plant table tomatoes for both the winter and summer U.S. export markets. As San Quintín is only four hours south of Tijuana, transportation costs are lower than in Culiacán, and cheap labor is available from Mixtecs recruited in the Culiacán Valley and from the slums of Tijuana. The main problem for San Quintín farms is the shortage of water, and indeed the vegetable boom began there only after a series of exceptionally wet years that recharged groundwater levels during the late 1970s. But with continual pumping, the pumps have begun to pull up salty water, not surprisingly, since the fields are located within a few miles or less from the sandy beaches of Baja's Pacific Coast. The growers hope to persuade the Mexican government to finance a dam in the region to guarantee fresh water supplies in the future (Kistner 1986; Drum 1987; López Ballieau 1989; direct observations, June 1985, June 1987).

The Baja growers include the Canelos brothers, who have extensive holdings in Sinaloa and the Culiacán Valley particularly and who are said to be one of the most politically powerful families in Mexico. Their ABC Farm label appears on vegetables sold winter and summer throughout the United States. The Canelos brothers make financial and marketing arrangements with Castle and Cook, one of the world's largest multinational agribusiness corporations, headquartered in San Francisco (Kistner 1986).

On a day in June 1985, a crew of Mixtec workers applied a mixture of fungicides and organophosphate insecticides to tomato fields on the Canelos farm. They wore no protective gear. Behind them, as they sprayed the acutely toxic mixture, walked a line of women and children, some of them no more than four years old. The women and children picked discolored leaves and yet living insects from the plants, working in the still visible mist of the poisonous mixture the men were applying.

The head of the emergency clinic at the nearby Seguro Social hospital told me that in the summer months it is typical to see eight to ten cases of organophosphate poisoning per day. Subsequent investigation by anthropologist Michael Kearney and independent journalists found that the growers were working hard to keep pesticide poisoning cases from reporting to Seguro Social, where the very high rates of poisoning might eventually cause political problems. They encouraged workers to go to private doctors under the pay of the growers themselves. They also instituted a practice that shows a remarkable degree of cynicism about the health of their workers (Kearney 1987).

The standard antidote for organophosphate insecticide poisoning is atropine, a belladonna derivative. It is a dangerous substance in its own right, particularly when administered with no regard to weight, physical condition, or medical history of the patient. It does, however, tend to block the effects of organophosphates on the human nervous system, although it may do so for a brief time only before the symptoms of organophosphate poisoning reappear.

Workers and doctors in the Baja region report that foremen supervising field workers have been issued boxes of prepared atropine injections, administering the drug to workers when the foremen begin to notice symptoms of organophosphate poisoning. This may allow the worker to continue to work, and in any case helps to keep him or her from reporting to a doctor or clinic. In this way, the growers introduce a new level of danger to workers' lives to avoid financial loss or embarrassment due to the routine dangers of highly toxic pesticides (Kearney 1987).

To the south of the Culiacán Valley several hundred miles, in the state of Michoacán, are the vegetable fields of the Apatzingan region. There, an epidemiological study conducted by Mexican doctors documented more than a thousand cases of pesticide poisonings in the agricultural season of 1981. Between Culiacán and Apatzingan are rich tobacco fields where Huichol Indian laborers are commonly exposed to roughly the same variety of chemicals as in Culiacán, with crews of workers frequently sprayed directly by aerial applicators (*Boletín de Epidemiología* 1981; Díaz Ramo 1986).

In the cotton fields of lowland Chiapas the extravagant use of pesticides has become notorious, with high rates of poisoning, and a growing public health problem created by the increasing resistance of disease-carrying insects. Malaria, yellow fever, and dengue fever are all on the rise in the humid tropical lowlands of southern Mexico, as in similar cotton-growing regions in Central America. When farmers clear tropical rainforest to open new land for agriculture, the malaria-bearing mosquitoes that live mostly in the high canopy of the forest where most of their animal hosts live are forced to live closer to the newly opened ground. In addition, in the cleared forest there are many more pools of standing water for mosquitoes to breed in than in the original forest. As a result, forest clearing typically means a large increase in malaria and other diseases carried by mosquitoes. The epidemic gains momentum as more people move into the region, providing a large group of hosts living close to one another (Desowitz 1981).

Because of the forbidding character of these diseases and because of the rapid florescence of crop pests, agricultural activities have been sharply limited in the humid tropical lowlands of the Americas. But with the advent of chemical pesticides, both of these problems at first seemed to be solved, and "The Insecticide Revolution" (Williams 1986) brought cotton and labor-intensive row crops into such regions, especially in southern Mexico and the Pacific Coast of Central America. Researchers discovered in analyzing the subsequent resurgence of malaria that agricultural experts involved in pesticide-intensive agriculture in these regions took it for granted that insect-borne diseases would increase as the combination of agricultural spraying and spraying specifically for disease control combined to make for especially rapid and thorough resistance among the key mosquito populations. Public health officials have no ready solution to this rapidly growing problem, a problem that has appeared at least as far north as the former thorn forests of the Culiacán Valley (Chapin and Wasserstrom 1981; Morales 1987).

In the mid-1960s, pesticide use had already created an economic crisis in the fields of northeastern Mexico and South Texas due to the resistance of cotton pests to insecticides. More than a decade of economic decline was partially resolved by the adoption of new Integrated Pest Management (IPM) programs requiring the use of smaller quantities of pesticides, but unfortunately the types of pesticides required are unusually hazardous, raising in a new form a series of questions about the degree to which public health is being sacrificed for agricultural production (Gips 1987; Adkisson et al. 1982).

Pest resistance and secondary pest outbreaks have created serious economic problems in the La Laguna region of north-central Mexico, and as a result private and cooperative farms have looked to diversification as a solution. Unfortunately, the most important form of diversification, dairy farming, has carried pesticide problems from the countryside directly into the urban household. Analysis of dairy products from La Laguna shows high level of pesticide residues. The concern created by the La Laguna residues led to further research, documenting high residue levels in a great variety of widely consumed raw and processed food products sold in urban markets all over Mexico (Albert 1980, 1981).

In El Bajío, one of the richest and oldest agricultural valleys in Mexico, a few hours north of Mexico City, farmers grow strawberries for export to the U.S. market, with enough left over to glut the Mexican urban markets. The transformation of this region into an extension of the U.S. economy has become notorious in Mexico as the result of a widely known book written by the late international agricultural expert Ernst Feder. Feder's 1976 book, *El imperialismo fresa* (Strawberry imperialism), warned Mexicans of the way their best land was being turned over to purposes mostly alien to the welfare of the Mexican people. Since strawberries are a very pesticide-intensive crop, it is not surprising that pesticide poisonings are common in the strawberry fields of Irapuato. But strawberries are not the most important crop in El Bajío—that distinction goes to sorghum and corn grown for livestock feed, displacing to a large extent the corn formerly grown for direct human consumption in El Bajío. While grain crops are usually less pesticide intensive than vegetables or cotton, the boom in livestock feed has been strongly dependent on pesticides, creating a variety of local health and pollution problems. But as with strawberries, the benefits of feed-grain production reach only a few, perhaps a quarter of the Mexican population, who regularly eat meat in quantity, and the consumers of the United States, who enjoy lower meat prices due to the import of large quantities of Mexican beef each year (Barkin and Dewalt 1984).

In the Puebla Valley of Central Mexico, peasants who have been reduced to day laborers through the corruption of the *ejido* form of cooperative landownership spray fields of iris and other commercially marketed flowers with potent mixtures of herbicides, fungicides, and organophosphate insecticides. In January 1984, I observed four men working in such a field, mixing parathion with fungicides and applying it with backpack sprayers. One of the tanks was leaking onto the back of the oldest member of the crew, and while he was trying to fix the leak by tightening the nut that held down a gasket, he told me something of the conditions under which he worked. He said that he was aware of the danger of the materials they were using and knew that it was absorbed readily through the skin. That's why he was worried about the leak. He also said that three members of their *ejido* had died of pesticide poisoning.

"Does that make you worry about working with these chemicals?" I asked him.

"No," he replied, "you just have to be careful."

I asked, "What would you do to be careful?"

He looked up from his work and, looking directly into my face with a pained expression, shrugged his shoulders.

The younger men working with him told me with great bravura that they weren't afraid of working with any kind of chemicals. Neither they nor the older man wore any kind of protective equipment, and all were either barefooted or shod in open huarache sandals.

In the upper, northern part of the same highland plateau as the Puebla Valley, near Huamantla in the state of Tlaxcala, landholders with fairly small farms but a good deal of capital invested grow potatoes for urban markets. They use a variety of the most hazardous pesticides, applied by hired laborers. One such farmer in the fall of 1983 said that they did experience problems with pesticide poisoning from time to time, "But, after all, isn't that the price of any technology? A certain number of injuries, a cost?" An extension agent and researcher with the federal Sanidad Vegetal reported persistent serious problems in the region with pesticide poisonings on potato farms (Trujillo Arriaga 1983).

Large-scale commercial farms in cotton, vegetables, and feed grains account for the lion's share of Mexican pesticide use and abuse, in spite of the conception so popular among Mexican bureaucrats and the public that pesticides are primarily a problem because they are misused by poor, ignorant, and desperate peasant farmers. But while it is untrue that small-scale peasant farmers account for most pesticide use and the most widespread or serious abuses, there are some very disturbing instances of dangerous and destructive pesticide use among poor peasants.

Peasants sometimes fail to understand the difference between use of a pesticide in the fields and in the home. Such was the case of a farmer who brought home the parathion he had used to spray his crops to spray his home for lice and bedbugs. In doing so, he killed four members of his family in the first night after spraying. Even rural doctors sometimes use agricultural or domestic bug sprays for delousing patients, arguing that the task must be done and that most of the drugs available don't seem to really do the job where people have a difficult time maintaining personal hygiene with no running water and precarious living conditions. It had not occurred to one such doctor that the variety of skin diseases and respiratory ailments common in his patients might in some cases have been derived from such pesticide use, in spite of the small print on the pesticide containers that warns against skin contact or inhalation (*Boletín de Epidemiología* 1981).

It is important to realize, however, that while pesticides are frequently misused in peasant farming and rural medicine, there remains an active resistance by some peasant farmers to the use of chemical toxins in the fields. Contrasting attitudes in this regard could not have been made clearer than on a fine summer day in the Oaxaca Valley, the air still fresh from rain that had fallen the night before.

On this day, in June 1986, I observed a young man applying pesticides to a bean crop a few miles north of the city of Oaxaca. He informed me that he was working for his grandfather, who was sitting at the edge of the field trying to puzzle out the label on what turned out to be a can of 45 percent parathion concentrate. The grandfather said that he frequently used such pesticides because "that's the only way you can grow beans around here. We all do it." A quick glance at the bean plants gave abundant evidence of heavy insect damage. I asked why he was growing the beans alone, rather than in the traditional Mexican way of interplanting them with corn and squash, a method that, among other things, provides a considerable degree of protection against pest damage. "That's the way we all do it here.

Look around you here in the valley, we all do it. The extension agent has told us that that's the way to do it, and has shown us how to control the insects with these chemicals."

I asked the grandfather if he had ever had any health problems with the chemical used to spray the beans. "Oh, yes, we all get problems with it. Makes you dizzy. We get crazy in the head. That's why I have my grandson doing it instead of me. But it passes."

Two kilometers north, on the same fine morning, another man of about fifty was cultivating his field with a wooden plow pulled by an ox and a holstein cow. The man wore an aluminum hard hat and was happy to chat for a bit while he rested the animals, though he had to move on soon because he was anxious to turn the soil across the whole field after the rain that had fallen the night before.

"I notice that you have planted corn, beans, and squash together here."

"Yes, I always do it that way."

"Why?"

"Why not? It doesn't hurt any of these three plants to do it this way, and so there is no reason not to. Besides, if the beans did not have the corn stalks to climb, I'd have to set out stakes, which is a lot of trouble, and expensive around here where wood is scarce."

"Are there any other advantages?"

"Oh, of course. Later, when the squash are well along, they will keep the weeds down by shading them out. This makes things easier."

"Are any of the plants good for the others in any way?"

"Oh, well we here think so. That is the way we do things. You see, the corn needs a lot from the soil, and the beans give something to the soil, so it works out."

"Don't you need pesticides to control the weeds, insects, and fungus?"

"Well, yes, I am prepared to use them. I know about them. One year, I even used some herbicides to control the weeds. But they are expensive." The man paused and reflected for a moment. "And, the truth is, I don't like these chemicals. I believe," and as he said this, he chuckled, as if laughing at his own eccentric thoughts, "I believe that these things take the energy out of the plants. It is not good for them."

"So how do you control pests?"

"Sir, the truth is that I don't seem to have much trouble with pests. Now, the problem is, if there isn't enough rain, then there is a problem. Then my family will not have enough to eat through the year and we can become very hungry, things get very difficult. I don't watch for insects in the field much, no sir, I watch the sky for clouds. That's my problem."

The man's fields were remarkably free of any visible pest damage, although certainly one observes pest damage in such traditionally interplanted fields on some occasions. As we will see later, however, many modern pest management specialists endorse interplanting and other traditional practices as having a strong rationale in current ecological theory and have proved their usefulness in field trials. But the sad fact is that even in some of Mexico's most traditional areas, the older wisdom is often abandoned in favor of the chemical approach.

In the mountains of Mexico's southernmost state of Chiapas live the Tzetzal people of traditional highland villages who for centuries have planted their fields in clearings cut from the high altitude pine forests. In the lower altitudes of the same mountain ridges live in the Lacandones, so isolated in the lush tropical rainforests that their existence was unknown to the rest of the world until the 1930s and so traditional in their ways that many scholars consider their culture the best surviving remnant of the classical Mayan culture that ruled the Chiapas forests more than a millennia ago.

Since 1960, upwards of 55 percent of the rainforest of Chiapas has been cut down. The Tzetzales, hemmed in by narrow mountain valleys, displaced by commercial cash crop production, and growing rapidly in population as a result of public health practices and medicine, have begun to move into the rainforests. Lowland peoples, for similar reasons, press in on the Lacandon rainforest from below. Lumber companies profit from the desperation, buying up the timber cut by peasant farmers, encouraging their movement into the forests by legal and illegal means. When the rainforest has been cut and farmed for one to three years, soils become exhausted and unsuitable for crops. The wild grasses and weeds that move into the fields are often unattractive to cattle because in the fiercely competitive tropical environment they have evolved a variety of defense mechanisms: waxy outer coverings to reduce water loss and discourage animals, poisonous chemicals that kill out competing plants and/or grazing animals, thorns, and hardy root and rhizome structures for reproductive vigor, making elimination of undesirable species very difficult. But if such land can be obtained cheaply enough, and if labor can be had for little, it is possible to make large profits on cattle production in spite of such obstacles. Big cattle corporations take advantage of such opportunities to move in behind the peasant farmers, usually getting the land in return for some combination of small cash outlays, legal maneuverings, and physical intimidation by thugs. Together, commercial cash cropping that dispossesses peasants in their traditional areas, timber companies, cattle companies, and the unrestrained growth of peasant populations are a powerful engine of forest destruction. Pesticides also play a part (Nations 1984).

One morning in early summer, 1986, I was on the way to the delightful food and crafts market of San Cristóbal de las Casas, the highland capital of the state of Chiapas. I noticed men carrying four-liter plastic jugs with labels filled with small print. On stopping one of these men, I discovered that he was a Tzetzal who was clearing plots in the transition zone between pine and tropical forest areas— he had found weed control a serious problem. The chemical in the jugs was paraquat, a powerful, nonpersistent, acutely and chronically toxic herbicide.

"Yes," he said in answer to my questions, "it is difficult to clear such plots and farm them without some method to control the weeds. We use this chemical a lot. It is a big help. It allows us to farm an additional year sometimes, before the plot must be abandoned. I don't know how we could do without it. This is a very hard way of life, sir."

"Are you aware that this is a highly poisonous substance?"

"I believe that I have heard that somewhere, yes, someone has said so. Can you tell me about that?"

I told him that paraquat was a very poisonous substance and that he should avoid all contact whether through breathing or swallowing it or through having it on his skin. I warned him that if he took in enough of the pesticide it could cause immediate illness or death, and also that if he used it frequently over a period of months or years, he might lose much or all of his ability to breathe as a result of tissue growth in the lungs.

"Nobody tells us much about it. I am glad to know this. I will be careful, because this product is very important to us, we don't have much money and this helps because we have to open new land."

Other farmers, a woman selling pesticides out of a farm supply store, and nurses in clinics in nearby villages confirmed that paraquat was being used heavily in weed control in newly cleared *milpas*. One old man near the market in San Cristóbal used his bare hands and paper tissues to clean out a backpack spray tank, tossing the tissues into the gutter where he worked.

An ongoing investigation by researchers at Stanford University has shown that farmers in highland Chiapas also use paraquat heavily on their traditional lands, and that the higher production possible in the initial phases of the use of paraquat makes it possible to reduce the amount of new deforestation for crops. However, serious localized health and environmental problems may arise because of the increased pesticide use (Collier and Mountjoy 1988).

There are many regions of Mexico where pesticides are an important part of agriculture but where the nature of the pesticide requirements makes for less obviously abusive situations. Such are the grain fields of Sonora and Sinaloa and the sugarcane fields in the Culiacán Valley, where federal agencies and private growers have worked out IPM programs that keep pesticide applications down to two or three per agricultural season. The careful planning of planting, mechanical cultivation, irrigation, and harvest schedules can interrupt the growth cycle of the pests. The use of parasitic wasps to prey on pest species and releases of sterile males reduce the populations of insect predators to acceptable levels. Selection of pest-resistant varieties minimizes the need for control measures. Through such classical control techniques, it is possible to keep pest loss down while using relatively small quantities of pesticides. This, the bright side of Mexican pesticide use, brings researchers from many nations to learn from proud Mexican agronomists who are noted for their successes in integrated pest management in many crops (Restrepo and Franco 1988; Ussher 1984; Aviles González 1984).

Unfortunately, cotton, vegetables, some fruits, and some grains present more difficult control problems, especially where they are grown in large-scale monocrops and where climatic conditions during the growing season favor rapid pest growth. Commercial farmers with good lines of credit and command of a large cheap labor force are able to put chemicals, people, and the land together in a profitable but unstable and dangerous combination.

It appears that there are few regions of Mexico where pesticides have not to some extent been incorporated into the lives of the people. In many regions, such as Culiacán, Apatzingan, El Bajío, and the cotton fields of La Laguna in the north and Soconusco in the south, agricultural pesticides have become an integral part of economic life and an ongoing public health and environmental problem. The resurgence of insect-borne diseases and resistant agricultural pests in these regions emphasizes that the combination of agricultural and public-health spraying can make for especially radical and disturbing ecological results.

The road to Culiacán promised great riches. The booming agribusiness of the state has made good on the vision. But boom and bust cycles have long played an ominous and relentless part in Mexican history, from mines of silver and gold to sugarcane and cotton plantations to oil wells. In the Culiacán Valley, the boom is new—commercial agriculture has only been important there for a matter of decades—but the terms of the bust are already discernible.

3
Doctors and Bureaucrats

Sitting at his desk in the emergency room of the rural clinic near Culiacán, Dr. Pérez said, "No, I can't recall a case involving a man named Ramón González. That name doesn't mean anything to me. But 1981, no, I wouldn't know about that. I wasn't here then. I don't think we could help with that. But it's not unusual. We have fatalities from pesticide poisoning, no question about that."

Dr. Pérez was interrupted by a call from the emergency room. "Yes, I'll be there shortly, but I'll wait until you've had time to prepare her. OK? Sure, let me know." He hung up the phone.

Dr. Pérez is a handsome man in his thirties, stocky but not fat, with heavy facial features and a self-confident manner. His dark eyes focus unerringly on the person he is talking to, seemingly penetrating deeply into the thoughts of the other.

"You were saying that fatalities were not unusual," I reminded him.

"No. For instance, last year, 1983, there were three fatalities here. That was while I was still working regularly in emergencies. But since, I've been the head of teaching here, and I don't have such close contact with the cases."

I asked what he could tell me about the three cases.

"Tell you? Well, three young women. They were brought here after working in freshly sprayed fields, or after they had been working as sprayers. There wasn't anything we could do for them, finally. The acetylcholinesterase level was severely reduced. We gave them the standard atropine injection and care. Because the cases were severe, we sent them on to the Culiacán hospital, where they deal with the life-threatening cases. All the young women died within three or four days after they showed up in our clinic here in Costa Rica."

"How old were these women? Where were they from? What else do you know about them?"

"They were about fourteen to fifteen. From Oaxaca. Indians. All three were pregnant, in their first trimester, in fact. One thought she was miscarrying. The other two didn't know they were pregnant, or at least didn't tell us. I discovered it doing the exam."

"Anything else about the state of their health?"

"Anemic. Malnourished, of course. But then, that's not really a particular characteristic of these women. All the Oaxaquitos [a diminutive of the term for someone from Oaxaca—the diminutive form is frequently used in Sinaloa to describe Indian people from the state of Oaxaca] who come in here, almost without exception, are anemic. We do the blood test for anemia the first thing, and it's almost always there. Well, they don't get enough to eat, or not the right things. So." Dr. Pérez stroked his forehead with his hand for a moment while looking down at the floor.

"What about their age? Is it typical of the poisoning cases?"

"Definitely. From about fifteen up to twenty-five. That's the average in the poisoning cases. It's a young person's problem in our experience."

"Do you know why?"

"No, I couldn't say with certainty. I suppose because they are put to work spraying more often. Maybe they aren't as careful. I don't really know. But that pattern is very strong."

"Anything else?"

"I can't say for sure in these cases, I don't recall specifically, but I think you could assume there were some intestinal parasites and some degree or other of intestinal infection. I seem to recall that. In any case, again, it's just to be expected among the Oaxaquitos."

I asked Dr. Pérez how I could find out more about these cases. He told me that I would have to go to the large federal hospital in Culiacán, where the young women died. He also suggested that it was there that I would be able to find statistics on pesticide poisoning cases in general.

"Talk to the director, Dr. Jorge Millan. He should be able to tell you," suggested Dr. Pérez (Pérez 1984).

The hospital in Culiacán is a beehive of medical and bureaucratic activity. For blocks around there are food and refreshment stands serving personnel and visitors. For such a large institution, the office of the director seemed a surprisingly open, friendly, and informal place. People with shoes that needed major repair and polyester shirts that had been washed and worn many dozens of times compete with doctors and hospital officials, with fair success, for the attention not only of the secretary in the outer office but of the doctors going in and out of the director's inner office as well. I could not imagine a similar situation in a large hospital in the United States, where the director would surely be protected by a more forbidding phalanx of receptionists and secretaries. Here, the middle-aged, fashionably dressed secretary seemed to manage traffic and tempers more than paperwork. She took my card into Dr. Millan's office, returned, and promised a short interview with the director in a matter of minutes. But as she took others into the office, my case was apparently discussed further; she asked me two more times to elaborate on my purpose in talking to Dr. Millan. Finally, I was taken into the office and introduced to Dr. Millan, a handsome, rugged-looking middle-aged man, self-assured and articulate.

I told Dr. Millan that Dr. Pérez of the Culiacán clinic had recommended I see him, and that I was looking for further information on the numbers of pesticide poisoning cases that occurred and on the severe cases, including deaths, that had been referred to the Culiacán hospital.

"I'm afraid there would be no way to capture that data, we simply don't keep that kind of information. And we wouldn't be able to share with you the details of any individual cases. As far as I know, there have been no deaths. But I couldn't assure you of that—I have only been director a few months." He put his fingertips together in front of his face as he leaned back in his swivel chair.

"Is there anyone who could tell me more?"

Dr. Millan stroked the large class ring on his finger with a thoughtful expression on his face, hesitating before he answered. "You see, you are really going after the wrong issue. The real problem is ecological, all the chronic and long latency period diseases like cancer as far as humans are concerned, and all the many unknown ecological effects for other species and soils, and so forth. That's the real question. As a matter of fact, we don't worry about the acute poisoning cases because there are so few of them."

I replied, "The rural clinics in Campo Gobierno, Costa Rica, and El Dorado seem to have twenty to thirty cases each every month, and perhaps more, according to their doctors and their own records. These, of course, are the records they pass on to you."

Dr. Millan gazed out over the city of Culiacán through the big window. "Well, perhaps so, but as far as serious cases sent here, we don't see many. You see, I've been interested in these problems, too. I have given a few papers on pollution myself, but they weren't well received here in Mexico. Everybody thought they were Malthusian, or a little catastrophe-mongering. So, that has passed."

"Getting back to the chronic diseases and the latent problems like cancer, do you have any studies or information on those problems with respect to pesticides?"

Dr. Millan smiled. "No, I'm afraid the epidemiology of it, all the statistical problems and time lags, make it impossible for us to know anything about that, much as we would like to."

After trying different approaches to getting more information out of Dr. Millan with no success, I asked where I might be able to find any further information on the question of pesticide poisonings in the Culiacán Valley.

"I'm afraid the only thing I can do for you that would be of any help would be to refer you to Salubridad y Asistencia. They are the ones who keep public health records. Everything is reported to them."

Dr. Millan gave me the name of a doctor to talk to at Salubridad y Asistencia. After three visits to Salubridad y Asistencia, always finding the doctor was not in, I was able to talk to the jovial and intelligent woman in a house dress who was in charge of gathering and computing the statistics on health for the state of Sinaloa. She was very helpful, but there was an insurmountable problem. The questionnaire that her agency circulated to hospitals, clinics, and doctors' offices, with several dozen listings each specifying a particular disease, accident, or condition, had no listing for pesticide poisonings. She explained that data from the clinics that showed a diagnosis of pesticide poisoning would therefore be listed under "other, unspecified," thus making it impossible to document the number of cases of pesticide poisonings in the state of Sinaloa. "There really should be a space for that on the form," the head of the statistical department told me.

The large hospital in Culiacán that Dr. Millan directs is part of a network of federal clinics and hospitals in the Culiacán Valley. As a prosperous region producing a sizable portion of Mexico's foreign exchange earnings and taxes, and with a corresponding degree of political influence in Mexico City, the Culiacán Valley enjoys more rural health care facilities than any but a few other places in Mexico. There is the Culiacán hospital, the hospital-clinic in Costa Rica with a multistory building and its own medical training program, and the smaller rural clinics in El Dorado and Villa Juárez, usually referred to locally as Campo Gobierno. All of these facilities are busy, crowded places, and all receive patients who have become ill or had accidents while living and working in the fields of the Culiacán Valley. In Costa Rica, El Dorado, and Campo Gobierno, pesticide poisoning cases are treated as a daily, routine matter. Many of the doctors working in these rural clinics are doing the mandatory social service required of them by Mexican licensing law before they enter private practice or regular employment. These young men and women tend to be hardworking and enthusiastic. They are given to commenting on the fact that rural people have special medical problems and traditionally suffer from a lack of adequate medical and public health services.

Dr. Luis Velázquez works in the emergency room in the Seguro Social clinic in Campo Gobierno, a settlement surrounded by vegetable fields in all directions. He is a short, slight, cheerful man, eager for conversation. The day before I arrived in his clinic, he had been quoted in the government-subsidized newspaper, *El Sol de Sinaloa* (February 15, 1984), saying that he thought the growers must take responsibility for the large number of pesticide poisoning cases coming into the clinic. "It is irresponsible on the part of the bosses to contract children as young as ten years old for agrochemical application tasks which require knowledge, great caution, and special equipment, but older people, who have already suffered the consequences, don't want to do it." The day after this statement by Dr. Velázquez appeared in the Culiacán daily, along with a photograph of him and a nurse attending an eleven-year-old girl poisoned by the organophosphate pesticide parathion, Dr. Velázquez seemed more reluctant to throw blame on the growers.

"No, we have their complete cooperation. There are no problems between us and the growers. There is excellent cooperation and they do what they can to avoid these problems."

Dr. Velázquez introduced a thin, intense young woman dressed in jeans and sandals who was in the emergency room with the doctor, his nurse, and another young woman when I was invited in. She was a social worker employed by the growers' association of the Culiacán Valley. The other young woman, less aggressive and dressed more conservatively, was a social worker employed by the federal medical system that runs the clinics.

"How many pesticide poisoning cases do you get a day, on average?" I asked.

"From one to four, more or less. Sometimes none. Here, take a look at our statistics for last month. Laura, go get them, will you please?" he said to the nurse.

The growers' social worker said, "It's hard to avoid. We try to tell them how to protect themselves, to wear the right clothes, to wash the tomatoes before they eat them, and to exhale when the airplane sprays over them while they are working the field. But they don't remember or they don't pay attention."

I asked if the workers were offered protective equipment and given proper training before they began working with pesticides.

"Yes, of course, the main problem is that the people won't wear the protective equipment offered to them because of the heat and discomfort. They would rather go barefooted, for example." (In interviewing twenty-five Mixtec workers on this point, I could not find one who said he or she had ever been offered protective equipment. Two in the San Quintin area of Baja California said that they asked for protective gear and that the foremen laughed at them and ridiculed them. In contrast, every supervising agronomist or foreman when directly questioned on this point answered that workers had been offered equipment but refused to wear it.)

Dr. Velázquez smiled. "You see, the people don't lack intelligence, but they have no idea of what they are working with or the dangers involved. Many of them hardly speak Spanish. They don't take precautions because of their ignorance."

I asked, "What are the usual symptoms you see associated with pesticide illness?"

Velázquez replied, "Just what you'd expect: headache, nausea, vomiting, difficulty breathing, sometimes convulsions, complete loss of appetite, tightening of the muscles around the thoracic region. Sometimes we see serious skin problems associated with it, sometimes not. We don't see many deaths, there haven't been any in this clinic since I've been here, almost a year. But then, I wouldn't know for sure about our cases, because the severe ones are passed on to the big hospital in Culiacán, where they have better facilities. They stay here only for twenty-four-hour observations, or, in a few cases, three days as an absolute maximum. If they came here and were passed on to Culiacán and died, we wouldn't be aware of the death."

The nurse, the social workers, and a couple of patients were standing in a semicircle around me and the doctor, listening and contributing their ideas when they liked. Dr. Velázquez joked easily with everyone.

"And tell me, what do you think of the people of Sinaloa?" he asked me with a grin.

"I'm saving my observation till I have enough to generalize with," I said.

"Good, safe answer. But don't you think we're different from other Mexicans? We have the reputation, a very strong reputation, of being braggarts, very egalitarian, no respect for hierarchies, even aggressive. Would you agree?"

"I can see what you mean. People here seem less formal to me, certainly."

"Yes, but more than that. You see, the theory is that because there were no big settled Indian populations here, no high civilizations like in central and southern Mexico, there never was a real Spanish conquest. There were no big mines or plantations here in the colonial period. So you don't have the long time of conquest, that mental conquest that changed the others. Here you get a lot of wild men."

"Do you see that difference showing up in the clinic?" I asked.

The social workers, Alicia and Mari, laughed. "You sure do," Alicia said, and Mari and she rolled their eyes and chuckled.

Velázquez laughed too, and added, "As far as pesticide poisoning, you'll notice that we hardly ever get anybody from Sinaloa. They know better than to do the dangerous work or to eat the contaminated vegetables. They've had the experience. So what you see are people from the other states, Zacatecas, Durango, usually Oaxaca, of course. And children. They don't know and they are more physically vulnerable to the poisoning. A high percentage are children."

"They just don't know what they are doing," Alicia, the growers' social worker, said.

Mari nodded agreement and added, "You can see that because people come in here sick, get some treatment, and go right back to fumigating work."

Alicia jumped in forcefully, "No, that doesn't happen, that's not allowed, there's no way that could happen."

Dr. Velázquez raised his eyebrows. "That may be. But look, we see it in here all the time, the very same people coming back the second or third time."

The nurse, Laura, returned with "all the statistics we have available at the moment." They covered the previous two weeks (the first two weeks of February 1984) and included all the cases treated in the emergency room. The most common diagnoses were gastroenteritis, colitis, "strange sensation in the body," "finger cut while working," in that order. There were eleven cases of pesticide poisoning during the two weeks. Dr. Velázquez excused himself while I looked over the figures and asked Mari to give me a tour of the clinic when I was done.

Mari showed me through the small building: the archives, small consulting rooms, observation rooms, and, in a separate building in the back, the pharmacy and small apartment space for doctors doing their year of service to live in, if they wish to live on the premises. The majority of people waiting to be seen, perhaps twenty-five of the thirty-five people, had the facial features, body size, and dress typical of Mixtec and Zapotec Indians from Oaxaca. Several babies sucked on plastic bottles, some filled with milk and some with juice. Many people, especially the women, preferred to sprawl on the floor rather than sit on the lines of blue plastic chairs. In contrast to their well-scrubbed and vigorous appearance in the villages in

Oaxaca, most of these people looked unkempt, dirty, and demoralized. But they were here, of course, because they were sick.

Back in the clinic in Costa Rica, Dr. Pérez had told me, "You see, the problem of pesticide poisoning interests me and my colleagues very much. But it is a difficult problem for analysis. There are the problems of diagnosis—pesticide poisoning symptoms are often easily confused with other diseases. One would like to sit down and design epidemiological instruments, statistical instruments that would give us a measure of things. But these people are migrants. They can't be studied over time. They won't hold still. And there is the problem of their illiteracy, and their reluctance to cooperate with strangers in the Culiacán area here, where they are never comfortable, never certain about what is going on. They are reluctant, for instance, to report any medical problem, because it very often means they will be dismissed from work for a time, maybe lose their job for the season. They are a people accustomed to a great deal of suffering—often they simply bear it rather than complain. Complaints can cause problems, after all. And look at the statistics of the clinic here. You will see such a wide variety of diseases, and you will note that pesticide poisoning is not among the top ten. But should it be? That is, what is the role of agrochemicals in the overall health situation in which these people live? We do not even begin to know how to study that question in a reasonable way" (Pérez 1984).

Epidemiologists have been asking such questions about pesticide poisoning all over the world for several decades now, and the accumulated results of hundreds of their studies are not very satisfying. Nowhere are the epidemiological questions simple or straightforward. Translating laboratory results into meaningful field studies among human beings in complex biological and social environments is a terribly difficult task. It is certainly absurd to believe that the underfunded and overworked medical personnel of the Culiacán Valley should be able to answer questions about pesticide safety that remain unanswered anywhere in the world.

The Mexican government is aware of many of the problems of pesticides in the Mexican landscape. Their response to criticism of pesticide abuses tends to be consistent and predictable: all technologies carry with them certain dangers and these dangers are likely to be most evident during the early years of the adoption of a technology in a region or nation previously unfamiliar with it. After a technology is transferred to a new social context, it takes time to put in place all the education, training, and control measures necessary to make the technology safe. Thus, current problems are merely a reflection of that transitional phase in the diffusion of chemical pest control techniques. The educational and regulatory efforts of the government will guarantee that abuses will soon be brought under effective control.

The primary agency responsible for the control of pesticides in Mexico is Sanidad Vegetal, a department of the Ministry of Agriculture. The ministry's full name in Spanish is Secretaría de Agricultura y Recursos Hidráulicos (Ministry of Agriculture and Irrigation, SARH). Sanidad Vegetal (the English equivalent would be "plant protection") has a network of regulatory personnel working in every agricultural region of the nation. Sanidad Vegetal is also responsible for extension work in pest management, which means that its personnel regularly recommend that pesticides be used. The extension agents are given control of quantities of pesticides they may give to farmers when it is considered in the public interest to do so. An obvious question that arises in Mexico again and again is whether Sanidad Vegetal is more interested in promoting pesticides or in regulating them.

The man in charge of Sanidad Vegetal in the Culiacán Valley in 1983–84 was Ing. Jorge Ricardo García Ussher. Ing. García Ussher (who is commonly known as Ing. Ussher) has a degree in parasitology from the National Agricultural University in Chapingo and had worked in Mexico City, the state of Jalisco, and, for the last five years, Sinaloa. He is a handsome, blue-eyed man, courteous and seemingly direct.

Ing. Ussher wanted a good deal of information about who I was and the intention of my study. As with others in the Culiacán Valley whom I interviewed, he wanted to be sure that this was a "scholarly" and not a "journalistic" investigation. He assured me that there were no serious environmental or public health problems with pesticides in the valley and that to his knowledge there had been no fatalities caused by pesticides in the valley (Ussher 1984).

"There are two very natural forms of control over the use of pesticides. One is the pressure from those who import the products into the United States, who make sure that there are no residue problems that would represent a threat to their ability to market their products. The second is the fact that pesticides are quite expensive, with prices rising very rapidly in recent years. This ensures that producers are never going to knowingly waste a product through excessive or improper use.

"Then, of course, there is the law, which is really very severe. Any damage done by pesticides must be compensated by the user, with double and triple punitive damages where any negligence or impropriety can be proven. For pilots who spray outside of the contracted areas, five years in prison is the penalty if there are demonstrable damages, two years if there is no demonstrable damage."

"Can you cite cases where these penalties have been exacted?"

Ing. Ussher looks surprised, thinks a moment, and replies, "No, I can't, precisely, I think, because the natural forms of control and the law itself would make it simply irrational for anyone to contemplate misuse."

"And yet," I say, "only yesterday, between here and Moroleón, no more than a few hundred yards from here, I saw very hazardous pesticides handled in complete disregard of all Mexican and international standards."

Ing. Ussher looked grave. "In our work here, we very seldom observe or are informed of such abuses. Of course, that doesn't mean they never occur. We are dealing with a relatively new technology here, and there is of course a very real necessity for a continuing and serious educational effort to see that people more and more realize the dangers and the potential penalties that the law provides." He pulled from his drawer a variety of publications issued by Sanidad Vegetal and showed them to me one by one. Some were technical in character, designed to inform growers and agronomists of the pesticides approved for particular crops and recommended by the local government experiment station as the result of recent tests. Others were in comic book form, with very graphic illustrations showing proper and improper use, with a text designed for barely literate readers. Similar technical and nontechnical literature issued by the federal experiment station and the Mexican chemical industry was among the official Sanidad Vegetal literature. (*Fitofilo* is a technical journal published by the agency meant to carry the most important news and articles on pest management problems in Mexico.)

"We use these publications, training seminars, and on-site visits in the field to try to improve the level of awareness and skills of people handling pesticides. One cannot expect perfection, especially in the use of such a new technology. In addition to such efforts, we have the reassurance that we simply experience few demonstrable health or environmental problems related to pesticides in the Culiacán Valley. Those that do occur happen as a clear result of the illiteracy and low educational level of the workers, and this must be seen, not just as a problem of pesticides, but as a problem of national development that can only be resolved by continued improvements in productivity and the educational efforts that can be afforded by such improved productivity."

"How do you explain the lack of problems given the low levels of training and skills you refer to?"

"The basic thing is that we are not authorizing the use of chemicals that present serious risks. In fact, we rely very closely on you in the United States. Our approval and standards follow closely on the decisions made by the EPA in the United States. If you believe we are running unacceptable risks, then look at your own standards and revise them. You will see that we follow very close behind. In the last five years, we have had no problems with our products entering the United States due to residue inspections at the border. Our coordination with you in the United States has been quite effective."

Ing. Ussher went on to explain the system of Sanidad Vegetal committees set up to determine what chemicals are authorized on which crops and what residue standards are to be applied to marketed crops. Each major agricultural region, such as the state of Sinaloa, has a regional committee made up of growers, personnel from Sanidad Vegetal, and other representatives appointed by Sanidad Vegetal from the agribusiness community. Each year, this committee considers what pest problems are most likely to be encountered in the coming year, what pesticides have been effective, what experimental results are available from experiment stations, what chemicals are available from sales agents, information on resistance problems, environmental hazards or health risks associated with particular pesticides, and various economic factors. On the basis of this information, the regional committees make recommendations on what pesticides should be authorized for what crops and what residue levels should be permitted. A national Sanidad Vegetal committee, with interest group representation similar to the regional committees, considers the recommendations of the regional committees and makes final recommendations for each region and for the nation. These recommendations then are meant to regulate pesticide imports, manufacture, and use; for instance, a pesticide that is not on the authorized list cannot be imported during the year governed by the list. Ing. Ussher apologized for the fact that a published version of this list was not available for the current or upcoming agricultural seasons due to budgetary problems of the federal government in the ongoing debt crisis. "But growers may consult with us and we can let them know what is on the list and what is not."

Ing. Ussher cited a number of successful programs of Integrated Pest Management recently developed in wheat in the Culiacán Valley and in cotton in the northern part of the state of Sinaloa. In wheat, reduced pesticide use was achieved by the simple insight that beetles encountered in the early season would succumb to natural predators before serious damage was done without spraying—before, the presence of the beetles in the young crop had been taken as a sign of a need for spraying. In cotton fields in the Fuerte Valley, the use of sex attractant chemicals, pheromones, and bacillus thuringiensis, a bacteria that can be applied to infect and kill caterpillar pests, had made it possible for some farmers to reduce pesticide applications from ten to twelve per season to three per season. Ing. Ussher was hopeful that similar successes might be expected for the use of pheromones and bacillus sprays in vegetable crops. (As noted above, specialists at the federal Instituto Nacional de Investigaciones Agrícolas [INIA] experiment station do not share his optimism in this regard, remarking, for example, that bacillus sprays were likely to be made ineffective by continued high rates of pesticide application within the field or by drift from neighboring fields, and that apparent successes with pheromones were really simply features of natural cycles within pest populations.)

Ing. Ussher arranged for a day of field visits in the company of one of his field inspectors, Ing. Vargas. Ing. Vargas took me along roads I already knew well, visiting farms and fields where I had observed dozens of pesticide applications and routine violations of Mexican law. But on the day's trip with Ing. Vargas, in the company of a young biologist from another government agency who later pronounced the trip "fraudulent," I did not observe a single example of a pesticide spraying operation of any kind. We visited a very large, well-fenced, and well-guarded farm to which I had previously been denied entry. On this farm, Ing. Vargas showed us traps hung from stakes in the fields. The traps were baited with pheromones, and each contained a number of insects stuck to the paper of the trap. There was some dispute between Ing. Vargas and a foreman about whether the traps were being used to monitor pest populations or whether they were meant to actually control the pests. Vargas, contradicting the foreman, insisted they were for actual control. Later, an agronomist on a neighboring farm told me that they were most certainly meant to monitor pest populations and that no one imagined that such traps could successfully control pests under conditions prevailing in Culiacán (Sánchez 1984).

The most remarkable thing about the field visit was that it was the only day out of forty that I have spent during the pesticide application season in the Culiacán Valley when I did not observe an application in progress—on a typical day during the winter months I had seen several applications, often ten or more. It was also the only day when my presence was previously known to Sanidad Vegetal.

Interviews with Sanidad Vegetal officials in Mexico City tended to emphasize the same themes as Ing. Ussher: environmental and health problems caused by pesticides were certainly minor, and to the extent they existed they were an unavoidable feature of the early stages of technology transfer; ongoing educational and training programs would resolve what problems existed, and Sanidad Vegetal was committed to advancing all available IPM programs as they became available in order to reduce pesticide dependency; and the U.S. Environmental Protection Agency rulings, procedures, and techniques are the basic guide that Mexico intends to follow (Corillo Sánchez 1983; Morgado 1983).

When pushed about abuses observed in Culiacán and the nearly total reliance on chemical pest control, officials admitted that the problems of vegetable and cotton production in tropical and subtropical areas tended to be especially severe, particularly where large areas were devoted to a single crop or crops sharing major pest species. In no case were they willing to discuss whether such chemical-dependent, monocrop agriculture was appropriate to Mexico's needs compared to other possible alternatives. The patterns of agricultural production are taken as a given by Sanidad Vegetal officials.

Many of the top officials of Sanidad Vegetal have master's and doctoral degrees from major North American universities such as the University of Wisconsin, the University of California at Riverside, the University of Florida at Gainesville, and North Carolina State University, or from Mexico's own internationally renowned National Agricultural University at Chapingo. Like Ing. Ussher, these men, and a few women, are clearly intelligent and articulate. They express great concern for the welfare of Mexico's rural people and for the urgency of the development task.

Their denials and reassurances with regard to conditions routinely observed in Mexican fields are, in a sense, hard to judge. Mexico City is physically far away from the areas of most heavy pesticide use and abuse, and when these officials visit the fields, they are apt to do so under the same conditions as my tour in the company of Ing. Vargas. One wonders, however, if it is not the psychic distance between Mexico City and the miserable *campamentos* of Culiacán that is not more significant than the physical distance—from the cacophonous streets, cosmopolitan cultural institutions, and intensely competitive bureaucratic offices of Mexico City, it may be difficult to imagine the experience of the miserable *campamentos* of the Culiacán Valley or Soconusco or to remember the *campamentos* even when they have been experienced. While the national debt may seem a remote issue to a farm worker with an airplane spraying deadly chemicals over him while he works, in Mexico City the apparent imperatives of the nation's competitiveness and credit worthiness in world markets may seem to be the more immediate issues.

But Ing. Ussher, who says the same things as the officials in Mexico City, must drive to his office in a rural area of the Culiacán Valley each day, and he certainly must observe the conditions he emphatically denies. And his superiors in Mexico City hardly seem the naive sort of people who would be fooled by Potemkin villages.

Ingeniero Javier Trujillo, a young agronomist working in Huamantla in the central highland state of Tlaxcala, provided a critical, insider's view of the workings of Sanidad Vegetal. In 1983 he was the technical director of Sanidad Vegetal in Huamantla, in charge of extension as well as of various research projects. He is a tall young man in his late twenties, with a warm smile, a ready humor, and a remarkable self-assurance. Although he has been doing controversial work, he was recognized in 1983 by the national directors of Sanidad Vegetal for innovative efforts toward adapting extension and research to the precise conditions of local agricultural systems. In 1987, he took a position as a professor in the National Agricultural University in Chapingo after receiving his Ph.D. from the University of California at Berkeley in pest management (Trujillo Arriaga 1983, 1985, 1987).

When I spoke to him first in Huamantla, he told me, "After my technical education, I began working here in Huamantla. I soon realized that with respect to the real problem raised by peasants in the region, we simply did not know how to fight pests. There was a lack of method, a lack of any quantitative technique by which to evaluate pest problems. When we realized that we didn't have the information, we also realized that the only way to get it would be to generate it by ourselves."

I asked, "Doesn't INIA do this kind of work?"

"Not really. What they do is to look at a specific pest and then try different chemicals on it. This perspective is much too narrow to be able to actually use the results on specific pest problems encountered in the field. With the INIA work all you can do is look up a pest and find a chemical recommended for it. This does not tell you, for instance, what population level constitutes a pest problem, it doesn't give you any clues about special local conditions, associated crops or plants or predator populations that influence the pest. There was not in INIA a political tendency favorable to a more ecological approach." He went on to tell a story of an INIA researcher who presented a paper at a national conference on pest control through ecological analysis of "agro-ecosystems," showing encouraging initial pest control results and very promising insights to guide future research. When Trujillo expressed an interest in the work, the researcher told him that the work was going to have to be abandoned—

not only had the funding been taken away from him but he had been threatened with job loss unless he directed his research to more conventional and "productive" areas. Trujillo concluded, "We had to develop a much more technical, scientific approach than INIA could offer.

"One advantage of Huamantla was that no one pays much attention to what happens here because this is a poor region with very little influence on the national economy, unlike a region such as Sinaloa or La Laguna, where national directives tell people exactly what to do because the results are so important to the national economy and so politically important as a result.

"Another advantage was an ecological one—the Green Revolution had never had so much impact here as it did elsewhere. There were farmers here who had never used pesticides, for example. Because of the needs we saw and the freedom we had to follow our own judgments, we tried, then, to create the beginning of a new tendency with respect to research and the application of pest control techniques.

"By coincidence, this work coincided with some changes in Sanidad Vegetal. In 1981 and 1982, Sanidad Vegetal published the *Plan Nacional Fitosanitario,* mandating an integrated pest management approach nationally. This plan was largely the work of two men who had long been critical of Green Revolution, chemical-dependent approaches: Dr. Alfonso García Escobar and Dr. Felipe Romero, both with Ph.D.'s from prestigious North American universities. This plan was very important to us because it supported our work and showed that what we wanted to do was not simply a personal whim."

Ing. Trujillo explained that research funding was practically impossible to obtain in Mexico, as the Sanidad Vegetal funds go almost entirely to extension work. Trujillo's major professor at the University of California at Berkeley helped find some small grants. Trujillo added with a rueful smile, "Also, people here used their own money, even though the salaries here are very low. People on the staff here in Huamantla would throw in a day or two's wages to buy equipment or supplies, because that was often the only way to continue with the work.

"The reason for political support of our work from Sanidad Vegetal is that there was a recognition that we had to solve various problems, particularly, pest resistance and the ecological balance between pests and predators. It was well recognized, officially, that these constituted problems that had to be solved. These problems had been identified in cotton in La Laguna, Soconusco, Sonora, and northern Sinaloa and in vegetable crops in Sinaloa and Apatzingan."

Ing. Trujillo added that in his private opinion, not a view shared by his colleagues, the fact that pesticides are produced mostly from petroleum and that petroleum supplies will one day be exhausted created an urgent need for a turn away from pesticide dependency.

"You have not mentioned problems associated with workers' health. Are such problems recognized as a matter worthy of attention?"

"Up to now, no. The position of Sanidad Vegetal is that if the chemical in question is on the authorized list and workers are hurt, then the problem is misuse, mishandling of the chemical. Therefore, it is not a matter of direct concern to Sanidad Vegetal but a concern between those injured and those responsible. [This position was indeed repeated to me on several occasions by Sanidad Vegetal officials.] As for tolerance for consumption listed on the authorized list, it is universally understood in Mexico that the tolerances there are exclusively meant for export products."

"What relationship does the manual of authorized pesticides have to reality as far as what chemicals are actually used for what crops?"

Ing. Trujillo answered without pause, "Almost none. It's largely something that the agronomist can use when somebody asks what they should use. He looks it up in the manual and says, 'These are the chemicals authorized for this crop.'"

I asked Ing. Trujillo how it was that chemical approaches had become so entrenched in Sanidad Vegetal. He replied with his own complex history of the Rockefeller-supported Green Revolution program in Mexico. The Green Revolution program, dating from a 1941 report to the Rockefeller Foundation that amounted to a research agenda that was to be followed for nearly two decades, focused on grain crops that are not now the most pesticide-intensive crops. But Ing. Trujillo, like many others in Mexico, believes that the chemical-dependent approach promoted as part of the Green Revolution had a generalized effect on the development of Mexican agriculture that went far beyond the technologies associated with the most famous of the "miracle seeds" promoted by the Green Revolution researchers. Among other things, it had a far-reaching impact on the agricultural and research establishments in Mexico, especially INIA, the network of experiment stations set up under the Rockefeller program and taken over by the Mexican government in 1961, and Chapingo, the National Agricultural University, whose faculty and research commitments were shaped by the Rockefeller program beginning in the 1940s.

"When I went to Chapingo," Ing. Trujillo continued, "I noticed that in spite of the enormous prestige of the place there were seriously flawed approaches to pest management problems. The main idea was to study what was done in the United States, studying text-books in English. We received solutions; that is, we would be given the name of an insect pest and the four chemicals which had at some time and place been used successfully against it.

"Fortunately, it was necessary for me to wait a year before I could enter Chapingo. I used this year to study biology. I had thought of agronomists as primarily biologists, so I prepared by reading and studying biology. With the biology fresh in my mind I entered Chapingo and was immediately conscious of something wrong there. There was a religious sense in Chapingo—what was taught in Chapingo you couldn't question. In addition, there was a constant fight for power at Chapingo that tends to preoccupy people there, be-cause what happens at Chapingo has such enormous national conse-quences. But the basic problem for me was that at Chapingo there is very little or no ecological or biological preparation. At the post-graduate level you find some consciousness of the biological and ecological problems, but the people at the postgraduate level have almost nothing to do with the undergraduates. Thus, you have hun-dreds of agronomists every year given their degrees without any real consciousness of what they are involved in. Those who do know better are not prepared to enter into the struggle for power at Cha-pingo, and so things remain the same. Chapingo remains the guide for the whole country."

Ing. Trujillo went on to discuss analogous problems at other agri-cultural universities in the country, concluding, "So what you have now is this enormous infrastructure composed of badly educated people."

Nonetheless, Trujillo remains optimistic about the potential for change in Sanidad Vegetal, believing that while INIA is too heavily influenced by its past dominance by the "agricultural moderniza-tion" strategies of the Rockefeller program, Sanidad Vegetal has shown signs of its responsiveness to change. He believes that the wave of publicity in newspapers and television shows with regard to pesticide abuse in the early 1980s has persuaded top Sanidad Vegetal officials of the need for change. While Trujillo is willing to privately describe some Sanidad Vegetal officials as "bought" by the chemical companies, as a whole he believes that the agency retains enough integrity and enough strength to avoid control by private economic interests. It is clear that the recent support by Sanidad Vegetal for Trujillo's ecologically oriented research and praise from the directo-rate have encouraged a favorable view of the agency as a whole on Trujillo's part.

Dr. Lilia Albert is a chemist who headed the pollution research

program of INIREB, the National Institute for Research on Biotic Resources (Instituto Nacional de Investigaciones sobre Recursos Bióticos). For a decade and a half, she has worked tirelessly to document
the nature of the pesticide contamination problem in Mexican foodstuffs and in the Mexican environment. She showed in the early
1970s that the waterways of Sinaloa were seriously polluted with
the persistent pesticides then commonly in use there. She has shown
that there are high levels of contamination in La Laguna in north-
central Mexico and in the dairy products that come from La Laguna.
She has demonstrated high rates of contamination in mother's milk
in Mexico. Her research has led her to the conclusion that most
commercially marketed food in Mexico's cities is contaminated
with pesticides, some of it to a very serious degree (Albert 1977,
1980, 1981). (At this writing, the Mexican government has announced
the closing of INIREB. Well-informed sources—not Dr. Albert—say
that the government's action was inspired by various politically irritating studies done by the institute and by the involvement of many
of INIREB's personnel in the fight against Laguna Verde, Mexico's
first commercial nuclear reactor, located on the Veracruz coast within
two or three hours' drive from INIREB headquarters in Xalapa.)

Dr. Albert has served on government commissions to study pesticide policy and problems, but "nothing has come of these studies.
We are beyond the point where it makes sense to just keep proving
that a problem exists." She laments that there are very few people in
Mexico studying pesticide problems and that there is little funding
available. "There are virtually no studies of the effects of pesticides
on flora, fauna, or fisheries, studies that are desperately needed" (Albert 1983, 1986).

Dr. Albert believes there is an urgent need for epidemiological
studies in all the areas where pesticides are used heavily. An obstacle to launching such studies is the fact that there is no comprehensive information available on what pesticides are used on what
crops in what regions. While the Sanidad Vegetal manual of authorized pesticides is supposed to provide a beginning point for such an
analysis, Dr. Albert says that the list "has nothing to do with reality;
without doubt, it has no relation whatever with reality.

"People in Sanidad Vegetal and other bureaucracies typically have
very little ecological, toxicological, or biological understanding of
the issues, with, of course, a few notable exceptions. The orientation of such people, due both to their education and to the way their
jobs have always been defined, is what we here in Mexico call *pro-
ductiva,* or directed at increasing production almost at any cost." As

a result, Dr. Albert maintains, the regulatory functions are ignored.

Dr. Albert's viewpoint seems to correspond with that of Sanidad Vegetal officials in the sense that both speak of better education as the primary need. But while Sanidad Vegetal officials speak mainly in the context of the need to educate growers, peasants, and farm workers about the elementary aspects of pesticide hazards, Dr. Albert believes that it is in improving the education of agricultural professionals who have been so important in promoting pesticides and supervising their use that change can come about. "The key is with the agronomists. The government is not going to change. But if we can educate new agronomists and have influence over those now working, we could change things."

Dr. Albert, with support from the Pan American Health Organization, has organized a series of workshops on pesticide hazards to retrain agronomists and public health workers with respect to pesticides. These workshops have ranged from one- to ten-day periods and have been given in different regions of the country. She has worked especially hard to see that at least a portion of those people attending are professionals who have already expressed an eagerness to break through some of the bureaucratic and political barriers to changing the situation with regard to misuse of pesticides—she is looking, clearly, to train people who are willing to be as determined and as combative as she has been.

There are others in Mexico who take a much harsher view of the situation than either Ing. Trujillo or Dr. Albert. Dr. Arturo Lomelí and his wife, Lili Lomelí, direct the Mexican Consumers' Organization. They maintain that the misuse of pesticides is simply an aspect of the undue influence of multinational corporations on the Mexican state combined with the corruption of the ruling party and state apparatus itself. They have been working in a Ralph Nader fashion for many years to expose various commercial practices harmful to consumers, using sympathetic insiders in government and corporations to obtain much of their information. They have circulated this information in their own newsletter, and both major and minor newspapers in Mexico City and other cities regularly carry news of their various exposés (Lomelí 1983a, 1983b, 1984, 1986).

While their view of the underlying corruption in the relationship of the Mexican state multinational corporations is more harsh than that of many other observers, discussions with the Lomelís reveal all the complex ambiguities that necessarily complicate the Mexican political scene. They describe the pesticide problem in Mexico as being in all essential respects identical to the pesticide problem of all other less-developed, economically dependent nations and in doing so, in a sense, remove the Mexican government from special culpability.

In an editorial in their newsletter, *La Voz del Consumidor*, they point out that it is "in the countries of the Third World that the most crassly commercial and irresponsible practices take place daily" that not only expose the peasant to every kind of danger but that

> also contaminate the agricultural products destined to con-
> sumers of the whole world, rich and poor. Thus, it is evident
> that we are dealing with a worldwide problem, but one which
> has special significance in countries like our own, such that it
> is in our countries that we see the greatest number of acci-
> dents, abuses, and deceptions. At the same time that our gov-
> ernments frequently minimize or plainly deny problems, on
> many occasions it has to be precisely the foreign researcher
> who puts his finger on the sore and whose research calls at-
> tention to the abusive and irresponsible practices. It is also
> the case that in our region and other underdeveloped ones, the
> problems of technology and the shortage of experts are every-
> day matters. (Lomelí 1983a: 1)

While they are frequently angered by what they and many others in Mexico regard as the pervasive corruption of Mexican politics and government, they readily admit that if anything, the Mexican state is more rigorous than most in Latin America in promoting the inter-ests of its people against large international corporations. "Of course, the situation in Central America would be even worse. One need only imagine!"

In 1984, a key question of tactics in the debate over pesticides in Mexico had to do with the possibility of completely nationalizing the pesticide industry, including pesticide retail sales, the five pesti-cide manufacturing facilities in Mexico owned by large multina-tional corporations, and the network of privately held formulation plants. This is no abstract question. With nationalized banks, elec-tricity, steel, railroads, telephones, petroleum extraction, and some 15 percent of the retail trade, the Mexican government is an active participant in sales, services, and manufacture. Fertimex, a parastate corporation with a guaranteed 51 percent share ownership by the government, is the largest manufacturer and wholesaler of pesti-cides in Mexico. Should it become the exclusive manufacturer? Should it retail its own products rather than selling them wholesale to the multinational corporations for relabeling with their own brand names, as it did up to 1987? The answers one receives to such questions are a kind of touchstone for underlying attitudes, but, interestingly enough, few people wanted to be quoted on the subject.

Some people would like to see more nationalization and perhaps complete nationalization of the pesticide industry, including sales. This, they believe, would relieve the pressures coming from the multinationals, and it would allow for a capture of a greater share of the profits, the bulk of which come from retail sales rather than manufacture or wholesaling. The authors of internal Fertimex studies have argued vigorously for such a position. People inside and outside of Fertimex also say that it would allow for better quality control and standardization of pesticides, claiming that the lack of such control has led to serious problems due to the inconsistent chemical content of identically labeled products (Fertimex 1981).

Others believe that while it would be possible to improve the situation with Fertimex as the government pesticide monopoly, the prospects are not good. They fear not only excessive bureaucratization and rampant corruption, but they also worry because historically it has been very difficult to publicly and effectively attack state-owned enterprises. State-owned enterprises tend to be political sacred cows—to attack them is to risk being seen as an agent of foreign interests—and even severely criticized operations like the petroleum monopoly continue to operate in arrogant disregard of criticism because of the protection they enjoy from high government and party officials. For example, when liquefied petroleum gas tanks exploded in the midst of a slum community in the northern Mexico City suburbs on November 19, 1984, killing at least 452 and injuring many thousands, a Pemex spokesman replied with surprise to a reporter's question about damage claims against Pemex, saying, "Damage claims against Pemex? But it was Pemex which suffered the damage!" Against such a background, it is not surprising to hear those involved in the pesticide debate say, "It is only the tradition of criticism of multinationals as agents of imperialist control that has allowed us to get so far with this issue. . . . Many people join the battle who are simply interested in combating the influence of the international corporations on whatever grounds, and without their help, the pesticide issue would not on its own attract great attention in political circles." And, from another observer, "It is only because of the role of the multinationals that we have sometimes had the cooperation of people within the government and the party itself." The potential for total nationalization is an important political tool, but many who wield the tool privately fear nationalization as much or more than they fear the multinationals.

It is a difficult paradox. Nearly all critics of pesticide abuse in Mexico hold that it is the multinational corporations who are primarily to blame, either through their aggressive sales tactics or through the pressures they exert on government policy. "Obviously, without multinationals, there would be no problem," said one. But the same person said, when asked about the advisability of nationalization, "It would be even worse with nationalization." The experience of nationalization in other areas of the economy and the political barriers to criticism of any state-controlled enterprise make many cautious of expanded government responsibilities in manufacture or sales.

All of this debate took a surprising turn in 1986 when the Mexican government, under pressure from the International Monetary Fund to divest itself of "inefficient nationalized firms," sold Fertimex to private stockholders. The debate over the politics of pesticides was overwhelmed by the debate over the disposition of the nation's national debt.

The Lomelís, in any case, tend to see the problem less in terms of ownership than in terms of the lack of independent political forces in Mexico. "Here in Mexico, we lack what you have in the United States, and what so many other countries have, and that is the existence of a whole network of civil associations—environmental groups, consumer groups, labor unions independent of political parties or government, clubs and associations of all kinds—that constitute a kind of check on the worst corporate and government abuses. Here there is hardly anything of the kind, or what there is tends to be pitifully weak. As for a political left wing in this country, it is hardly an exaggeration to say there is none. Certainly there are plenty of people with fine rhetoric in the universities and in political parties, even in government, but when it comes to a commitment to real change, it can't be taken seriously. The loyalties are to family, to little groups, to patronage networks, not to wider programs or principles."

Political scientists have indeed frequently described the Mexican political system as big city machine politics on a national scale. Political power is based on the operations of a patronage system, with the key to power being the ability to endow a multitude of people with large and small favors. Combined with the ability to use brute force whenever necessary, the dependency on patronage networks is a powerful incentive to stay in line. With respect to the workings of the agricultural economy, this has some very particular implications for pesticide use.

As in any rural economy, Mexican farmers rely on credit to get from one harvest to the next. Some credit is issued by private firms and some by public institutions. In regions such as Culiacán, with large-scale, privately run farms, most credit comes from private sources. In the Culiacán Valley, according to the head of the Mexican Union of Vegetable Growers, confirmed by the informed guesses of many other observers, about 90 percent of the financing is from North American firms and individuals. The private credit is usually extended under conditions that provide the creditor with some means of control over the production and marketing of the crop. In some cases, this takes the form of contract farming, where the creditor extends a large share of production costs to the farmer in return for interest, a commission, and specific commitments with regard to production techniques. The creditor promises to buy all or a stated portion of the farmer's crop at the end of the season at a specified price. The creditor believes it legitimate to set out production techniques in order to ensure that the money he has loaned and the commitment to buy result in a reasonably predictable quantity and quality of produce which he will undertake to market. In most such contracts, the use of pesticides and fertilizers are mentioned, and in some contracts the types and amounts are established, with room for response to special circumstances during the year. In other contracts, the grower promises to work with a representative of the creditor, taking his advice with regard to such matters as pesticides. Where contract buying is not involved, specification of chemicals used or submission to advice from the lender's representative will still often be included in the loan contract, in order to ensure the lender that the harvest for the year will result in income sufficient to cover payments on the debt. For example, Frank Maggio, who has extensive farm holdings in California, works through his firm, American Products Overseas Services, to finance Mexican growers. In early February 1987, he traveled to Mexico to give production advice to the growers, telling many of them to cut back production expenses and have workers pull flowers off tomato plants because of a glut of tomatoes in the North American market (González 1987; Maggio 1987).

With the stated purposes of lessening dependence on foreign or large domestic private financiers, of encouraging operators of small farms, and of improving Mexican rural life, various Mexican government agencies also offer credit to Mexican farmers. The amounts of credit offered and the methods and institutions used change frequently in Mexico and are often a matter of intense public debate. The most important institution in recent years has been Banrural, a government-owned rural credit bank, and its predecessors under other names. In theory, Banrural is to favor land reform beneficiaries, small-scale cooperative and private farmers, and farmers with less access to such basic advantages as public irrigation water. In practice, however, Banrural often makes loans in order to reward political friends and punish enemies. Partly because of criticism of such politically inspired practices in the past and partly due to increasing economic pressures on the government, the criteria used by Banrural have recently tended to resemble the profit-oriented strategies of private lenders. In either case, whether the stated purposes are distorted for reasons of patronage or in response to financial pressures, the effect is to favor larger-scale commercially oriented growers in direct contradiction of the supposed objectives of the institution.

The pressures to produce income from the bank's operations were loosened during the 1970s when the government was working with high revenues from petroleum sales, but with the drop in petroleum prices and the skyrocketing debt of the Mexican economy, commercial norms more and more predominated. Even where there is resistance to this trend in the Mexican government and from within Banrural, the International Monetary Fund and major international banks with large portfolios in Mexico have each year succeeded in exercising more pressure on government rural credit institutions to use narrowly commercial criteria (Hewitt de Alcántara 1980: 65–71; Millan Echeagaray 1974: 103–128; Gutelman 1974: 150–151; Yates 1981: 193–211; Reding 1988). Like the private creditors, the government credit institutions then have an increasing interest in the supervision of production techniques. The typical Banrural contract specifies given sums to be spent on pesticide purchases, presumably in order to guarantee the borrower's productivity. In many cases, Banrural actually purchases the pesticide for the borrower, deducting the costs from the loan amount (Yates 1981: 208; González 1984).

Individual farm productivity, however, is not the only reason government-owned credit institutions may require specified levels of pesticide use. Banrural was instrumental in financing an ambitious network of pesticide formulation plants, where farm cooperatives and growers' associations mix pesticide ingredients to produce a commercial product under their own label. Fifteen government-financed formulation plants typically work at only an average of 10 percent of their production capacity, and three dozen more privately financed plants work at less than one-third capacity. Thus, Banrural must be concerned that these rural industrial investments will be a burden on both the associations that nominally own them and on the government when the loans used to finance them cannot be repaid. The government would like to see greater utilization of the installed capacity, and thus credit institutions have an extra incentive to push pesticide use, arguing, of course, that additional use will mean increased productivity all around and greater rural income (Fertimex 1981: vol. 2, 29–39).

The government's anxious efforts to sell more petroleum in whatever form (pesticides are mostly petroleum products), its long history of ownership of Fertimex's pesticide and fertilizer manufacturing capacity, and its investment in chemical-dependent agricultural research over more than four decades give it further incentives to promote and even require higher levels of pesticide use. Most pesticides are used in export-related production, and both the government and the International Monetary Fund are pushing for more exports to provide foreign exchange earnings to pay off the debt, providing yet another incentive for pesticide use. The Mexican government may be as hooked on pesticide production and use as private farmers.

Another indication that this may be so is the fact that the government supplies Sanidad Vegetal with quantities of pesticides to be offered free or at very low prices to Mexican farmers. In one rich state in the El Bajío region, the director of Sanidad Vegetal reported in a confidential interview that he was unable to keep most of the pesticides he had been given for free distribution to farmers from falling into private hands for sale at commercial prices to farmers. Another Sanidad Vegetal official, also off the record, reported that in an effort to reduce pesticide use in his central Mexican highland state, he had refrained from distributing free pesticides through Sanidad Vegetal and was faced with a major problem justifying the fact that the agency still held them and had to pay the corresponding storage costs. While the Mexican government justifies pesticide subsidies and give-aways by claiming that pesticides stabilize rural income and welfare and that the availability of cheap agricultural inputs encourages investment in agriculture, it appears more likely that they are trying to convince the cart to pull the horse by encouraging agriculture to absorb products that encourage industrial growth and/or growth of nationalized firms. The government-owned or financed banks and factories needed farmers to need the factories; otherwise, these firms that represent the milk cows of patronage and corruption would no longer have been able to grow, nourishing the political system as they did so.

An obvious question to ask at this point is, If more narrowly commercial criteria are in vogue for government institutions that offer credit and services to farmers, how can the stoking of artificial demand for pesticides and fertilizers be justified? The answer is that it can't in economic terms, but there are abundant political reasons for the inconsistencies, and there are very few people in the Mexican public who will ever be aware of the contradictions. The International Monetary Fund and private international bankers have been pressuring the Mexican government to sell nationalized firms on the grounds that they are economically inefficient, encourage corruption, and discourage private investment. But international bankers are as wedded as the Mexican government to the largely unexamined idea that agricultural growth requires more chemicals, that growth feeds exports, and that increased exports are necessary to earn more foreign exchange, which is necessary to make debt payments. With this firmly fixed idea, they raise few questions about the folly of pushing pesticides and fertilizers.

Now that nationalized firms play a reduced role in pesticide manufacture and promotion, having helped to form the patterns of pesticide use in Mexico, private firms like Castle and Cook and Mr. Maggio have their own reasons for requiring chemical use, as they do now in Culiacán. Because the Mexican economy is a mixed one, with a variety of both private and public firms, the motives involved in any economic phenomena, like pesticide use, are mixed as well. In many cases, and certainly with respect to pesticides, the public firms have operated to channel and encourage private investment— the cheap pesticides offered by Fertimex, the promotion of pesticide use by government agencies, and the use of rural credit to require pesticide use all have operated to create special opportunities for bureaucrats, politicians, chemical manufacturers, sales agents, and agribusiness farm operations. There is then a special community of economic and political self-interest finely woven into the Mexican system to encourage continued pesticide use.

This situation is typical of Third World nations, whether or not their economies are heavily state owned or directed. A study by Robert Repetto (1985) of nine Third World countries reveals that local government subsidies to pesticide use amounted to an average of 44 percent of the retail price of the pesticides, and in some cases went over 90 percent. The mixture of preconceptions, patronage, and corruption has produced an enormous bias in favor of pesticide use around the world, far beyond the play of market forces.

In 1982, the Mexican government recognized that there was a wide range of environmental problems that were not being sufficiently examined by existing agencies by establishing the Ministry of Urban Development and Ecology, or SEDUE (Secretaría de Desarrollo Urbano y Ecología). While SEDUE has made some efforts at finding out what pesticide problems exist in Mexico, its chances for early success at effecting change were undermined from the beginning by a number of factors.

The leadership of SEDUE came largely out of the older ministry concerned with public housing and related public works projects— they were architects, urban planners, urban transportation experts, and public administrators experienced in public works construction and management. This reflected both a money-saving effort to create a new agency by recycling personnel out of an old one and a certain mind-set about what constitutes environmental problems. The view from Mexico City is that the environment is primarily threatened by air and water pollution and the multiple effects of uncontrolled urban growth. Mexican cities have been growing at about 7 percent per year—doubling in size every ten years—and it has been very difficult to plan or provide for this growth, especially in a country with substantial and persistent economic problems. It is not surprising that those politicians and bureaucrats in Mexico City who conceived SEDUE would think primarily in terms of stultifying air pollution, gridlocked urban traffic, sprawling slum settlement, uncontrolled raw sewage, water pollution, noise, and inadequate protection against fires, earthquakes, and floods. These are the stuff of their daily experience and of countless editorials, cartoons, jokes, books, articles, and songs. The countryside tends to be seen as a refuge from urban madness, and one that is emptying its human problems into the cities. About 19 million people live in Mexico City, nearly a quarter of the population of the country, and the environmental problems they face now and in the future are truly overwhelming. Almost two-thirds of Mexico's population is urban. Deforestation, erosion, farm chemical contamination, wildlife depletion, and the collapse of fisheries are abstract problems to those working inside the glass towers of Mexico City, while the traffic and smog must be faced daily.

With regards to pesticides specifically, there are very few people in SEDUE qualified to study them knowledgeably, and those few have no administrative authority to set priorities. Perhaps most serious of all, pesticides have already been staked out as a bureaucratic turf. The leadership of the Ministry of Agriculture (which is the home of Sanidad Vegetal) and their friends in the world of agribusiness want to see as little of SEDUE as possible, and they have the means to see that their authority is by and large respected. (This conflict is strikingly similar to ones in the United States—the federal Environmental Protection Agency versus the Department of Agriculture and the California Department of Health Services versus the California Department of Food and Agriculture.)

In February 1984, the young architect in charge of pollution problems for SEDUE in the state of Sinaloa became concerned about a widely publicized incident in which pilots of pesticide spray planes were washing out their tanks in the canal that provided the water supply to the town of Montelargo. Arq. Calderón (the abbreviation "Arq." meaning *arquitecto* is used as a title like Dr. or Ing., for engineer or agronomical engineer) prevailed upon his superior in SEDUE to talk to people in Sanidad Vegetal and arrange a joint meeting of the two agencies to discuss the Montelargo problem in particular and, more generally, to explore possibilities for cooperation. The Montelargo problem apparently provided an urgent excuse for the meeting, which had been requested on more general grounds earlier and continually postponed. The meeting as finally arranged was between Ing. Ibarra, Arq. Calderón, a chemist, a biologist, and an agronomist on the SEDUE side, and Ing. García Ussher, members of his staff, and growers representing the Sinaloa regional committee of Sanidad Vegetal. I was invited to attend the meeting as an observer by officials of SEDUE.

After a short and cordial introductory statement from Ing. Ibarra saying that the main purpose of the meeting was to get to know the people from Sanidad Vegetal and the regional committee, representatives from Sanidad Vegetal eagerly presented all the evidence they had that they were working hard on Integrated Pest Management programs and that they had had some success in some grain crops and in citrus. They said that it had to be accepted as a given, however, that heavy pesticide use was necessary for vegetable crops, soybeans, and safflower. While they mentioned a list of IPM programs that had been successful in Sinaloa, when Ing. Ibarra asked if such a list was available, Ing. Ussher said that there was no such list actually ready for distribution in any form and changed the subject to discuss the severe penalties available under law for abusive use of pesticides. He added that Sanidad Vegetal was considering the use of economic penalties for pesticide misuse.

When Ibarra and Calderón advanced the Montelargo case vigorously as an example of lack of enforcement, citing a petition they had received from seventy Montelargo residents, Ussher said that the real control over pesticide pilots was in the hands of the Civil Aviation Authority. He was quick to add that they had always been very responsive to Sanidad Vegetal complaints and had suspended pilot licenses until reassurances were given of proper practices in the future. "The problem of improper use of roads and lanes as run-ways is always with us, for example. We would like to see a program of government provision of proper runways throughout the valley." Then Ussher mentioned that he was also aware of improper container washing and disposal and that his agency was at work on the problem. Ibarra said that he had seen the "Oaxaquitos" fighting with each other over who would get cast-off pesticide containers. Ussher replied that nonetheless the runway problem had to be seen as "numero uno."

Calderón mentioned that they had been using the drug-testing lab of the federal prosecuting attorney to test for pesticide contamination of water and soil but that the facilities were inadequate. Ussher offered the use of the Sanidad Vegetal labs, to which the SEDUE chemist, Sánchez Lara, said, "Yes, your people have been very generous in offering us the use of these facilities, but it seems that they are always so overworked that they are never able to actually accept our samples for analysis." Ussher shrugged his shoulders. Calderón mentioned that they were being pressured by fishing and hunting cooperatives about toxic red and yellow water draining into swamps and lagoons, making it impossible to go on harvesting fish and waterfowl. Ussher said they would take care of it without giving any specifics. Calderón said that the rat-poisoning program was killing other fauna. Ussher said this was impossible—that the poisons now in use only killed rats, that the existence of such a large number of birds that they were a serious agricultural pest proved his point. As further points were discussed, Ussher either denied the existence of the problems, claimed that his agency had no authority over them, or promised to take care of the problems without giving any specifics about when or how. He politely brushed off Ibarra's suggestion for a written agreement on methods of cooperation.

One of the growers representing the Sanidad Vegetal committee took the floor for a time, introducing himself with the stock phrase, "Yo soy nada mas que un pobre ejiditario" (I am nothing more than a poor *ejiditario*—*ejiditario* meaning that he was a member of an *ejido*, precisely the land reform units that had been corrupted or displaced to allow large growers like this man to flourish). One of the staff members from SEDUE commented after this absurd introduction, "You know, you are the first *ejiditario* I have ever met who owns his own private plane." There was a shocked silence for a moment, and then everyone broke down laughing hilariously for a few moments, before the blushing *"ejiditario"* went on talking. Although the laughter was at the grower's expense, the effect of the incident was to bring to the surface for a moment the awareness of the bold-faced political power that lay behind the Sanidad Vegetal committee and its operations.

After about two hours, during which the Sanidad Vegetal side had managed to refuse nothing but to give very little away in terms of authority or commitments for definite action or future cooperation, Ing. Ibarra of SEDUE commented, "We need to take into account the aggressiveness of the press. They love the eight-column headline, as in the Montelargo affair. And since agriculture is the number one thing here in the valley, that is going to be the target."

García Ussher was eager to agree, adding that "this presidential administration [*sexenio*—the six-year presidential term] is the time for environmental questions. And worldwide, everywhere, there is this new concern that we all have to take into account."

Ibarra said, "And in some countries there are even political parties, like the Greens in Europe, that are founded on these concerns . . ." as Calderón interrupted to say, "Look, the point is that we have to get moving on these problems."

García Ussher said, "Well, who knows in the long run what the effects of these rapidly degrading pesticides might be?"

Ibarra replied, "Certainly, we don't know. The important thing to decide here today is a schedule of meetings to insure that we continue to work together." Ussher and the others from Sanidad Vegetal simply let this suggestion drop, and in the silence, a staff member from Sanidad Vegetal said, "Well, yes, official talks are alright, but we also have to talk as human beings, people who live in a certain environment that we can't change, recognizing the realities." And on this note of political caution, one of the growers launched into a speech about the desirability of getting experts from the United States to come to Sinaloa to give advice on the possibility of reducing spray drift and cost—"The big thing for us is that we have to lower the costs." The meeting was adjourned with no agreement for any future meetings or any methods by which the two agencies might work together.

The net result of this meeting was clear. Sanidad Vegetal intended to maintain, if possible, a polite working relationship with SEDUE, but not at the expense of any commitments to the new agency, even for a program of scheduled meetings, and, most of all, with no surrender of authority over the question of pesticides. Throughout, Sanidad Vegetal officials and their growers spoke with the assumption that SEDUE could ask for meetings and remedies, but that Sanidad Vegetal would maintain the right to make any decisions. The presence of wealthy growers arrayed on the Sanidad Vegetal side and officially connected with the agency through the Sanidad Vegetal regional and national committees and the absence of anyone other than staff members from SEDUE served to remind everyone of where the power resided. Such a relationship between agencies located in agricultural departments and ministries and their agribusiness allies on one side and environmental agencies on the other is typical of pesticide politics in many countries, certainly including the United States. There is nothing to suggest that this relationship will be different in Mexico, in spite of the superficially fashionable interest in pollution problems in the press and political circles.

It would be misleading, however, to leave the description of the Mexican bureaucratic reaction to pesticide problems without recognizing that there is an extraordinary diversity of viewpoints within the Mexican government. The government is a member of the fourth Socialist International and describes itself as a revolutionary force. Internationally, Mexican presidents usually align themselves officially with other revolutionary movements in Latin America, or at least speak against interventionism by the United States. In Mexican universities, Marxist thought is usually taken as the basis for all serious social science and political theory. At the University of Sinaloa, for example, the Mexican Communist party (under different names as coalitions with other parties have been fashioned) has been the most powerful influence on internal university politics, having a substantial degree of control over faculty hiring, firing, and assignments. Conferences at "la UAS" (Universidad Autónoma de Sinaloa) sometimes begin with a speech from the rector celebrating "international proletarian solidarity in the struggle against capitalist imperialism," with hammer and sickle emblems decorating the hall. Even highly placed politicians and businessmen are likely to use the terminology of class struggle in candid discussion of the Mexican political scene, even though this is given a special twist by the assumption of a paternalistic relationship by which it is argued that Mexican capitalists and politicians will advance the interests of the Mexican proletariat against imperialism.

As should be obvious, the essentially conservative and authoritarian Mexican state and its supporters are playing some very complicated rhetorical games here, but such games have their consequences. One consequence is that scattered throughout the Mexican bureaucracy are people with genuinely radical political viewpoints and commitment. Among these are a fair number of political exiles from other Latin American countries, such as Chileans exiled after the coup against Salvador Allende in 1973, Argentines exiled during that country's "dirty war" against revolutionary movements, and Central Americans fleeing violently repressive regimes over the last several decades. The tradition of harboring such exiles and taking advantage of their often impressive professional qualifications dates back at least to the sheltering of Spanish Civil War refugees and the ill-fated Leon Trotsky in the late 1930s. Such exiles join with the substantial number of Mexican professionals with left-wing sympathies in the Mexican bureaucracy to make up a kind of dissident and somewhat self-conscious group able at times to influence policy and develop critical perspectives.

There are several agencies in the Mexican government set up at different times that are supposed to promote rural development, each of these agencies having its own special history and viewpoint. A Chilean economist in one of these agencies has done much to document the conditions of farm workers, including their exposure to pesticides (Astorga Lira and Hardy Raskovan 1978). Others in the rural development agencies have worked with farm labor and peasant organizers on various occasions. An Argentine epidemiologist working for a federal public health agency documented the strange case of affluent people in Mexico City poisoned by organophosphate insecticides in tamale wrappings—the detective work of his team led to the discovery that an entire valley in the state of Hidalgo had been heavily contaminated with parathion due to the wreck of a Bayer Chemical Company truck carrying parathion. The wreckage of the truck had left many of the parathion packages ruptured with the contents spilled along the roadside. The insecticide had been carried by wind and water through the valley. Intact packages had been stolen by peasants anxious to acquire free what they had been told was a miraculous product, and after a guard was posted, the guard himself became an illicit sales agent for the chemical, selling at rates about one-twentieth the normal retail price. The truck wreckage was left for several months until the public health team discovered it, and by that time numerous mysterious deaths with all the symptoms of organophosphate poisoning had occurred in the valley. The residues in tamale wrappings reached relatively well-to-do people in Mexico City because the valley produces an especially large corn husk prized by housewives and cooks as a wrapping for fine, large tamales served at very special occasions. The sickness of groups of friends and family at parties and the death of one person led to an autopsy and investigation that would not have occurred had the victims been poor. Dr. Damante Bilello, the Argentine epidemiologist, commented that this investigation had led him to the realization that pesticides may be a serious public health threat in Mexico and that there was a pressing need for more study. Mexican physicians who attended their colleagues' presentation at a national public health conference said afterward, in whispers, that Bilello could afford to take these relatively strong stands because he was planning to leave the country soon—they believed that there was a systematic government cover-up of the pesticide problem by high government officials and did not want to stick out their own necks (Damante Bilello 1983).

University professors and researchers sometimes collaborate with people in rural development agencies in research or service projects that are meant to advance independent organizations of landless rural laborers and peasants. They may provide support services to organizers and they bring media attention to particularly abusive situations. In many Mexican universities, students must write a kind of senior thesis, the writing of which is meant to give service to communities and groups who have been neglected or mistreated by private or public authorities. Sometimes complex networks of collaborators are involved in the completion of these projects and in seeing that the findings are used to bring attention or remedies to the problems analyzed. In cooperating with such projects, bureaucrats can sometimes quietly create leverage on their own or rival agencies to address issues that otherwise would not be recognized. For example, students have conducted surveys of public health problems among farm workers in the Culiacán Valley, working with university researchers, labor union organizers, and public health personnel.

The complex personal and institutional patterns involved in such efforts is both a strength and a weakness of the Mexican system, where conservative authority may be undermined by the actions of individuals using informal networks. Positive change can result from such work. But it may also be that the opportunity for critical or discontented people to work in such informal fashion helps to avoid direct confrontations and responsible, forthright changes in policy. In the process of doing the work for this book, nearly every Mexican with whom I had serious conversations complained about the confusion and inconsistency of government policy, sometimes attributing it to incompetence, sometimes to corruption, and sometimes to deliberate manipulation of the public's consciousness of social problems and the government's failure to resolve them. It often seemed to me that Mexicans were over-critical of their governing system, especially in their belief that the corruption and confusion of the Mexican system is fundamentally different or worse than that found in other countries. But whether the critical perspective so common among Mexicans is overdrawn or not, a sense of dissatisfaction and malaise is so pervasive as to be a serious obstacle to the solution of any social problem. The lack of confidence in the system generates failure in its own right.

The state director of Sanidad Vegetal in a rich agricultural region of central Mexico epitomized the sense of frustration I found so universal among Mexicans when they could speak candidly. Ing. Torres (not his real name) is a charming, middle-aged man who offered to give me a great deal of detail about pesticide use in his region. While he acknowledged that there were some problems of pesticide poisoning, he asserted that they were minor. He also argued, as did Ing. Ussher in Culiacán, that if people in the United States were worried about dangerous pesticides in Mexico, they should lobby the EPA to ban such pesticides. "You are the older brothers and we follow your lead." He maintained that his extension agents were well qualified to be sure that pesticides were properly used in the countryside and that they had developed model programs for biological control of pests, particularly in response to pesticide resistance problems in pests. In the midst of our interview, however, this glowing report on the work of his agency was interrupted by a telephone call from home. There was a crisis that seemed to require an immediate decision. First, he talked to his eight-year-old son, who insisted that he needed a new pair of tennis shoes. "But you already have three pairs!" The problem seemed to be that the boy's school was requiring white tennis shoes for a parade. "But your new ones *are* white!" Alas, there was a stripe on the new ones. "Don't you think this school is being rather demanding?" Ing. Torres asked with rising irritation. After a long and inconclusive discussion of the topic, first with the boy and then with his wife, Ing. Torres promised to be home soon with a definite decision on the topic.

Upon hanging up, Ing. Torres was a changed man. He asked that we be able to speak without my quoting him by name, to which I agreed. He began by saying that his children simply didn't understand the value of a peso or the meaning of work but soon moved into saying that this was the problem with the whole country. After describing the modest poverty of his own upbringing on an *ejido*— "Well, if we needed a pair of shoes, fine, but if we *had* a pair of shoes, there was nothing to discuss"—he brought the topic around again to pesticides. "But, you see, no one is direct or honest anymore, and nothing works right. Look, I have to tell you something. It's hard working with the peasants. They don't have any confidence in the representatives of the government, they just don't want to talk to us. Why? Well, they feel betrayed. All this stuff I was giving you about how we tell them to use this pesticide or that pesticide, they don't listen and they don't need that. They know well enough what to do, but they need the real economic and political support of the government—put the tractor on the land, credit so they can pay with the harvest—they would work, work hard, work well. But let me tell you. Last year we were supposed to get eight hundred thousand pesos worth of pesticides to distribute here. It was supposed to be divided up, not sold. But what happened? Well, it got sold here, sold there, nobody knows what's happening to it. And the peasants got disappointed again. I tell you we are reaching a limit. I've been meeting informally with peasant groups every two weeks. I've been hearing a lot. People are suffering! There is a lot of poverty out there, a lot, and I think there is a limit. And that limit is hunger, that's the limit. Hunger." He went on to say that he and his associates had been talking to people in the army about the situation. "There's no point in these political parties, these little groups. . . . We start from this point: The government has all the power. What should we do? A revolution? The army's too powerful. Not a chance. Forget it! Do we want a socialist country? Forget it! In Mexico that would mean a collective farm with four guys ordering everybody around, and the rest living in the worst kind of poverty, with no say in anything.

"No, I have grown up close to you, to the United States, I'm with you all the way. Mexicans are, in the depths of our most intimate beings, capitalists. Capitalists! What we want is not socialism but a fair system, with the same rules for everybody, but let us decide what to do with our money. That's what 98 percent of Mexicans, that's what they want. This Miguel de la Madrid [president of Mexico at the time]! What does he mean criticizing the United States [over Central American policy]? Doesn't he realize that if the United States didn't stop them in Grenada, they'd soon be invading us? He must know. But he's not a socialist either. No, these men in government—I know them personally—they are neither socialists nor communists nor capitalists. They are just a bunch of guys robbing the rest of us of everything we've got. How do they think? Apparently, it's just 'Let's see, I'm going to be president for six years, I'll take everything I can. As far as resolving problems, to the devil, I'll make mine and be gone. . . .' There is an absolute lack of morality, it hurts me to say this but among 80 percent of us there is no morality."

He went on to explain that often when he tried to hire laborers on his own farm at the going rate, people refused to work. "I don't blame them. Here they are feeding eight to twelve people. They say, 'What can I do with 400 pesos? There's no point in working for this.' Better to work on their own, look for something else, just stay at home. What can they do with 400 pesos? And yet, look at the army, a soldier who doesn't know how to read or write makes 800 pesos a day, plus benefits, and he does nothing! Absolutely nothing! And a peasant working his eight hours a day and more, real work, he's lucky to get 400 pesos. It doesn't make sense, this kind of system."

Ing. Torres leaned back in his swivel chair and was quiet for a moment before speaking again. "So the peasants, they have a hard time trusting him. Not me, I can work with them because I understand them, I come from them, they recognize me. But, well, now it's happening that a lot of the agronomists come from higher economic levels, and they can't establish trust. It's hard for them, and hard for the peasants. There is no understanding. So, pesticides, and all that, well, we don't really do much of anything here, speaking honestly. . . . So, I need to go deal with this matter of the tennis shoes." Much as I disagreed with some of the man's sentiments, he clearly spoke from the heart of his own experience and sincere concern.

Expressed in this case by a man whose thinking is clearly carrying him in a right-wing direction, such frustrations and anger are almost universal among people working close to the situation of rural Mexico, of all political persuasions. The sense of powerlessness against a vast and well-defended political system constitutes the background against which all that has been said here about pesticides, public health, the environment, doctors, and bureaucrats must be understood. For all the volumes that have been written about the ills of the Mexican system, the pessimism of Mexicans about their government sometimes seems out of proportion to circumstances, because political corruption and mismanagement are so common in modern states nearly everywhere. Mexico has done relatively well compared to dozens of other nations at providing a decent life for a large portion of its people. It seems that Mexicans have revolutionary expectations of themselves even if they have failed to create a genuinely revolutionary society. It also may be that they realize in ways they cannot always fully articulate that the solutions of the last decades will no longer satisfy the needs of the future. There is a sense of having traveled a long distance on a pleasant road that finally turns out to be a dead end. Can another road be found?

4
Going Home

*R*amón González grew up in the village of San Jerónimo Progreso in the rugged mountains of the northwestern part of the state of Oaxaca. Through the lively and surprisingly accurate informal network of information maintained by the Mixtecs along their migration trail, I was told that in January 1984 I would be able to find Ramón's father at home in San Jerónimo. Since he had been with Ramón at the time of his death I thought that Ramón's father might be able to tell me more about what happened. In any case, I would surely learn a good deal about Ramón, his family, and his community. Most of all, perhaps I would learn more about why tens of thousands of Mixtecs every year travel thousands of miles in search of grueling, low-paid agricultural labor.

Anthropologists have written a number of excellent studies of Mixtec communities. Historians and archaeologists have recently begun to pay serious attention to the story of the Mixtecs both before and after the Spanish Conquest. It is still nonetheless the case that, considering the importance of Mixtec civilization in the central and southern Mexican highlands and western coastal region, we know relatively little about them. Ordinary people in Culiacán and Mexico City have a whole series of stereotypes about the Mixtecs and other Oaxacan Indian groups—words such as "simple," "ignorant," "backward," "innocent," "sly," "primitive," "clannish," "closed," "somber," and "fatalistic" are applied to the Mixtecs. The Mixtecs have existed as a people with a clear identity for at least a millennium, and probably more than two millennia, and they were noted in pre-Conquest Mexico for a high degree of perfectionism and inventiveness in arts and crafts—jewelry, pottery, sculpture, and architecture—and for an unusually refined and successful diplomacy, through which they exercised great influence over their neighbors with a minimum of warfare. They were also recognized for their agricultural ingenuity in an extremely forbidding mountain environment. It is not surprising, then, that a visit to a Mixtec village reveals a complexity and sophistication that makes all labels and simple generalizations irrelevant. (Some of the basic works on the Mixtecs include Benítez Zenteno 1980; Butterworth 1975; Secretaría de Programación y Presupuesto, Centro para el Desarrollo Rural Integral 1984; Cook 1949, 1958; Flannery 1983; Kearney 1984a, 1988a; Paddock 1966; Ravicz 1965; Spores 1969.)

In the context of this work, it would also be all too easy to see the Mixtecs as little more than victims, perhaps robbing them, in the mind of the reader, of their diversity and creativity. They are philosophers and theologians as well as peasants and workers. They are witty or irresponsible at times, just as they are, more generally, hardworking and serious. But all labels, even those that account for contradictory qualities, fail to communicate who these people are and what they are like. I will tell the story of my first encounter with a Mixtec community in the hopes that it at once tells something of the generalized problems and character of the community and also something of the variety of human personality and individual drama that may be found in a Mixtec town. Following the story of this encounter is a discussion and analysis of the major problems of the Mixtecs, how these problems came about, and how the larger world influences the Mixtec land and people. I hope it will be clear after both these approaches that the future of the Mixtec people and their homeland will be determined by complex interactions of the Mixtecs with the wider world—the Mixtecs alone cannot shape their destiny, nor are the Mixtecs merely passive victims. They are active agents in a larger drama that involves us as well as them.

It is only about three hundred miles by road from Mexico City to San Jerónimo Progreso, but the social differences between the Europeanized, industrial Mexico City and the disintegrating ancient agricultural towns of the Mixtecs make the effective distance far greater. For just that reason, it is important to pass through the stages of the physical journey to get some sense of how the urban ways of central Mexico give way step by step to a different world.

I started my drive from smart, fashionable downtown Mexico City, out along the crowded throughways past miles of slums and slum-clearance housing projects, past soccer fields littered with blowing paper and plastic bags, by traffic islands where transients and new migrants to the city camp under the eucalyptus rows, through the vigilant gauntlet of a small army of traffic cops stationed at stoplights, and alongside mountainous garbage dumps with dozens of people clambering over them in search of salvage. Mexico City in the mid-1980s counted 19 million inhabitants, and by the end of the century at current growth rates will have 31 million, meaning that the city must accommodate more than 1 million new people every year. This growth rate is typical of Mexican cities, except for those along the border with the United States (Tijuana, Mexicali, Ciudad Juárez), which grow even faster. The high rate of natural increase combined with the exodus out of the countryside demands a herculean effort of the Mexican economy and government to simply maintain living standards—by the mid-1980s this effort had begun to fail, and the many improvements in Mexican life in previous decades were visibly deteriorating. A journey through the periphery of slums that surrounds most Mexican cities is a clear reminder of the massive problems created by such rapid urban growth.

The Mixteca region has one of the highest rates of natural increase of population in Mexico and an exceptionally high rate of migration out of the area into Mexican cities. The slums of Mexico City are home to hundreds of thousands of Mixtec people. Without traveling to rural Mexico, it would be difficult to understand what attracted people to forsake their homeland in preference for these disheartening urban settlements stretching for mile after mile around Mexico's capital.

Reaching the edge of Mexico City and beginning the climb out of the basin, the smog begins to clear enough to see the great snow-clad volcanic peaks, Popocatepetl and Iztaccihuatl, rising from the seven thousand feet of the valley floor to nearly eighteen thousand feet. The pass over the northern shoulder of the peaks runs through fifty miles of pine forest and down into the Puebla Valley, with small, lush fields of corn and vegetables surrounding hundreds of colonial churches scattered throughout the countryside. The traditionally conservative Catholic city of Puebla has been overtaken by automobile factories and urban sprawl. South from Puebla to Izucar de Matamoros the small, diversified farms of peasants give way to sugarcane plantations. Then the road rises sharply into barren-looking mountains, stripped of most of their natural vegetation for centuries, pines and oaks displaced by shrub and cactus, little eroded fields on steep hillsides and goats picking through what passes for pasturage. This is where the Mixteca begins.

The little rented Datsun, manufactured in Puebla, ground through one hairpin turn after another. Most of the big, long-distance buses and trucks favor a newer road to the city of Oaxaca, farther east, so on this road the older second- and third-class buses and local truck traffic dominate, pouring out clouds of black diesel smoke. A surprising number are broken down by the side of the road with orange warning markers carefully laid out and someone flagging down cars

near the breakdown to avoid accidents on the narrow two-lane as-
phalt highway. This was January, the dry season, when the land
looks most forlorn, without the bright green patches of fields fed by
summer rains. In January, the adobe mud of the houses and thatched
clay granaries, the corn stalk fodder tossed up into tree crotches to
keep it from the animals, blend into the tans and greys of the steep
fields. Each little herd of goats picking through the shrubs and dried
grass was tended by a child or old man; even where one cow roamed
across a hillside, there was always somebody following, keeping
close track of the few assets the family owns. I got out of the car to
take a photograph and found that the string of children walking
home from school had taken a fifty-yard detour up through the
thorny brush to avoid me, but they couldn't resist furtive glances
and giggling as they went.

There are bright spots from time to time. The road will dip into a
little valley with a hundred acres or a few square miles of deep soils,
intensively cultivated, and a creek bed, usually dry in this season.
Most of the towns in these valleys have some craft specialty: baskets, pottery, or hats, offered for sale in roadside stands, expressive of
the notorious rural underemployment that plagues the region. In
rural Mexico, unemployment and underemployment waste about 50
percent of the potential work effort of the adult labor force. In the
Mixteca the situation is a good deal worse than in Mexico as a whole.

The green trees and fields of the valleys contrast sharply with the
sunstruck hillsides surrounding them. Before the Spanish Conquest,
the difference between the green valleys and the hillsides defined the
distinction between the Mixtec nobility of the bottomlands and the
commoners forced to make a living on the slopes.

Even before I reached San Jerónimo, hitchhikers told me stories
that helped portray the struggles and ways of thought in the Mixteca. A man in his fifties, about five feet four, dark and wiry, got in
thirty miles short of the turnoff from the main road in the small city
of Huajuapan de León. His leathery face under the brim of an old
straw hat was one great smile as he greeted me formally and asked
me what I would charge for a ride into Huajuapan. He was delighted
to hear I would be happy to have the company.

Although one often reads of the laconic character of Mexican Indian people, I have usually found the opposite, as with this man,
who seemed comfortable only during conversation. Loquaciousness
seems a comfort, something anyone, even the poorest person, can
own and keep. This man talked along steadily in a relaxed way, with
unexaggerated emphasis from his hands, his eyes, and his smile. His
story is typical of the never-ending struggle for land and employ-

ment, the intense conflicts within communities over these issues, combined with a sharp awareness of a larger political and legal context.

"I was born south of Huajuapan. My family was very poor. As a child, I worked as a shepherd to herds of goats, like you see along here everywhere. They were that breed of goats that are very small, the ones so common in the Mixteca. There, like the ones you see there to your left. See?

"I wasn't able to go to school, and I had no education at all until I married. I was lucky, because my wife knew how to read and write, and she taught me something of those things. As I learned more, I began to come into conflict with my community, because I saw that the leader of our community was damaging not just myself but the community as a whole. Because of the position I had taken, I began to be attacked. I had to put up a fight against the leader's thugs. They tried to throw me out, beat me up. But I took my case all the way to the Supreme Court, which took several years, and I won in the Supreme Court. I received my piece of land. That was my right. I worked that land and began to study again, with the time I had because I didn't have to keep fighting all the time. I was given a job with SAHOP [Secretaría de Habitación y Obras Públicas, Ministry of Housing and Public Works] and later with Communications, working roads. I had won some friends and some respect in my struggle, you see. I began to conform, because I saw there was no point in continuing to fight. The work was good and things settled down. Now I work as a watchman in the machinery yard, not out on the roads, because I already have twenty-seven years of experience."

"Do you still own your land? Do you work it?"

"Oh, yes, I plant every year. I have just been working my *milpa*."

I was to think of this man and his *milpa* again when I learned that the Mixtecs of San Jerónimo make every effort to maintain and plant their land, even in years when they travel thousands of miles and are gone for nine to ten months. It is the land itself which in many ways attaches a person to the community, and the community gives identity and purpose.

Huajuapan de León, one of three major Mixtec market towns, the informal capital of the Mixteca Baja, looks like dozens of other small Mexican cities of several tens of thousands of people. During the day the streets are jammed with people, walking, on bicycles, pushing carts, carrying loads, waiting for buses, going to market, standing having a chat with a friend. Dust is everywhere. A small central plaza boasts a loudspeaker system with American and Mexican rock and roll blaring for the benefit of those who sit on the benches under the big trees. The trees were filled on this day with cattle egrets, coming and going. The town has still not recovered entirely from the earthquake of 1981, with buildings that have tumbled down and some that are being rebuilt. Mestizos in business suits, freshly pressed dresses, or polyester pantsuits and knit shirts go in and out of banks, stores, and government offices while Mixtecos, in cheaper and sometimes traditional clothes, hurry on their own errands or sit in family groups on the ground, waiting or resting. There is an overwhelming presence of teenagers and children. In Huajuapan, the new, urban Mexico mixes, uncomfortably and imperfectly, one senses, with the older, rural, Indian Mexico.

On the side road toward San Jerónimo, my first hitchhiking companions were two short young men with massive arms and shoulders and a thin, unshaven elderly man with a traditional straw sombrero and huarache sandals. One of the young men said nothing and got out after about three miles. The other two stayed with me, the old man anxious to chat and the younger one gazing out the window. Occasionally, in the rear view mirror, I could see him sneering at the old fellow's words.

Many people in the Mixteca are living their lives and organizing their families and communities on the basis of beliefs and assumptions that are very old and very distinct from the rough, secular rules of modernized Mexico. The old man had a story to tell that exemplifies the cultural distance between the politicians, businessmen, and bureaucrats who run modernized Mexico and the magical world of the old Mixteca.

As we approached the valley and town of Tonalá, the old man said, "You must understand, sir, that Tonalá is a very lucky community, because a living cross came to live with us many years ago. One of our men found it up where the canyon is very narrow, the canyon of the Río Balsas. He had to cut it to bring it to Tonalá, where they set it up in a cave on the bluffs. There, each year, it produced a new branch, which was cut for veneration. After four years, it produced no more branches. While it was in the cave, the *viejitos* [little old men—often a term specifically referring to men with a special ceremonial or counseling role in a village] said that an eagle would come to pull a thread from it. When the eagle came, many men derived tremendous power from the thread that he extracted from the cross. The cross was eventually put in a church, where it still remains, but now it's dead. The priest had always opposed it. But because of this cross, Tonalá has been a very lucky community." The old man told this tale with great animation, and with each point I noted the younger man's discomfort—contempt or perhaps embarrassment—

as I glanced in the mirror. Both men asked me how much they owed me for the ride before I let them out on the outskirts of Tonalá.

Tonalá indeed looked like a lucky community, spread along good black bottomland for several miles, with lots of trees and solid, prosperous-looking plastered and brightly painted adobe houses. Later, a rider with me on my way back to Mexico City who was carrying the alternator from a tractor to be fixed in Huajuapan told me that Tonalá was organized as an *ejido,* a form of landholding defined by the Constitution of 1917, after seven years of bloody revolutionary warfare. The term *ejido,* which had been used to describe a variety of community-based indigenous landholding systems (*ejido* is a Spanish word derived from medieval Spanish land law which also recognized community-based forms of landholding), now means that the land is held by the community through the authority of the federal government and may be worked either individually by peasants holding a revocable life-time permission to the fields or, more rarely, collectively by the community. Most often, the land is worked individually, with various forms of cooperative arrangements for credit, tools, and small public works through the *ejido* organization. Tonalá, as an *ejido,* contrasts with more traditional Mixtec communities like San Jerónimo, organized under Mexican law as communes, where community ownership and governance are meant to be stronger forces than in most *ejidos.*

From the town of Tonalá, with its special luck—its tractors and potable water system and paved streets—it is another great cultural distance to the more remote communes like San Jerónimo, where an ox is considered a great luxury, potable water is still a dream, and no street is more than a mud track. Shortly after leaving Tonalá, the road becomes a long, steep grade leading to the bus shelter that marks the turnoff on a dirt road toward the little town of Nieves. Nieves sits high on the shoulder of a ridge. Looking from the ridge across a barranca are strange patterns on the hills, straight lines along the contours of the mountain, unmistakable, but overgrown with brush, cactus, and an occasional oak tree. From a distance, the lines look like the evidence of sedimentary layers of rock, but the way they contrast with the other patterns of the hillside makes that seem unlikely. Archaeologists say that these are terraces of the ancient Mixtec people, built before the Spanish Conquest. They are unusual in Mixtec country because ancient Mixtec terraces are found normally only in bottomland that was controlled by the Mixtec nobility and worked by a special class of laborers obligated to the nobles. Most of the Mixtec people were forced to work the highly erodable mountainsides without the control over their own labor to be able to afford terraces. As the steep fields of the commoners eroded downhill, the nobility were enriched by fresh deliveries of good forest soils, captured in the bottomland terraces. The commoners were forced further uphill in a process of progressive erosion and deforestation that has been speeded up, as we shall see, in the modern era. The hillside terraces of Nieves are interesting as evidence of a puzzling exception to the ancient Mixtec ways.

Near the town of San Luis de los Reyes, I stopped to check on directions with two young men working on a Volkswagen Beetle. They warned me that my car might not be able to negotiate the remaining kilometers to San Jerónimo and then invited me to share a soft drink and chat. It developed that they assumed that as a foreigner traveling in this remote area I must be looking for drugs to buy and offered to find a good deal for me. When I told them I wasn't interested in drugs, they could not seem to apologize enough and elaborately called on God's protection for me on the rest of my journey.

After crossing a new concrete bridge, the road deteriorated badly. Recent rains and traffic had raised high banks of mud in the middle of the road in many places. With some difficulty, I reached Silacayoápan, a pretty little town in a creek bottom. A man reassured me that the muddy track to the left of the statue of Lázaro Cárdenas, the Mexican president, 1934 to 1938, who went furthest to carry out the promises of the revolutionary Constitution of 1917 for land reform and nationalistic economic policies, was indeed the road to San Jerónimo. He said that the road was "average."

I tried to imagine a road that would be worse than average. The road ascends sharply from Silacayoápan for several kilometers before reaching San Jerónimo. Soon, I had four riders, a young, giggly married couple squeezed together in the front bucket seat, and two old men with hoes and machetes, dressed in loose homemade white cotton trousers and shirts and huarache sandals.

"What are you doing here?" one of them asked with not a shade of hesitancy.

"I am looking for a family, a family that I know through a friend of mine who has visited you here in San Jerónimo."

They beamed with delight. "Oh, how very good! How good! You are looking for a woman! You are very lucky, señor, because there are two fine young señoritas who happen to just now be looking for a man."

"How good! How good!" the second man agreed. They laughed uproariously. (In fact, as in most Mixtec towns, people in San Jerónimo rarely marry outside their village or families from the village. Kearney said of this incident, "The old scoundrels were pulling your leg.")

"No, I'm not looking for a woman. I'm looking for a man to talk to him about some things. His name is Irineo González."

"You didn't know?" one of them asked in surprise.

"No, I don't think he can know," the other one said. "He must not be aware of it."

"Oh, well, I would think you would know. But well, no. Because Irineo González died."

"Yes, he died about three weeks ago, after Christmas, no, maybe more like ten days ago."

"Yes, he died in an accident. A truck they were riding in. Three people died. Irineo and two other men, friends of his. Very bad."

I thought of the photograph Michael Kearney had given me as a present for Ramón's mother and father. It was a photograph taken in Riverside three years earlier, before Ramón's death in Culiacán. Ramón, his father, and a friend were sitting together, smiling, waiting for a dinner cooking on an outdoor fire to be ready to eat. The friend, who in the photograph wore a round terrycloth hat and a warm smile, had been killed by a gang of teenage *cholos* (hoodlums) in Tijuana. Ramón had died in Culiacán. And now, Irineo and two other friends were dead as well.

"Is the mother, Guadalupe, alright?"

"Oh, yes, just fine. Sure, there's nothing wrong with her."

As we came over the shoulder of a ridge, with the little cornfields freshly plowed, running down the heavily eroded slope to the right, San Jerónimo lay at the top end of a little valley directly ahead. The older part of town was in the bit of flat land at the bottom, with more recently built houses running up the mountainsides in three directions. There was a church sitting prominently in the middle of the village, with one bell tower brightly painted and the other, a different size and shape, half-built, with a clock. Still on the ridge line, we passed a lone, newly remodeled chapel.

The road into San Jerónimo and the view of the town and the surrounding fields reminded me of Kearney's phrase that much of the land looked like "a moonscape." Soil erosion, primarily sheet and gully erosion caused by rainfall, with some help from the wind during the long dry season, has devastated San Jerónimo. The UN's Food and Agriculture Organization classifies the Mixteca as one of the most severely eroded landscapes on earth, with about 70 percent of the potentially arable land no longer able to grow crops due to soil loss. Looking at the scarred hillsides that surround San Jerónimo, a similarly gloomy figure seems reasonable as an estimate of the damage done by erosion (Secretaría de Programación y Presupuesto, Centro para el Desarrollo Rural Integral 1984; Cook and Borah 1968).

I recognized the municipal hall, *el juzgado,* from photos Michael Kearney had shown me. The old men pointed out the *juzgado* and assured me that though the doors were shut at the moment, the *agente municipal* would soon come to open them because it was now time for him to do that. They also pointed out the González house, a couple of hundred yards above the *juzgado* on the same straight lane. The giggling couple had already gotten out to take a road leading to another village, and at a particularly rough spot on the way into town, the *viejitos* got out so that the car would have a better chance of making it. They jumped back in to ride to the middle of town, and then set off walking after wishing me well, thanking me for the ride, offering to pay, and inviting me to come see them while in San Jerónimo. I drove directly through town and up the hill to the *juzgado.*

Soon, a short, vigorous man with a grave expression and a big straw cowboy hat came walking quickly up the hill. He was carrying an enormous old-fashioned key about ten inches long, with which he opened the *juzgado.* I began to explain that I was a friend of Michael Kearney's and that I had come especially looking for Irineo González, and why. The *agente* introduced himself as Quirino Torres and brought out a little, worn wooden table, set it in front of a bench along the front of the building, and sat down to read my letter of introduction from Kearney. He read pronouncing the words slowly and softly, puzzling out each Spanish word phonetically. He nodded his head solemnly as if to say that he understood.

We talked for a while aimlessly, avoiding any serious topic or direct questions. Then I mentioned that I had just heard of the death of Irineo and that I had come partly to talk to Irineo about his son's death in Culiacán. I asked whether Ramón's mother, Guadalupe, might be able to tell me anything. Quirino said that he didn't think she would be helpful because she spoke only Mixtec. (Most of the women of San Jerónimo do not speak Spanish. The men usually speak Spanish, with varying skill, and sometimes a bit of English, because of their travels. The children all speak Mixtec and some speak some Spanish learned in school or in travels with their parents.) After chatting for a few minutes more, Quirino invited me to come to his house for lunch. As we walked, I asked Quirino about the big rolls of plastic pipe stacked against the wall of the *juzgado.* He told me the pipe had been given to them by the government so that the men of the village could build a potable water system, but it turned out to be worse than useless because it cracked and broke so easily. They were waiting for new and better pipe to begin construction. The town has a discouraged and pessimistic attitude toward cooperation with the government on community projects, engendered in part at least by such problems as the unusable plastic pipe. For example, several years ago, the National Indigenous Institute proposed to San Jerónimo that the town cooperate with the institute in building a boarding school in San Jerónimo for children from more remote villages. The town council turned the project down, calculating that the cost and risks to San Jerónimo would not compensate for the benefits of having the school located there—as a consequence, the poorer, neighboring village of San Andres took over the project and is pleased with its model school (Kearney 1984a: 55).

We walked down the mud streets past the church with the clock and off to the right to Quirino's solidly built new house made of concrete blocks reinforced with steel, plastered and painted inside and out. There was washing on the line in the front yard and dozens of pieces of goat meat hung out to dry on other lines. The yard was planted with peach and papaya trees.

Quirino told his smiling, plump wife to bring me something to eat, and we sat down at the formica-topped table. She soon brought a big basket of store-bought wheat flour tortillas surely sixteen inches in diameter, along with a plate of chopped egg and warmed-up dry goat meat. Looking around the room, I noticed two beds at opposite ends of the big, single room. A pile of yellow and purple corn ears about four inches long lay in one corner. On another table were a small radio and cassette player, with a few tapes of Mexican popular music, and a record player with two speakers. The record player and speakers were covered with a carefully hand-sewn and embroidered cloth cover. As I ate my lunch, Quirino explained that he had drunk too much beer one night and had broken the record player. Could I fix it? On fooling around with it a bit, we agreed we could do nothing for it, and he said he'd take it into Huajuapan to be fixed. The well-built house, the radio-cassette player, and the phonograph had all been bought with the earnings from Quirino's migrations to work in the fields of northern Mexico and California in recent years. Some of these investments are very carefully considered. The reinforced block houses, for example, are much more earthquake resistant than the traditional wood and straw houses, and they are also tight enough to keep out most rodents, an important matter because the family's supply of grain for the year is stored inside the house. Grain conservation is an important matter. The people of San Jerónimo have little good-quality land, virtually no inputs such as chemical fertilizers, and little time to cultivate crops when migration takes people away from their fields a large part of the time. The median landholding is about six-tenths of a hectare, less than one and a half acres, much of it unusable. Corn yields are about 300 kilograms per hectare, or one-fourth of the national average. In sum, the people of San Jerónimo are able to grow only about 20 percent of the corn they consume in an average year. A good, tight house to store grain in makes the meager supplies go as far as possible and holds cash purchases to a minimum. When Kearney and Stuart surveyed households in San Jerónimo, they found that of fifty households, all but three had members who had migrated within the previous three years, and all but nine had members of the household who were involved in migrant labor during the year of the survey. Eighty-five percent of the men over fifteen have worked outside the community and 55 percent of the women have done so (Stuart and Kearney 1981).

Quirino began to talk some about his first of three successive trips "al otro lado" (the other side—the United States). He had spent most of his time in Portersville and Farmersville near Bakersfield. As he talked, children came and went and I asked how many he had. He answered that he had five and that he thought that was too many, because life was so expensive and it was very difficult to support so many. He asked how many I had. I answered that I had two and showed him my billfold pictures. He said that I should be sure to talk to his wife about my two children, because he was trying to persuade her that they shouldn't have any more. "She wants more, of course, because she loves them so."

Quirino's perspective is interesting because it is often said that men in male-dominated cultures such as Mexico's are anxious to have more children as a sign of virility and that women are more likely to want to restrict family size. If women are admitted to want large families, it is generally ascribed to the notion that women are only valued as childbearers and child rearers, not to the possibility that many women may simply love children and enjoy caring for them.

More generally, demographers tend to believe that both men and women in peasant societies value large families because they can productively use their help in the fields or in other opportunities to earn income, that the children cost little out of pocket to rear, that children represent the only available form of disability and old age insurance, and that children are in any case regarded as gifts of God or nature, gifts one questions or prevents only at great risk of cosmic retribution. In apparent contradiction to these views, Quirino's attitudes deserve close consideration.

Some demographers believe that economic motivations are both strong and finely calculated in peasant cultures, explaining differences in attitudes toward family size by fairly precise analysis of the particular economic factors pressing in on a family. Quirino's attitudes could then be understood in the context of several facts: Quirino's landholding is so small and poor that additional labor applied to it will yield little additional income; Quirino no longer regards his land as his chief source of income but rather relies mainly on migrant labor for wages in the United States; and it is difficult to cross the U.S. border with children and even more difficult to find well-paid field labor for them once across the border. Combined with the ambitious effort of the Mexican government since the early 1970s to provide information and birth control to the population, it may be fairly easy to understand Quirino's viewpoint, the viewpoint of a man held between peasant life and the life of a wage laborer. Anthropologists and sociologists have usually assumed that a more secularized world view likely accompanies the move from peasant ways, changing fear of cosmic retribution into fear of economic failure. Of course, for parents two or three generations back, the very high child mortality rate in rural Mexico turned such questions around—it was difficult to bring enough children to adulthood to satisfy the family's economic needs. It is relatively cheap and widespread public health and medical improvements that have raised the issue, creating a Mexican population where nearly half the people are under age fifteen, still to enter the reproductive period of their lives.

In one sense, though, the reproductive patterns of San Jerónimo in recent decades, like most of peasant Mexico, have clearly conformed to a major insight family planners use in analyzing population growth patterns. The high degree of generalized insecurity in the village correlates as one would expect with a high birth rate, and the introduction of some of the most important elements of modern public health and medicine has ensured a high survival rate relative to the experience previous to World War II. Thus, the population growth rate is remarkably high. Very rapid population growth is the result of the combination of poverty, insecurity, and inexpensive public health measures. The poverty and economic subordination of the Mixtec villages predates rapid population growth, but the present rates of population growth, as Quirino Torres understands, often make the problems of poverty more difficult to manage.

I noticed that the family altar was extraordinarily simple and unkempt. There was a murky religious picture at the center, two votive candles in cracked cups, and plastic flowers stuck in Pepsi bottles at either side. An electric drill and skill saw lay nearby.

Quirino wanted to know my religion and I explained that I was not a church member but had my own religious ideas. He let it pass with a bit of a quizzical look. When I was done eating we went back to the *juzgado*, where about a dozen men were gathered, sitting on the benches, talking. As we approached, I asked if this was a meeting and Quirino answered in a distracted way that, yes, it was.

Quirino told me to sit on one of the two rows of benches that faced each other under the unpainted adobe arches of the *juzgado*. The men assembled ranged from about twenty to their sixties. Some were dressed in jeans, cowboy hats, western shirts with pearly snaps, and heavy work boots, while others wore huaraches, serapes, loose muslin pants, and blouselike home-made shirts. Every combination of these elements was represented by one man or another, giving an impression of varying degrees of identification with the outside world. While the conversation proceeded, several of the men braided ropes and wove hats and baskets, products that would later be sold for less than a tenth their retail value to middlemen. When I came, there were two Pepsi bottles filled with a clear fluid that turned out to be mezcal. I was asked to take a swig when the bottles were passed, and the home-made mezcal turned out to be mild and delightfully flavored. Later, a young man showed up with a case of Superior beer, and we each had a couple of bottles during the afternoon. Every time someone took a swallow, he offered a word in Mixtec, later changed to "Salud" in apparent deference to me, and all drank together. It took awhile for me to realize that whenever I took a sip, I was forcing everyone else to drink with me.

Quirino introduced me in Mixtec, and a very long conversation ensued in that language, with only an occasional Spanish word I could pick out. After perhaps an hour of conversation, I asked to excuse myself for a moment, but Quirino directed me to stay seated, "Because we are about to come to a decision in your case." Then, a man in his fifties with a long, sober, and pained-looking face read my letter in the same slow, deliberate way Quirino had. He then explained or translated it in Mixtec to the other men. While he talked, a woman appeared and asked permission to sit down, which was granted. I guessed correctly that this was Guadalupe, Ramón's mother. Her eyes were open unnaturally wide and focused fitfully. She was clearly in great distress. She frequently bowed her head and brought her rebozo up to her face as she restrained sobs. There were large, unrepaired holes in her rebozo. She sat beside me and stared at me a good deal while the conversation among the men continued. Quirino explained to me that the older man, the translator, had something to say to me.

The old man looked intently and unflinchingly into my eyes as he spoke. "We are poor people. Whatever you see here that we have, we obtained with much sacrifice. When the government comes to help, nothing is as it should be. As with this water pipe here, this plastic stuff you see piled against the wall here, useless, a gift that serves for nothing. There is very little that we get from them, and they take a lot from us. We are poor and small and there is little that we can do." As he paused, men around the circle shook their heads and grumbled in agreement.

"You who have come here to visit us, to come to know us, also with so much sacrifice, we do not need much from you. But what we must know is, when the government deals with us, or when the powerful men deal with us, with all these things that happen to us that are so difficult, we need to know, sir, yes, what is your opinion in these cases, we must know, what is your opinion?"

I began a little nervously, "I have come here to know you a little, to come to know how you live here. I have come to do this, because as Michael said in his letter, I am studying problems of sickness caused by pesticides, chemicals used in the fields of Culiacán, Sonora, and California. I came to talk to people here who may have had this kind of problem, and also to simply know you better so that I can understand the problems you have when you go north. As far as your problems with the government are concerned, it is difficult for me to give an opinion, because I cannot say yet that I am familiar with these problems here in San Jerónimo. I know that the campesinos, the *ejiditarios,* and *comuneros* in Mexico have had many very serious problems and that often the government has not helped as it should, or has hurt the poor people of the countryside. When these things occur, I am on the side of the campesino, the *comuneros,* the poor people."

There was a general nodding of heads in what seemed approval, various Mixtec words that also sounded approving in tone.

Quirino said to me, "Now we must talk about whether we can be of help to you and about what form your cooperation with us will take." What followed may be expressive of the fact that while the men of the San Jerónimo town council and I are all fluent in Spanish, for none of us is it our first language. And while Dr. Michael Kearney's letter in Spanish had certainly expressed clearly the purposes of my trip, none of the men of the village are fully literate. I also suspect that it typifies many of the interactions that traditional peoples with their own languages have with outsiders, except that in many cases the confusions are never clarified.

Perhaps ten minutes of conversation in Mixtec took place before the man who had assumed the role of translator turned to me again. During the Mixtec conversation I heard the Spanish words "lower prices" mentioned several times.

"The problem is that we are not sure how to arrange things. We are not sure about how to help you in the best way, and we haven't decided on the way you will be able to help us. So that everyone can know what you are offering, we will have to draw up a document with your proposal, so that we can all decide. There will be a wedding tomorrow afternoon, and perhaps, if the document could be ready, we could let everyone know there. The problem, señor, is that we don't know if we can afford these chemicals you are selling, and we don't know if we want to use them, because we have never used them here before."

I felt very disconcerted and replied, "I have not spoken clearly. I am not selling chemicals. I have no chemicals to sell. I am here to talk to any of you who became sick while working in the fields of Culiacán or California, because these chemicals have been used in the fields and some are very poisonous. I want to try to help, in working with others, to be certain that people do not become sick from these chemicals. It would be helpful to me to talk to anyone who has become sick while working in the north, especially if the person was in a place where these chemicals were used."

Several replies came at once, all in Mixtec, as the men looked around at each other. Several of the replies contained the Spanish word for medical doctor. Quirino said to me in Spanish, "Now we understand that you are a doctor. Now we will discuss how you can help us."

I replied, "I have still not spoken well. I am not a medical doctor. A doctor could tell you that if you are sick in your stomach that it is perhaps because you drank some water that was not clean or that you ate some bad food. He could give you some medicine, perhaps. But he does not work to see that your water is clean and your food is good. The chemicals used in the fields of Culiacán can be very poisonous. My colleagues and I are working to see that the poisons that make people sick are not misused in the fields. If we can do this, then the workers will not get sick from these chemicals. I need to talk to people who have become sick from these poisons to know more about how these poisonings happen, so that we can avoid them in the future."

A man sitting next to Quirino answered. He was a lively looking man, in his forties, with luxuriant salt and pepper hair and a bushy mustache, named Samuel. His face was handsome because of the way his expressions constantly reflected his response to whatever was being said or done at the moment. His eyes twinkled. He had been treated with special respect by others, and he often seemed to fill in when Quirino was at a loss. He had been at the center of nearly every outbreak of laughter. He was dressed in a cloth jacket with fake fur sheepskin lining, jeans, heavy, high-heeled laced boots, and a broad-brimmed felt cowboy hat with a red tassel in the back. Samuel said to me, "Look, I know what you are getting at. But you have to understand what it is like in Culiacán. We don't get sick while here in San Jerónimo, not often. And we usually don't get sick in California. There's usually good food, good water, a clean place to sleep. But in Culiacán we only drink from the irrigation ditches. The water is all foul. One person washes in the ditch and a meter away someone draws a bucket of water for the family. And yes, the chemicals, the liquids they use in the fields, they drain into the water and we drink them. We work with these liquids in the field. So, we get sick. But there is sickness everywhere. Culiacán is a land of sickness. Sickness is in the water, the food, the camps, and who knows when the sickness may be in these liquids, these chemicals, you are talking about? I will only be able to worry about pesticides when I can be permitted to deal with disease, and hunger, and the endless miles of the road."

An old man seated next to me, dressed in a brightly colored cheap serape, muslin pants, and huaraches with rubber tire soles, pushed his face into mine and said, in broken Spanish, "I want you to know, they say the Indian eats bad food, drinks bad water. Yes, they say that. It is not true. The Indian makes clean food, we drink clean water, *potable water* [pointing at the plastic pipe leaning against the *juzgado*], everything clean. We know about cleanliness. But when we travel, they feed us bad food, bad water, very bad, they always give us bad things to eat. And then we get sick. Then we ask for the doctor. And then they say, the Indian eats bad food, drinks bad water. Then they talk about the dirty Indian. No, señor! We make clean food, drink clean water, we do not get sick. But when we are with them, they sell us dirty things and then say we are dirty. Señor, we are clean! We need no doctors here, señor." He shook his head in apparent satisfaction with his words, smiled a toothless smile, and patted me gently on the shoulder.

At this, everyone looked a little nonplussed and uncomfortable. Someone picked up a beer bottle, and all followed. "Salud!" All drank. The man in the serape bowed, excused himself, and went shuffling off.

"Look," I said, "I think this can be very simple. I want to talk to anyone who has become sick in the fields. I came to talk to Irineo because I believe that his son died because of the liquids, the chemicals, the poisons they used in the fields. I wanted to know more about the death of Irineo's son, so that perhaps we could avoid other such deaths. But, unfortunately, Irineo now has died, too. It is difficult to discuss these things in Culiacán because there is not the liberty to speak freely there. [A few words and grunts of agreement, vigorous nodding of heads all around.] So, perhaps we can speak here.

"This is what I want. If anyone has become sick in the fields or the camps of Culiacán, Sonora, or California, I would like to talk to them. Ask your friends and relatives if this has happened to them, and if so, tell them I would like to talk with them. If you have become sick, I would like to talk to you. If someone does not want to talk about this, that's fine. I have come to know you better here, and this is enough. This is all we need do."

There was another general round of nodding, and the men looked at each other for signs of agreement. Quirino said, "Fine, now this is settled. Thank you for coming to know us. Later we will discuss the form of your cooperation with us. Now, we have another matter. Salud!"

At this point, Guadalupe was given permission to speak. She spoke at length, very forcefully. When she was done, an old man dressed traditionally, now sitting in the spot vacated by the man who had spoken to me about cleanliness, began to speak. He spoke with obvious anger and irritation, and for a long time, as much as twenty minutes each time he was given the floor. The other men listened without comment, some of them showing occasional expressions of sympathy with the speakers, most of them impassively, as this exchange between Guadalupe and the old man went on very nearly two hours. Often, while the man was speaking, Guadalupe would cover her head and face completely with her rebozo and sob heavily. Her dark eyes, set in a squarish, small face, seemed enormous. She appeared, in turns, outraged, dismayed, and imploring. Quirino seemed somewhat at a loss at times in trying to preside, and Samuel, the man who had told me about conditions in Culiacán, often intervened with a soft, sympathetic voice that helped calm things down.

At one point, I leaned across to ask Samuel quietly if Guadalupe was asking for help from the community. He said yes. I said that since I had come to visit Guadalupe's husband, perhaps an appropriate form for my cooperation with the community would be for me to give Guadalupe some money. He said yes, that might be possible, but asked me to be quiet and wait until the present matter was settled. The exchange continued for some time. It ended with Quirino and Samuel speaking to Guadalupe and the old man for a while as the others listened. The old man bowed and left, and Guadalupe seemed to indicate her assent as he had, but remained seated. Quirino would explain to me later that evening what had happened between Guadalupe, the old man, and the community.

Samuel introduced me to Guadalupe formally, explaining that she was Irineo's wife. He said that I could ask her a few questions if I liked, and that he would interpret. He suggested that since Guadalupe was curious about my interest in the family, it would be good to do this immediately. I explained through Samuel why I was there, and I gave Samuel the photographs Michael Kearney had taken of Irineo, Ramón, and the friend in Riverside, and a photo of Ramón alone. She grabbed them from me, gave them a quick glance, and stowed them away somewhere under her rebozo. She thanked me perfunctorily. I explained to Samuel that most people who knew of her son's death said that it was caused by pesticides, but that others said it was caused by a kick he had received while practicing karate with a friend. She replied through Samuel that she had heard both things said, that she did not know whether it was a blow or a poison—that she knew that he had died, and knowing that, it really didn't matter to her how he had died. I asked whether she knew the date of her son's death, and she answered without hesitation that it was February 9, 1981. She covered her head again, sobbed a couple of times, recovered, and looked me in the eyes again. I asked if there was anyone else in San Jerónimo who knew the circumstances of the death and got a negative nod. I thanked Samuel and Guadalupe. Guadalupe excused herself.

The group of men began another discussion. Guadalupe had only gone around the corner of the building, where she was surrounded by her five children, who began to peer around the corner and beckon to me. I excused myself and went to speak with Guadalupe, with her ten-year-old son serving as interpreter. She wanted to know if I could give her money. I said that I would speak to Quirino and the others and would come by her house before I left to speak with her. She thanked me and turned to take her five children up the hill to her house.

Soon after I rejoined the men on the benches, where they still talked, wove hats and baskets and braided ropes, and took ritualized swigs of beer until the meeting was adjourned. The little wooden table was taken back into the *juzgado*, the beer bottles put back into the case, and the lights turned off. The time was nearly 7:30, more than four hours after the meeting had begun. Quirino would later explain that meetings of this length were necessary five or six days a week—they were, after all, the complete administrative decision-making mechanism for a town of nearly two thousand people.

After a lot of handshaking, Quirino and I headed back down across the back of the churchyard to his house. On the way, I picked up some things I had brought along in the car to eat, including a six-pack of Tecate beer. Quirino's wife was in the kitchen, a separate room of the house with its own entrance. There was a raised fireplace, with a shaped adobe hood and chimney, but the children had built the fire this evening among three pottery jars on the cement floor. Quirino's wife, with her seemingly permanent look of amusement, was sewing and nursing her youngest son, while the oldest child, an eleven-year-old daughter, took care of the fire and the other children. Quirino explained when asked that they didn't use the fireplace when the nights were cold because all the heat went up the chimney—"one must conserve heat on such a night." He would order the children from time to time to open or close the door and wooden window shutters, trying to keep a balance between heat and the amount of smoke in the room.

We broke open a pomegranate I had brought with the help of my Swiss army knife. "Oh, those are really good," Quirino exclaimed, pointing to the knife. "They have everything you need." The children were delighted with the juicy red seeds of the pomegranate. Quirino and I popped open a couple of beer cans.

Quirino took up again the story of his first trip to the United States, four years earlier. He had crossed the U.S. border somewhere in the mountains between Tijuana and Mexicali. Some of the men he was traveling with got caught by La Migra (the U.S. Immigration and Naturalization Service), and in running from them the rest of the men got separated from each other. He spent several days walking in the mountains, staying away from towns and people. This meant generally staying as high as possible, and he found the cold at the high altitudes almost intolerable, sleeping on the ground at night with nothing but the clothes he wore and a jacket. He went three days with no food. One morning he woke up with one leg locked in a cramp, bent at the knee, and had to spend several hours getting it warmed up and flexible again.

In Ventura County, some fifty miles north of Los Angeles, he met up again with some of his *compañeros* from San Jerónimo who had crossed with him. They went together to a labor contractor (a Mexican man from Michoacán) who told them that there was no work in Ventura County but that he could probably put them to work over the Coast Range in the southern San Joaquin Valley near Bakersfield.

In the meantime, he invited them into his house and brought in some beer and groceries. The next day they were taken by truck to a farm in the Portersville area in the San Joaquin. There they got work for the season, and by the end of the summer had accumulated "bastante dinero" (plenty of money) and felt very satisfied with their work. By now there were forty to fifty Mixtecos from San Jerónimo working in the same area.

One night as they sat around discussing how well they had done, Quirino proposed that they should each contribute to some kind of gift to the village of San Jerónimo in memory of their good season. He proposed that each man make a suggestion for what the gift should be. They would discuss the suggestions and come up with an agreement about which idea was best.

Not everyone had an idea, but some did. One man suggested that they pay the priest in Silacayoápan to say mass in San Jerónimo every day in the month of September, since usually the priest visits the village no more than once a week and since September and October are the months in which most people have returned to San Jerónimo from their migrations. Quirino spoke against this suggestion, because it would be a gift that would last only one month, and it would cost an amount that one man could pay for alone, if he so desired. Quirino said the gift should be something permanent and should be something that no single person could ever buy for the village. He said that they should all put in $40 U.S. apiece and buy a clock for the church tower in San Jerónimo. After some discussion, this idea was accepted. They raised nearly $3,000 for the clock. Miguel (Michael Kearney) was asked to contribute, too, as were the two "gringas," Juanita and Juliana (missionaries from an evangelical sect who had lived off and on in San Jerónimo for five years). When Quirino arrived in San Jerónimo after the summer season in the San Joaquin was over, he began to look for a clock. They spoke with a representative of a factory that makes such clocks and made a deal. When the factory representative came to look at the church tower, he advised them that the tower would have to be reconstructed to be strong enough to support the clock reliably. Quirino organized the men of the village to reconstruct the tower and install the clock.

The purchase of the clock is an excellent illustration of a point made by Kearney with regard to the effects of migrant wage labor on Mixtec communities. Mixtecs are not simply becoming drawn into a world outside their village and region, either physically or culturally. As Kearney says, the migration experience intensifies community and cultural values as well as attenuating them. Mixtecs are putting money back into their villages, not only in the form of private economic investments such as oxen and concrete block houses, but also in public improvements specifically intended to express a sense of gratitude and concern for the community. Douglas Butterworth, in his study of Tilantongo, a village in the Mixteca Alta, noted that migrants to Mexico City had formed an association, one purpose of which was to make contributions toward public improvements in the home villages. At an annual meeting, members of the association made a precise reckoning of these contributions and the progress they represented. Butterworth also makes it clear that this effort went on in spite of the considerable tensions that existed between those who had migrated and those who remained at home, those at home seeing the migrants as a species of traitors who "lacked sincerity" and the migrants seeing the stay-at-homes as hopelessly backward. Kearney notes that the villagers of San Jerónimo have set up a kind of parallel village settlement in Colonia Obrera in Tijuana, located next to another parallel settlement composed of people from a town that borders the San Jerónimo commune. They have also set up another daughter community in Colonia Oaxaca in Nogales, Sonora. Ravicz, in his study of five Mixtec villages in the 1950s, noted that Mixtec political and religious organizations and such traditional values as respect for elders were strongest where mestizos were present in significant numbers to constitute a local ruling class, while Mixtec organizations and values were weaker in exclusively Mixtec villages. In my own work with educating farm workers about pesticide hazards in the Central Valley of California, I attended meetings where the Mixtec workers came into the meeting as a group and sat as a group. At the end of the meeting, they put away the chairs and tables and straightened up the hall as a group, without help from other Mexicans, Chicanos, and Anglos attending and without being asked to do so. It appears that there is a strong tendency among the Mixtecs to strengthen community bonds and values when under pressure from the outside world (Kearney 1984a; Butterworth 1975; Ravicz 1965; Kearney and Nagengast 1988).

During our conversation, a handsome woman of about fifty came by with her eleven- or twelve-year-old son. She asked permission to enter and sit with us, which Quirino immediately and ceremoniously granted. She spoke some words of Mixtec to Quirino in a voice barely above a whisper, whereupon Quirino turned to me to explain that his neighbor was fond of beer and would be pleased if I could give her one of the Tecates still left from the six-pack. When I handed the neighbor lady the beer she beamed gratefully and popped open the tab on the can expertly. The slightly awkward feeling among us that had arisen with her entrance vanished with her first long draw on the beer.

The neighbor lady wanted to know if I could do her the favor of delivering a letter to a relative in Colonia Obrera in Tijuana, where people from San Jerónimo have established a permanent barrio. I explained that I would make an attempt to do so through Michael Kearney, who sometimes visited there, but that I couldn't promise it would arrive or when. She decided to keep her letter and look for another way to send it.

Quirino began to explain what had happened during the meeting between Guadalupe and the old man. The old man was the owner and driver of the truck in which Irineo had died. There are only three trucks and no cars in San Jerónimo, and in the dry season the only bus leaves San Jerónimo before dawn each morning, so transportation is always a problem. Juan, the old man, had offered Irineo, along with some others, a ride in the back of his truck into Huajuapan. During the trip to town the brakes on the truck failed and it went over the edge of the road into a barranca. Irineo and his two friends had pitched over the cab onto the ground. Quirino explained that Juan had paid out a lot of money as a result of the accident. First, he had to pay a fine for hauling passengers without a license. Then, he had paid for the costs of an ambulance into Huajuapan. Then, he had paid for caskets and for the transportation of the bodies back to San Jerónimo. He had gone to Guadalupe to pay his respects to her and offer his apologies for what had happened.

Guadalupe was not satisfied because she said that Irineo had been carrying 12,000 pesos (about $75 at current rates of exchange) in his jacket pocket to buy some things for the household. When his body was returned to San Jerónimo he had the jacket on but the money was gone. The ambulance attendants swore that they did not take the money, although Juan suspected they had. Guadalupe had come to the council meeting to say that she held Juan responsible for the money and that she wanted him to give it to her. After the long discussion, the council decided that Juan could not be held responsible for the lost cash, that he had done all he could be expected to do to make up for the accident, and that Guadalupe could not be blamed for being upset, but that she would not be able to make further claims on Juan. They further asked that Juan and Guadalupe come to the *juzgado* the next day when they would sign a legal document to be drawn up and typed by noon, in which each of them promised not to defame or gossip about the other in the future. They had both agreed to this. Quirino wanted to know whether I thought this was the proper thing to do. "Is this the way you would have settled it in your country?" he asked.

I explained that where I lived such a matter would be fought between lawyers and insurance companies, and a little about the laws with respect to auto and liability insurance, ending with, "And so, the laws are so different, and so much more impersonal, that it would be impossible for me to say how they would be applied to such a case. Certainly the two people would not be confronting each other before a town council for a decision."

"Yes, I see. I understand, and it reminds me of things that happened when we were working in California. But do you think what we did was fair?"

"With what little I know of the situation and of your rules, it seems fair to me."

Quirino seemed satisfied with that reply and moved on to question me about my religion, with about the same inconclusive results as earlier in the day. It was very difficult, apparently, for him to make the distinction I made between my own religious and moral ideas and a church as an institution.

Quirino said that Juliana and Juanita, the two gringas who had lived in San Jerónimo for five years, were very popular in the village, although they were not Catholics. "Still, they cooperate with us when we need something, they always contribute, as does Miguel. Everybody who comes here contributes something."

I suggested that I make a contribution by giving some money to Guadalupe. "Times must be very hard for her with her two oldest sons and husband dead now."

"Yes, times are very hard for her." I sensed that there was something wrong with my idea of giving money to her. I suggested that I divide the money I had to give between Guadalupe and the community. That seemed to be better but still presented, I sensed, some kind of problem. Then I suggested that I give all the money to the community, with the understanding that half of it would be passed on by the community to Guadalupe.

Quirino brightened immediately. "Yes, that is the right way. You give me the money and tomorrow we will go up together to Guadalupe's house where I will explain to her that you have given this money to the community so that she could have half of it. That will be fine. And the rest we will discuss with the council tomorrow so that they will understand. I will check with some of them first before we go to Guadalupe's, but I'm sure that this arrangement will be very good. And you will be present at all these discussions to know that I am not keeping the money for myself."

During the conversation, the children and their mother had been talking a little, eating the various things I had brought in with the beer and listening. They smiled at me and Quirino almost constantly and warmed the room like a second smoky little fire.

We talked on for a while, then decided it was time to go to bed. Quirino offered me one of the beds in the big room, but seemed relieved when I showed him I could sleep comfortably in the car.

I awakened about four o'clock the next morning, and in the darkness could hear the sounds of the households around me beginning to stir. I took Quirino's advice about defecating in a dark place under a tree (he explained that there were no toilets or outhouses) and in a bit noticed that the pigs that range freely in the village were eagerly scavenging for the night's remains. I had read that in Mixtec villages, children were often taught to relieve themselves in the fields in order to return the nutrients to the soil. In San Jerónimo, apparently it is the pigs' feces that are returned to the fields.

The clock, as it had all night, sounded its bells every fifteen minutes. The noise of people with small hatchets chopping the morning's firewood was everywhere. Chickens and turkeys joined wild birds in the trees in a cacophony of avian conversation. Donkeys brayed and radios played quietly within the houses. At four-thirty, still in the dark, the bus driver blew the horn repeatedly to warn of departure, and shortly after the bus began to growl up the grade out of town. I climbed a ridge at the edge of town to listen and watch as the town awakened.

As the sun came up, its rays were filtered through the grey smoke of dozens of cooking fires slowly rising from the valley and slopes. The light on the higher mountain behind me showed an oak forest much like that on the foothills in some parts of the Central Valley of California, except that the regularly shaped oaks were smaller and more closely spaced here. Wherever the forest had been cut, great gashes of erosion ran down the hillsides. Within the village, houses had their own small fields, recently plowed, usually along the contour to reduce erosion, sometimes with a little effort at terracing.

In the earliest light of dawn, a young man approached me, introduced himself, and inquired why I was in the village. He said that he had been to the United States and spoke a little English that he had learned there. "Good morning, how are ya?" he said. "I was in Riverside, but not more," he said slowly in English. Then he shifted to Spanish and asked me what my religion was.

Once again, I explained that I did not belong to a church but that I had my own religious ideas.

"Of course, I know what you mean. Isn't it the truth that all of us

have our own religious ideas, our ideas of good and bad, whatever we
say our religion may be. Take me, for example. I say that I'm a Catholic, like the others here. But what does that mean? I have asked the
priest, I have asked different people. They have their own ideas. I
don't know what it means to be a Catholic. No one seems to know
what it means to be a Catholic. But I certainly have my own ideas
about religion, about God. And so I am a Catholic, but I have my
own ideas, like you. I understand. I don't know what it means to say,
'I am a Catholic.'"

He explained that his job was driving the big red flatbed truck
with the stake sides parked below us on the hill, hauling whatever
there was to haul on contract with the truck's owner. They had a
contract to haul all the soft drink bottles.

It was Saturday morning. Women and children hurried about
carrying water, charcoal, and food. Some little girls about three years
old carried enormous armfuls of lilies, peeking around them to see a
stranger pass. Men worked building houses, especially the new concrete block houses that had become so popular. By ten o'clock, some
men were already returning from their plowing in nearby fields.

A middle-aged man sitting outside his house near the center of
town called to me and asked me to come sit and talk to him. He introduced himself as Vicente Morales. As soon as we had sat down
outside the open kitchen at the back of his house he asked me, "Tell
me, sir, do you believe that God lives in sticks?"

I was not certain what he meant and looked confused, muttering
something or other.

"Well, look, if I put eyes and a nose on this stick with my knife,"
he said, pulling out a piece of firewood from a stack, "have I made a
god? Does God live there?"

I began to catch the drift. I answered, "No, I don't think so."

"Of course not. God does not live in sticks. You are right. God is
here between us as we talk. God is among us and in us. God is here
and everywhere. Do you agree?"

"Yes, I like that way of thinking about it," I hedged.

"And another thing. I have not been in this new religion very
long, just seventeen, no, eighteen months now. And I am still learning. But I have learned that the love of God is the love of self. And
love of others, of all creatures. But what I didn't know before was
that it is the love of self. Do you agree?"

"Yes."

"Yes. That means that if you love yourself, you don't spend your
time getting drunk in cantinas, wasting your money, getting in fights,
ending up in jail. I didn't understand that, but now I do understand

that the love of God is the love of self, I know that, and I don't do those things anymore. Because if you love yourself, you treat your own body and your own family with respect."

"I see."

"No, God does not live in sticks, sir, this I have learned. You know, when I was a child the old people taught me that the gods lived in the mountains, in the wild places, and then this Catholicism came along, with the church here and everything, and they said that God lived in sticks, in these saints they put up in the church. But now I know that God does not live in sticks, that they are wrong. That God is right here between you and me as we talk." (Catholicism had certainly been present in the region long before Sr. Morales's birth. It is not clear just what he meant here by "then this Catholicism came along.")

He went on to explain that he had been taught these things by Juanita and Juliana, the evangelical gringas. Later, he showed me some books that had been printed by the Summer Institute of Linguistics with the Mixtec creation legend and other Mixtec stories printed in the San Jerónimo Mixtec dialect, along with the New Testament, all of which Juanita and Juliana had translated. During that same week, Mexico City newspapers were carrying stories about the federal government's National Indigenous Institute throwing the Summer Institute of Linguistics missionaries out of Mexican Indian communities on the grounds that they were undermining Indian culture and exploiting the people for the sake of the growth and aggrandizement of fundamentalist religious sects. Some of the newspapers accompanied the news stories with accounts of the fundamentalist Summer Institute's alleged involvement with the U.S. Central Intelligence Agency and conservative political forces in various parts of Latin America (*Uno Mas Uno*, January 15, 1984). Vicente showed me the Temple of the Evangelicals of the Americas— Filadelfia that he had helped build just above his own field. I was struck by the prominence of the "Property of the Federal Government of the Mexican Nation" sign in the same letters as the name of the church (the anticlerical revolutionary Constitution of 1917 made all churches federal property and forbade church ownership of real property, but I had never seen it stated so prominently on a church building). Knowing little of this particular Protestant sect, I asked the meaning of "Filadelfia." Vicente didn't know. I suggested that it might refer to the city in the United States, and in the interests of keeping the conversation going, I told him that the city had been named the City of Love by the Quakers who had founded it. I mentioned some of the Quaker beliefs, including pacifism. Vicente was quite interested, and I told him I had known Quakers who had gone to jail rather than serve in the army.

"That is good. I like this 'pacifism.' That is a good belief. Of course, some atheists might be pacifists, too, although atheism, as you know, is not a religion but simply a belief. But I like pacifism because it would stand against all the destruction, all the killing that happens in war. But, not because you are afraid of death, no, not for that reason. It is foolish to fear one's own death. People often believe that they can avoid their own death, when they know nothing of it, when it is not in their hands, when their fear makes them cowards and useless to God. But this pacifism would not be the fear of one's own death, but to avoid the destruction of war. Is that right?"

"Yes, I think so."

"Well," he said, "there are not many of us here, we evangelicals. There are one hundred and eighty in this temple. But it is enough. Even if we are ridiculed."

After a pause, he continued. "Do you read the Holy Bible? Well, you might remember the story of Noah. God told Noah to build a boat because there would be a great flood that would destroy the wicked. Everybody laughed at Noah and asked why he was building a boat, and he said he didn't know why, that God had told him to do it because a flood would come. So they made fun of him. And the flood came. It began to rain, and rained, and kept on raining. People got up on top of their houses. Those who had two-story houses, they got up on the roof of the second story. And then they said to Noah, 'Let us on your boat,' and Noah said, 'No, it's already too late for that. Too late. I'm sorry.' They were afraid, but Noah had obeyed the will of God."

I asked to take a picture of him, and he posed expertly, lifting up his face out of the shadow of the big-brimmed hat. "What is that camera there? A Canon? That should be good. I have a couple. The Pentax is the good one. Come, have a cup of coffee." (Kearney later explained that Morales had been an itinerant street photographer in Hermosillo and Tijuana.)

We sat in the smaller room of Vicente's two-room house. His much younger wife brought in coffee and rolls and explained that she was too busy to sit down with us and that she hoped I would forgive her. Vicente showed me a photograph of his second house, one he built in Hermosillo, the capital of the northern state of Sonora, just four hours south of Tucson. He used the house when traveling "al otro lado" or while working in the fields of Sonora. He told me that he had gone to the United States for the first time as part of the official bracero program, to Texas. He seldom went to the United States now. He also wanted to talk about the bags of fertilizers stored in piles in the front room. He said the fertilizer seemed to work well on the deep soil of the field next to his house here in town, but up on the hillside, in the *tierra blanca*, it didn't seem to do much good. (*Tierra blanca* in the Mixteca usually means barely decomposed subsoil rock, granite, or limestone, the only thing left in many areas where sheet erosion has stripped away any other soil that existed there when the hills were forested with pine and oak.)

"Perhaps the chemical fertilizer does not do well where there is so little topsoil and organic material capable of absorbing and holding the nutrients and the water."

He thought about it and said, "Maybe so."

After some more conversation, I took my leave and began to walk in the mud streets again. Sr. Morales had reminded me of the work a former teacher of mine, anthropologist Robert Hinshaw, had done on Guatemalan villages, showing that becoming a Protestant was one of the ways to protect personal savings and become more wealthy than other members of the village. As a Protestant, one cannot serve as Quirino Torres was serving as an officer of the village and so cannot be asked to spend personal time and savings on the village celebrations and other community needs. The *cargo* system of obligation to serve the village serves as an economic leveler, keeping everyone more or less in the same state of shared poverty. Protestantism is a way to leave the obligations of the *cargo* system behind without completely losing the respect of others. Sr. Morales, with his two houses, two cameras, and bags of chemical fertilizer seemed to epitomize this process (Hinshaw 1975).

Soon after leaving Sr. Morales's house, I was stopped by a handsome young man named Adolfo who wondered what I was doing in San Jerónimo. He said that he remembered my friend Miguel (Kearney) quite well, and that Miguel and a friend had rented a house up the hill a ways and he had talked to him a number of times. I explained that I was interested in how the Mixtecs did when they were in Culiacán and especially whether they had any trouble with pesticides.

"Sure, lots of trouble. That place is terrible. The houses are just made out of that black cardboard or out of sheet metal. Everybody in those big camps. The water and everything is filthy. You work six and a half days a week. Everybody is scared there. If you get sick, they fire you."

"Have you ever known anyone sick from the pesticides?"

"Yes, lots of times. But it's hard to do anything about it. If you take off work to go to the doctor, they fire you, like I said. And if you make a complaint about the poisoning, they fire you, or they shoot you."

"Shoot you?"

"I've seen it twice. They just gun you down. The *pistoleros* who work for the growers. They'll just shoot you dead. [Another Mixtec with whom I spoke in Fresno, California independently made the same charge against the Culiacán growers, saying that he had seen a man gunned down for complaining too vigorously about pesticide illness.] They don't care about anybody and they don't want anything going out of the camps about what it's like there. But here, everybody knows that's what it's like in Culiacán."

I asked if he had been in Culiacán recently.

"I was there last year, and I'm going to be there next month. I've heard there's not much work in California now, so we'll be going to Culiacán."

"Good, I am going to be there too. Maybe I could find you. Where will you be?"

"Look for me in Campo Patricio, Campamento San Luis. I should be there, or ask for me there. We can go have a talk. I can help you. No one will talk to you in Culiacán. Too scared. But I know you from here, so we can talk."

I agreed to try to find him. (I did try, but failed to locate him in the teeming camps with thousands of people.) He changed the subject to Juliana and Juanita.

"Do you know them? Everybody here in San Jerónimo is very happy when they are here. We feel ill at ease then when they are gone, like now."

"Why?"

"Because no matter what we need, they'll always contribute. If some of the men are working on the *juzgado* and need to take a break and have a bottle of Coca-Cola, they'll buy a round. They contributed to the clock up there in the church tower. Whatever we need, they help. They bring medicines and sell them to us very cheaply, antibiotics and other things. That's a big help. They speak our dialect. They've been hiring me to help them learn our dialect, and they speak like one of us now. They are good, even though they are not of our religion. But nobody really cares."

I slowly worked my way back toward Quirino's house, but every few paces someone would stop to ask me what I was doing in San Jerónimo and to talk. One eight-year-old boy asked if I had three dollars in American money and what the current exchange rate was. When I told him I did have the three dollars and that the rate was about 160 to 1, he raced home and brought back 480 pesos to exchange for my three dollars, with many thanks.

A thin, short young man named Magdeleno stopped me to talk. We sat on an embankment as he told me of his experiences in Culiacán, about which he was very bitter. He had been in various *campamentos,* including one near Morroleón (a few kilometers from the state office of Sanidad Vegetal along the same main road), where there had been several thousand people. The workers had staged several strikes while he was there. There had been a great deal of sickness. The growers built a bathhouse, but there was no water. After protests, water was brought from the irrigation canal, with a filter system. "But the filter system didn't work, and the pump gave out, so we had no showers or drinking water except directly out of the canal after all."

Magdeleno picked loose threads from a seam in his jeans as he continued to talk. "A lot of people died that season, especially the children. They said it was a cold or the flu. It was very difficult for the mothers to take care of them. The women had to awaken at one o'clock in the morning to prepare the lunch that would be all the men had to eat in the field. The outhouses were locked up at night. There were no beds, just whatever blankets or clothes we had to put on the ground. The company store [illegal according to the federal legal code] charged very high prices, and we were given script for some of our wages that was only good in the company store."

I asked if there had been any special trouble with pesticides. "Yes, of course. After being in the fields that had been sprayed, we frequently had terrible headaches and dizziness. Sometimes there were bad stomachaches, like cramps. The liquids would hurt the skin and cause sores like acne. When you tried to wash off with water, it would burn the skin like very hot chilis. And when you complained of illness, they wouldn't take you to the Seguro Social clinics, or they wouldn't give you the papers you need to get in. Only a few ever got to the clinics when they were sick. It was terrible, a lot of people died and were beaten."

"Was there anything you could do about all this?"

"No. They treated people like slaves. They would order us around, making us go into deep mud in the fields to work, and beating people who refused to work under such conditions. They fired me when they realized I picked tomatoes that were too mature—you see, I am color blind and can't tell the difference between the light green color you are supposed to pick and the pink and red that are too mature. Of course, the workers themselves eat the mature ones."

"Why do people choose to go to Culiacán under such conditions?"

"Mostly when they are not familiar with the other side, with how to cross the border and find work there. But, of course, during the winter months there frequently isn't any work on the other side. And everybody needs the money. People go to get the money to buy land, to buy a team of oxen, or to have enough money to get married."

As I walked the streets of the town, women often hid their faces from me as I approached. I noticed that if two or more men were talking and a woman approached, she sometimes would hold her face to the side, ask permission to speak, and when granted it speak clearly and boldly about whatever was on her mind.

When I got back to Quirino's, he was talking with the men building a new house next to his. We talked about the way to arrange things with Guadalupe and the community, and Quirino went in to his house to get some American dollars he had to make it possible to split my dollars evenly between the community and Guadalupe. We walked through the churchyard and up the hill to Guadalupe's house, to find that she was gone. One of the younger children was dispatched by a twelve-year-old daughter to go look for Guadalupe and send her home to talk to us. We sat on the dirt veranda in two kitchen chairs, looking over the village cemetery and the church. I took the opportunity to talk to Quirino about village government and the land.

"So you have these meetings nearly every day?"

"Yes, nearly every weekday. There are a lot of things that have to be settled. This is my year when I have been elected to serve as *agente*. If I didn't have to do that this year, I would be in California working to make money. It's very hard here. There's no money, and you can't grow enough to eat. You have to go to work. Everybody does who can. You will notice there are only three or four dozen of us working men here now—those of us who are on the council, some with trades, like house building so they can make money here in the village. I'm just here this year because I was elected to be *agente*."

"How many times will you do this in your lifetime?"

"Usually it falls on a man about three times. And after he has done it three times, he is excused. He doesn't have to do it again. But it's a hard job to keep things going in the commune. There's a lot of work every single day."

"Do you have a *milpa* planted?"

"No, I haven't planted for three years. First, I was traveling north, and now, as *agente*, I don't have the time. Just living on what I saved from traveling. That's it, that's all the family has now."

"What are the qualifications for *agente*?"

"Any man thirty years or older."

"What about Protestants, can they serve?"

"Oh, no, no! They can't be allowed to serve because many of the duties have to do with taking care of the church, with the building and saints and ceremonies, and so a Protestant couldn't do those things. It has to be one of us."

"The community is organized as a commune?"

"Yes, for some time now. I wouldn't remember exactly when, but maybe the fifties. After World War II, anyway."

"So how is the land divided?"

"After a series of meetings, it was decided what a man needed, and the commune gave ownership as was needed. Then that's what a man has."

"And what is the average amount of land a family works?"

"Oh, I don't know."

"About how many hectares, more or less?"

"We talk in terms of square meters. Maybe a man has ten square meters, maybe twenty-five, thirty. Maybe ten here in town, another fifteen or twenty on the mountain. It's hard to say." (While Stuart and Kearney's median of six-tenths of a hectare from a household survey contrasts fairly strongly here, their figures were literally on the size of holdings, while Quirino spoke of the amount available to actually farm.)

One of the children ran up to tell us that Guadalupe would be returning soon, to be patient.

"I hope you don't mind my asking you these questions. Is it alright?"

"Oh, certainly. Why not? We are just talking to pass the time. We have to talk about something. Whatever you want to know, that's alright."

From Guadalupe's veranda at the top of the ridge you can see the oak woodlands stretching out on the mountainsides to the west of town.

"These woodlands, do they belong to the commune?"

"Yes, they all belong to us. Some of the other villages don't have any woodland, but we do. We have to watch it carefully to see that they don't come in and steal from our woodlands. We keep a constant vigilance."

"What do you use the wooded land for?"

"Firewood and building materials. When somebody is going to build a house, they come and discuss it with the council and get permission to cut what they will need. For firewood, everybody just takes as they need it." Quirino ran his gaze from one end of the oak-studded panorama to the other.

"Can you farm that land?"

"Yes, you can farm it. Sometimes a young man will go out and clear a field and plant. Once in awhile somebody does that."

"How is the soil? How long can you plant in such a field?"

"It's not bad. You can plant for two, three years. No more than three, though, and it's already gone bad. You just leave then. It's a lot of work for three years of planting."

"What about your own experiences in Culiacán? You have worked there, haven't you?"

"I was there once, for a few weeks, the first time I went north. But we don't go there if we can avoid it. There's better money on the other side. And the life there in Culiacán is a very hard way to live."

"Have you had any troubles with sickness from the pesticides?"

"No, I haven't. I guess that some people say they have trouble with that, with the medicines they put on the plants, but I don't know anything about it."

Guadalupe's daughter drew circles in the dust of the veranda. There were flash cards scattered on the ground with arithmetic problems on them. A big stereo radio and cassette player sat on the ledge of the only window in the adobe house.

Guadalupe came trudging up the hill with three children working to keep pace with her. When she got to where we were sitting, everyone shook hands and Quirino and I sat back down. Guadalupe remained standing, leaning against the house, with her rebozo held over her mouth. Quirino explained our business and gave her the little handful of bills. She thanked us both and then set into a long speech to Quirino in Mixtec. At the end of it, Quirino tried to speak to me a couple of times but would turn his head and look at his boots.

After some moments of such discomfort, he said to me, "She wants me to ask you to do something. I don't think you can. Just say whether you can or not. I'll ask for her."

"Fine, we'll see," I answered. Guadalupe looked very intently at me.

"She thinks that the owner of the truck doesn't know where the 12,000 pesos is. She thinks the doctor who examined her husband's body took it. She would like you to go into Huajuapan and tell him to give her money back. She couldn't do it, but maybe you could. What do you think?"

I thought of all the difficulties of this plan and its likelihood to succeed and said, "No, I'm sorry. I can't do that."

Guadalupe understood the Spanish and immediately began to sob again. She recovered and thanked me and said good-bye. As we walked down the hill, I said to Quirino that this was a very hard time for Guadalupe.

"Yes, hard enough, hard enough," he agreed.

5

King Eight Deer, the Plagues, and the Devastation of the Land

The first Mixtec people were born from the earth, the traditions say. But the traditional stories, in Mixtec codices preserved by the Spanish, tell the story of a new group of kings born from trees in the north who came around A.D. 1000 to rule over the Mixtecs. The most prominent of these kings was a man named Eight Deer, who lived from A.D. 1011 to 1063 and who came from the great city of Tula, to the northwest of the present Mexico City. The difference between the Mixtec lower classes and the rulers is symbolized by the distinction between those born of earth and those whose parents werc trccs (Spores 1967: 67; Wolf 1959: 121).

Modern anthropologists studying the Mixtecs say that before the Spanish Conquest the Mixtecs had perhaps "the most highly stratified society in Meso-America" (Flannery 1983: 218). The same anthropologists have detailed the agricultural methods of the pre-Columbian Mixtec people, methods founded on the sharp difference between rulers and ruled. As discussed briefly in the previous chapter, this system was based on the control of the bottomlands by the nobility. About 30 percent of those excluded from the nobility were people who worked directly for the noble families under a relationship that may have roughly resembled European serfdom, with their labor available to the nobility in corvées, or obligated labor. The rest of the commoners made their living by farming the steep slopes of hills and mountains. At some point, the nobility noticed that as the slopes were denuded of the natural oak and pine forest by the efforts of the commoners, the rich forest soils came down into the bottomlands carried by the runoff from rainstorms. To take advantage of the rapidly eroding hillsides, the nobles put their laborers to work building check dams in the valleys and canyons to catch the runoff and capture the soil. As these check dams grew into elaborately designed systems of soil and water retention, they took on the character of valley-land irrigated terraces. Valleys became broader and soils deeper, and the system was so effective that even V-shaped canyons completely unsuited to agriculture were transformed into valley lands with fine soil. There was steady progress in the desperation and impoverishment of the commoners on the slopes as their laboriously cleared fields washed out from under them, forcing them to ever higher and steeper ground. With each such impoverishment the nobility was enriched. Those born of the trees were gathering the soil of the forests using the labor of those born of the earth (Flannery 1983). One can speculate that they may well have gained new obligated workers as well, as the commoners failed to provide adequate support for all their children on the rapidly eroding hillsides.

Known as the *lama y bordo* (literally, mud and edge, or, in Mexican usage, mud and dam or terrace) system, it is still practiced in the Mixteca, to the consternation of college-trained agronomists who try to persuade the Mixtec farmers to stabilize the slopes first rather than concentrate on capturing the eroding soils on the bottoms. Of course, it is still the case that it is not the more desperate farmers of the slopes who control either the bottomland or the efforts of the community. Archaeologists working with geological techniques consider that the *lama y bordo* system has been in use in the Mixteca for about one thousand years, or, roughly, since the coming of the Toltec ruler Eight Deer (Flannery 1983; Mouat 1986).

The *lama y bordo* system was founded on the maintenance and effective control over deep social divisions in the society and on the tight control of gangs of laborers under the supervision of ancient Mixtec agricultural engineers. It was a way of life that must have been burdensome, and surely some considered it oppressive—the ancient text that tells of those born of the earth and of the trees seems inspired in a kind of protest, even though it must have been composed by a member of an elite. In any case, it is clear that a sharp disruption in the effective control of guided labor, or unbridled dissatisfaction among those who worked the slopes, would threaten the system with collapse. The arrival of the Spanish, for a variety of reasons, tore the system asunder and set in motion the modern tragedy of the Mixtecs. This modern tragedy has been both oppressive to the Mixtec people and disastrous, perhaps beyond repair, to their land.

The story of modern Mixtec history is not exclusively tragic, however. The Mixtecs are sharply aware of their distinction as one of the relatively few indigenous American people to preserve their language, much of their culture, and a degree of control over an extensive homeland in the face of European conquest and nearly five centuries of domination. This is no small achievement. If it can be said that this is partly due to the forbidding character of their territory and the relative lack of interest Europeans have had in exploiting it, it is also true that the Mixtecs can and do take pride in preserving their way of life in such a difficult arid mountain environment under heavy social pressure. The qualities of toughness, perseverance, and community solidarity that have held together a culture under extreme stress are also, perhaps, the qualities that may allow the Mixtecs to shape a more hopeful future than otherwise could be imagined. For that reason, it is worthwhile, before turning to the story of the tragic devastation of the Mixtec landscape in the wake of the Spanish Conquest and more recent insults, to try to understand the origins and character of the Mixtec people.

A distinctive Mixtec language appeared in Mesoamerica about three thousand years ago. This was a time of the proliferation of distinctive languages and cultures throughout Mesoamerica. Each major linguistic branch current then was to further divide and subdivide, as with the Mixtec, for example, that developed into Ixcatec, Mazatec, Popoloca, Chocho, and Mixtec. Linguists have identified 260 living languages from the northern border of Mexico to the Isthmus of Panama, and many of these have distinctive dialects. Today, various Mixtec dialects make it difficult for even Mixtec speakers to understand one another. The villagers of Ramón González's hometown speak a San Jerónimo dialect and do not take it for granted that all other Mixtec speakers will understand them (Wolf 1959: 34–47; Kearney 1984a).

The florescence of languages was expressive of the rapid cultural development that was occurring in Mesoamerica at least a thousand years before Christ. The hunters and gatherers who had worked their way south for millennia from the bridge of land and ice that had spanned the present Bering Strait during the last ice age were settling down into more permanent communities with a more or less strong reliance on agriculture. Archaeologists have unearthed Mayan house sites four thousand years old, identical in every essential respect with the house designs followed by Mayan people building their homes in rural Yucatán in the late twentieth century (Hammond 1982: 114–117). By the time of Christ, many of the peoples of Mesoamerica were erecting enduring monuments, adorning them with friezes, sculptures, calendric dates, and writing. By the time of Eight Deer, a millennium later, high civilizations with complex social organization, long-range trade, and massive cities had arisen and fallen into decay throughout what is now central and southern Mexico and the northern part of Central America. Each culture interacted with many others, but each developed unique characteristics and defended its existence as a political and cultural entity (Hammond 1982: 33–104; Wolf 1959: 1–151).

While it is relatively easy to locate and define the achievements of Mayan culture during this period, as with a number of other major and minor cultural groups, any interpretation of the Mixtec record must be more speculative. Readings of Mixtec history tend also to be provocative from the point of view of various schools of thought among those still trying to piece together the incredibly complex history of pre-Columbian Mesoamerica. It is certain that Mixtec people were cultivating valleys and slopes in the present-day Mixteca. They were building towns, making pottery and jewelry, and recording dynastic successions among their rulers. But some scholars think they were doing a great deal more.

The city without peer during the period approximately A.D. 300–900 was Teotihuacán, with the great pyramids of the sun and moon, now adjoining the northern industrial suburbs of modern Mexico City. Teotihuacán reached a probable population of 300,000. It may have had direct or indirect contact with people as far away as the northeastern Iroquois to the southern peoples of Tierra del Fuego. No one knows for sure who the people were who built this city or where they went after its rapid collapse. There was a neighborhood of Mixtec people in the great city, and some archaeologists believe it was the Mixtecs who built and administered the city. During the sharp decline of Teotihuacán, it appears that Mixtecs leaving the city went to live in the newly flourishing Tula, to the north about fifty miles, where they became involved in the intense political struggles of that city-state (Paddock 1966: 174–210; Jiménez Moreno 1966: 69–82).

In Tula, central Mexican traditions have it, Quetzalcoatl, a man-god, ruled for a time but was driven out by a faction dominated by military men. Quetzalcoatl was the god of peace, of the sacrifice of small animals and plants rather than humans, of the refinement of agriculture and the cultivation of fine crafts. Perhaps he was the human tutor of such arts rather than a god. He may have been the forbidden son of one of the sacred rulers of Tula, a ruler who was chosen by a parallel group of secular authorities. Under law, the sacred figure was denied the right to have children or choose a successor. In any case, when the man-god was driven from Tula he promised that he would one day return in the form of a white, bearded god traveling from the east and that he would depose the militaristic dynasty that had wrested authority from him. (This long-accepted explanation for the demoralization of Moctezuma II upon the arrival of Cortes has recently been called into question as a possible invention *after* the Conquest.) He left with a group of followers, and some modern archaeologists speculate that these followers established themselves as nobility in the Mixtec highlands, where they could not be conquered by the central Mexican city-states and their rulers. Secure in the rugged terrain of the Mixteca, the followers of Quetzalcoatl may have felt free to pursue their interest in fine crafts and the development of agricultural technique, things for which the Mixtecs were admired at the time of the Spanish Conquest (Jiménez Moreno 1966: 77−82; Paddock 1966: 380; Wolf 1959: 120−122).

It is probable that Mixtec speakers from Tula did rule Cholula, another large and prosperous city-state in the modern state of Puebla, an area where Mixtec and central Mexican Toltec and Aztec rivalry would be played out at a later date before the Spanish arrival. (The pyramid at Cholula is larger than the Egyptian pyramid "that assured Cheops his immortality" [Wolf 1959: 95].) Such was the reputation of Cholula that the Spanish covered its great pyramid with earth and built a church at its summit (Paddock 1966: 201).

About the same time as the rise of Cholula, the Mixtec people in their mountainous redoubts ("Mixtec" means "cloud people" in the central Mexican Nahuatl language) of the present-day Mixteca, well south of Cholula, were exercising increasing influence over the people farther south, especially the rival Zapotecs of the rich central valley of Oaxaca. The Zapotecs, beginning not long after the time of Christ and perhaps before, had built the impressive city-fortress called Monte Albán. The pyramids, great public buildings, and tombs of Monte Albán are built on a mountain that rises from the center of the valley of Oaxaca, granting both great symbolic power and military advantages to Monte Albán. At various times in the city's history, the Mixtecs had exercised influence over it, perhaps ruling it, but by the fourteenth century they were clearly in control. It is remarkable that their increasing dominance over the Zapotecs came more from economic and cultural influence and diplomatic maneuvering than from outright warfare, though they did engage in some military campaigns against the Zapotecs. As the Aztecs rose from being subject people and mercenary soldiers of the Toltec city-states of the central valley of Mexico, they began to pressure both the Mixtecs and the Zapotecs. The Aztecs exacted tribute in the form of craft items, feathers, and precious stones from the Mixtecs, but they never militarily or politically controlled the Mixtecs beyond keeping the main trail to the valley of Oaxaca clear. In one great battle, the Aztecs suffered one of their most disastrous losses at the hands of the Mixtecs (Wolf 1959: 151).

The many debatable and speculative elements of the ancient history of the Mixtecs need not be resolved—and they probably never will be—in order to appreciate some important characteristics of ancient Mixtec civilization. The Mixtecs managed to control a central homeland throughout two millennia or more before the Spanish arrived. The military security offered by the rugged mountains must have had great value to them as other great civilizations around them rose and fell and rose again. The Mixtecs were very active participants in the great events of their age, living in the largest and most influential cities and participating in political struggles. They exerted their influence and authority over regions and peoples who evidenced far more dedication to military strength and skill. The sharp division in social groups and the slogging, unending labor of most Mixtecs may have been the basis for the elite's abilities to cultivate craftsmanship, the fine arts, and the arts of diplomacy, but a strong cultural identity and coherence may have had sustaining value for commoners as well as nobility. We can easily imagine that cultural coherence had its virtues for the Mixtec people both in the tumultuous centuries of rising and falling city-states and empires before the Conquest and in the disastrous years after the Spanish arrived.

The most dramatic result of the Spanish Conquest was the devastating wave of European diseases introduced to New World peoples. So powerful was the effect of these diseases, to which New World people had never before been exposed and to which they had evolved no immunities, that a century after the Conquest the aboriginal population was one-sixth of that before the Conquest. Put another way, 80 to 90 percent of the Mesoamerican population were eliminated by European conquest. The effects of disease were of course amplified and complicated by war, forced movements of population, introduction of new animals, plants, and crops, and the rapaciousness of the conquerors. But while such factors had their predecessors or close analogies in the pre-Conquest history of the Americas, the European diseases were unique. There were probably no serious epidemic diseases established in the New World before the Conquest, and certainly none to compare with the smallpox, malaria, yellow fever, influenza, black plague, and childhood epidemic diseases brought by the Europeans. Of course, in the not-so-distant past European populations had been devastated by such diseases. The Black Death in two years (1349 to 1351) killed about one-third of the people living in a swath running from the Ganges plain to Ireland, including all of Europe. In the rest of the fourteenth century, continued waves of the plague added another 15 to 30 percent to the death toll. The plague had much to do with the unraveling of medieval feudal society. But by the time of the conquest of the New World, Europeans had been in contact with Asia and Africa for long enough, and had been living in populations dense enough to support epidemic disease for a sufficiently long time, that a large share of the population had acquired full or partial immunities to a wide range of diseases. In contrast, New World peoples were completely unprotected (Cook and Borah 1968).

The effects in the Mixteca were at least as devastating as they were in Mesoamerica generally. Demographers estimate that Mexico did not reach a level of human population equal to the pre-Conquest population until about the year 1900. Among the Mixtecs, it appears that the pre-Conquest population levels were not equaled until 1950, although they were to double between 1950 and 1975. (Many regions of the Mixteca remained far below the pre-Conquest level due to the relative depopulation of the countryside as a large share of Mixtecs spent most of their time outside the region while usually maintaining some identity with the Mixteca.)

The effects of European domination on the *lama y bordo* agriculture practiced by the Mixtecs can be easily imagined. Without sufficient labor to maintain the valley check dams and terrace walls, the conserved soil of the bottomlands developed great gulleys and the best, most productive land was seriously and rapidly degraded. Although the lighter population levels might have given some rest to the steep slopes of hills and mountains, the Spanish introduction of domesticated grazing animals and the plow assured that the destruction of these soils would proceed apace. (Mesoamerica had neither plow nor domesticated grazing animals before European conquest, so that in areas like the Mixteca the surface of the ground was disturbed intentionally only by terrace construction or by the *coa*, or digging stick.) The Mixtecs were briefly denied the privilege of raising *ganado mayor*, cows, horses, and mules, in order to protect Spanish landowners from competition, but the denial was lifted throughout the Mixteca by the end of the first century of Conquest. Meanwhile, the even more seriously damaging *ganado menor*, sheep and goats, were allowed to graze the slopes, with the goats attacking trees and brush as well as the grasses. As a whole, the results of Conquest for the Mixteca were that far fewer people were using a much more damaging set of technologies with no limiting social controls. The effects on the Mixtec landscape were horrific.

Ingenious solutions were not entirely absent, although we cannot say whether their specific local adoption was due more to Mixtec or Spanish initiative. The Spanish contact with the Far East had given them an appreciation for silk and the profits to be made from silk, and sericulture and industry grew up in many areas of the Spanish empire during the sixteenth century, most notably in the Philippines and the Mixteca. Silk brought an economic boom of major proportions to the Mixteca. Silkworms were at first set out on the native white mulberry trees of the Mixtec river margins and later on black mulberry trees imported from China. The Spanish interest in long-range trade based on capitalist agriculture merged with the Mixtec penchant for skilled craftsmanship. The Mixtecs became world-famous weavers of silk as well as growers and processors of it. But, as they were forced to recognize, the Spanish empire was governed with more than local interests in mind. Soon the silk growers and weavers of other areas in the empire, particularly the Philippines and Chinese trading enclaves, out-competed the Mixtecs in colonial politics, bringing down restrictions on the Mixteca silk industry and the subsequent bust of the great silk boom in the region. In 1596, the colonial government prohibited the planting of mulberry trees for silk production in the New World. Nothing since silk has lifted the Mixteca out of the economic depression that followed on the loss of the silk industry (Lira and Muro 1976: 400–401).

Fortunately, cochineal dye production provided a somewhat more reliable source of income suited to the ecological conditions of the Mixteca. Cochineal, a red dye highly prized in Europe before the nineteenth-century substitution of synthetic products, is made from an insect that lives on a particular variety of cactus, both insect and cactus being natives of Mesoamerica. The cottage industry techniques of cultivating the cactus, encouraging and harvesting the insect, and producing the dye from grinding the bodies of the creatures resisted the attempts of the Spanish to turn cochineal into a plantation product grown on a scale of interest to Spanish landowners and investors. As a kind of backyard agriculture and industry, cochineal was the basis for a certain degree of independence from Spanish control among Indian peasants and communities. Cochineal production was similar to the idea of small-scale rural industrial development using slack-time peasant labor that is so often favored by Mexican federal bureaucrats looking for ways to stabilize or revitalize the modern Mixteca. Unfortunately, cochineal could never more than cushion the worst effects of Spanish domination, and competition from synthetic dyes in the first century of Mexican independence wiped it out (Lira and Muro 1976: 401–402).

The Spanish remained strongly focused on gold, silver, and lucrative export agricultural products such as sugar, except, as in much of the seventeenth century, when generalized economic depression involving both colony and mother country discouraged all but local and regional trade and industry. Much of New Spain was put to provisioning the mines hundreds of miles from the Mixteca in an arc north of Mexico City in Zacatecas, Aguascalientes, San Luis Potosí, and Guanajuato. Peasants were uprooted and shipped to the mines, fields were turned to raising mules or food for mules, forests were cut for mining timbers, and a large share of food production was devoted to assuring the miners could continue with their labors. There is no evidence that the Mixteca was directly affected by the mania for mining, because the mines were mostly at great distance. But the indirect effects were important because the focus on mining and exports meant that the Spanish had little interest in cultivating or investing in the Mixteca outside of the silk and cochineal industries. Intensive farming of high-value perishable crops was out of the question because of the distance from mines and cities and the difficult terrain and poor roads. Aside from the silk boom, Spanish overlords were content to rest on the laurels attendant on the prestige of large landholdings and on the income gained from extensive grazing and grain production lackadaisically managed. The neglect was hardly benign, as the bottomland continued to be plowed and gullied into ever higher rates of erosion and the uplands suffered the unending depredations of donkeys, cows, sheep, and goats (Florescano 1971; Lira and Muro 1976; Cook 1949).

There were those in Mexico who hoped that independence from colonial bondage would mean a stronger Mexican economy and the recuperation of such especially poverty-stricken regions as the Mixteca. They were to be largely disappointed in the first century after Independence. The Wars of Independence went on sporadically from 1810 until 1822, followed by nearly two decades of scattered but locally devastating conflict between political and regional factions. The wars with Texas and the United States exhausted the nation and left it with only half as much territory in 1848. The civil war leading to the new Constitution of 1857 (La Reforma) brought more disorganization and destruction in its wake, and the French invasion of 1862 followed hard on its heels. Five years of war with the French invaders and their Mexican supporters were followed only by more years of intense regional and political feuding involving much of the nation at one time or another. Only with the hard-handed dictatorship of Porfirio Díaz beginning in 1876 did something resembling peace come to the Mexican countryside, and only at the price of a brutal, disruptive program of rapid political and economic change.

It is worth noting that the two great political figures of the nineteenth century in Mexico came from poor peasant backgrounds in Oaxaca. Benito Juárez, often compared in Mexico with Abraham Lincoln, author and champion of La Reforma, was a Zapotec who came from a family and circumstances not unlike those of Ramón González—both his parents died when he was young and he had to leave the village of subsistence farmers to find another way to survive. Porfirio Díaz, one of Juárez's most capable generals and dictator of Mexico from 1876 to 1910, was born in a poor mestizo family. His Indian ancestors were Mixtecs. In their separate ways, these two poor men of Oaxaca unfortunately chose dubious doctrines and treacherous allies and in so doing brought further calamity to the communities and land of rural Oaxaca.

La Reforma Constitution of 1857, while ostensibly dedicated to modern ideals of democracy and shared economic welfare, was a disaster piled on previous disasters for most surviving traditional Indian communities of the sort still found in the Mixteca. La Reforma, consistent with the laissez-faire capitalist economic theory of the nineteenth century known as liberalism, outlawed "corporate" ownership of land. In the context of nineteenth-century Mexico, this was understood to mean first and foremost that the Roman Catholic Church, monasteries, and convents, all of them corporate bodies, could not own land. But it also meant that the Indian communities, legally defined as corporate bodies, could not hold land. The theory was that corporate ownership by the Church or by a community dampened the individualistic motivation for economic gain that fueled economic growth and development. The effect was to dispossess the politically powerful Church of some minor parts of its holdings and to almost completely eliminate landholding by the politically weak Indian communities. Because Indians as individuals could only very rarely hope to purchase the land they had once farmed as a usufruct privilege granted by the community, most formerly Indian land was purchased by mestizo and white ranchers, commercial farmers, and real estate speculators. As a result, wherever Indian people had managed to maintain some degree of independence after four centuries of conquest, both economic privation and cultural disintegration increased rapidly (Díaz 1976: 835–837, 909–918; Vanderwood 1981: 39–50; Hale 1968, 1976; Sinkin 1972).

During the Mexican Revolution of 1910 to 1920, many Mixtecs joined revolutionary armies in the hope of improving their lot, and they were especially attracted to the forces of Emiliano Zapata. Zapata was from Morelos, between the Oaxacan Mixtec mountains and the capital in Mexico City. General Montero, Zapata's officer in charge of operations in the whole south of Mexico, was an able military leader from the Mixtec village of Tilantongo. But the Mixteca was badly split during the Revolution, with some villages following Zapata, others the more conservative Carranza, and still others a variety of national and local leaders who raised a revolutionary banner from time to time. Predictably, the towns and fields of the Mixtec suffered from the fighting, pillaging, and looting as loyalists of the various factions warred among one another. Most unfortunate was the fact that while the Revolutionary Constitution of 1917 promised thorough-going land reforms and political change, most of the hopes of Mexico's peasants, Mixtecs included, were unrealized seventy years later. Why this was the case is the subject of a later chapter, but for the moment we can summarize the recent conditions of the Mixteca and justly regard them as symptoms of the failure of the peasant-inspired provisions of the 1917 Constitution (Butterworth 1975: 53–56).

A recent study undertaken by the Centro de Investigaciones para el Desarrollo Rural Integral (CIDRI, Center for Research on Integral Rural Development) concludes that of every ten Mixtecs, three will emigrate permanently, four will be temporary migrants, and only three will remain permanently in the region. In Ramón González's district of Silacayoápan, as in two other major Mixtec districts, the rate of permanent emigration from the Mixteca was calculated by CIDRI as 50 percent—one out of two will leave the region. In the 1960s 115,000 people left the Mixteca permanently, and in the 1970s nearly 240,000 emigrated. Because earlier waves of emigration, after decades of a tradition of temporary migration to harvest sugar in lowland Veracruz, tended to be to relatively nearby large cities like Oaxaca and Mexico City, the largest concentration of Mixtec migrants may be found there. According to the CIDRI study, it is estimated that in Mexico City there are more Mixtecs than there are in their own land. More recently, Mixtecs have been attracted to the areas like Culiacán, which along with eleven other agribusiness valleys in northern Mexico accounts for more than half a million Mixtec migrants a year. Mixtec numbers in California are rising fast—some observers estimate that Mixtecs make up 30 to 40 percent of the migrant agricultural work force in the San Joaquin Valley. All of this is perfectly understandable when we read in the CIDRI study that "agricultural land in the Oaxacan Mixteca has been reduced by fifty percent in the last twenty years . . . with only 73 thousand hectares representing six percent of the territory, worked predominantly as *minifundias* (landholdings considered economically inefficient due to their extremely small size)." Seventy-two percent of the population are primarily agriculturalists, and agriculture is overwhelmingly dependent on uncertain rainfall rather than irrigation. In the district of Teposcolula, the one area where agriculture production has climbed due to the successful introduction of new seed varieties and mechanization, with 9 percent of the Mixteca's economically active population producing 74 percent of the Mixteca's agricultural produce, mechanization has produced a massive dispossession of former landholders and the highest emigration rate of all Mixtec districts (Secretaría de Programación y Presupuesto, Centro para el Desarrollo Rural Integral 1984).

In June 1986, I visited the Nochixtlán district of the Mixteca Alta, learning firsthand many of the daily problems that bedevil those responsible for trying to change the desperate condition of the Mixtecs. The fields of Nochixtlán are, if anything, even more discouraging to gaze upon than those around San Jerónimo or Silacayoápan. Sheet and gully erosion scar every hillside, and everywhere one looks there are permanently abandoned fields. At the same time, there are some very striking, massive terraces that for a moment raise hopes that something constructive is being done. But over several years of brief visits to the Nochixtlán area, it became apparent that a large proportion of these terraces were not being used and that many that were being planted were yielding very disappointing harvests.

I asked the soils experts at the local federal experiment station, run by INIA (Instituto Nacional de Investigaciones Agrícolas—the same federal agency that runs the experiment station in Culiacán), to tell me about the terraces.

"Oh, yes, those are constructed by huge machines as part of a federal program. A great deal of money has been invested in them. Very expensive. Very impressive." The man talking was a dark, bearded fellow in his late twenties, Ing. Sergio Campos, educated at Chapingo. When I encountered him in his office, he was reading Gabriel García Márquez's *One Hundred Years of Solitude*.

"Yes, they are very impressive. But I seem to notice that many of them are lying fallow."

"Good observation. Yes, of course. You see, most of those terraces are sterile, or virtually sterile. The problem is that when they began to make the terraces, they were starting with hillsides with very thin topsoil—the only thing there is on hillsides around here—and the machine digs very deeply, like a plow, in forming the terrace. The result is that the rocky, calcareous subsoil is brought to the top and the topsoil is buried, with what few nutrients the soils have put deep beyond the reach of the plant roots. Thus, the terraces are mostly sterile, or very unproductive."

"But surely somebody has noticed this and suggested that it is self-defeating, no?"

Ing. Campos smiled a rueful smile. "Of course. I have, my colleagues have, my predecessor has. But the men who make the terraces work for another agency, and they say to us, 'Our job is to make the terraces, it is your job to make them fertile.' And it's an expensive program that lines a lot of pockets, so it goes on. That's how it is."

"Is it possible to make them fertile?"

"We have selected some and experimented with them, with some fairly promising results. Any significant amount of chemical fertilizer is out of the question, because the farmers around here can't afford it, and besides, given the nature of those soils, chemical fertilizers alone won't solve all the problems. So, what can be done is to incorporate manure and the stalks of the corn plants from other fields into the terrace over a period of years, and things definitely improve."

Another of the soils experts at the INIA station said in another conversation on the same topic, "But the peasants around here are not like the ones in Jalisco, where I come from. They are just out of it." This even younger agronomist wore a permanent frown and seemed unhappy and restless. He was skinny and wore a baseball hat with "Guadalajara" printed across the front.

"What do you mean, they are out of it?" I asked as we talked at a bus stop in front of a famous, massive sixteenth-century cathedral at nearby Yanhuitlán, constructed on the site of an ancient Mixtec city.

"They won't change, they won't innovate. As Ing. Sergio told you, they could make those terraces fertile with manure and cornstalks, as we have shown, but they won't do it."

"Why won't they do it?"

"They're stupid and backward."

"Do the cornstalks and manure have some other value, perhaps?"

The young agronomist sneered in a way that gave up nothing to the obvious inconsistency between the remark that followed and his harsh judgment on the mentality of the Mixtec peasant. "Of course they have value. The manure is all committed to maintaining the fertility of existing fields and the cornstalks are fodder. In fact, around here the agricultural economists calculate that the fodder from the cornstalks is more valuable than the corn itself. It's essential to keeping the few animals around here alive."

"Then perhaps they are not so stupid to refuse to put their most valuable product back into long-term fertility restoration when they have to use everything possible to simply make it from year to year."

"Right, that's it," he said with another sneer.

When Ing. Campos was told of his colleague's approach to this problem he said, "Well, he's right about the value of the fodder and the manure, and that is certainly the problem. The peasants are caught, they're not stupid. But you see, his attitude, it's very common among the agronomists here. Their sympathy with the local people and their comprehension is of a very limited sort—the agronomists posted here sometimes can't see beyond their disappointment at being stuck here in one of the poorest and least promising regions of Mexico. They are angry, and it makes it hard for them to work effectively here. They feel like they're being asked to carry the impossible burden of all these insoluble problems on their own backs, sacrificing their own careers, perhaps because they didn't get the best grades in school or didn't have the right connections to get a better assignment." Sergio smiled his rueful smile again and said, "Well, I'm here partly because I find it interesting and challenging, and partly because of my own personal problems. So, we just keep doing what we can, with the peasants and with our colleagues." He chuckled quietly. "Really, this is Siberia."

As it happened, I visited the INIA offices on a day that one of the INIA agronomists was giving a seminar for his colleagues at the experiment station. There had apparently been a good deal of talk at the station about the desirability of introducing Mixtec farmers to the advantages of synthetic pesticide use, which had been used rarely and sparingly in the region. The agronomist who was speaking that day had volunteered to make a presentation on the subject of "Good Modern Pest Management Techniques," discussing mainly the principles of pesticide use with a few comments on the concept of Integrated Pest Management. Unfortunately, the young man seemed to have confused, at one point or another, basic concepts taken from old lecture notes or textbooks. For instance, he said, "When using a pesticide, you should try to use it only when you are able to hit a number of pests at once, to ensure the efficiency of the product," when the recognized principle is that pesticides should be targeted as carefully as possible at well-defined pest problems to avoid creating broader ecological problems and consequent pest outbreaks due to the differential response of populations of various species to the chemical. Similarly, he told his colleagues, "It's important to get the pests at the first identification of them in the field," which directly contradicts the recognized rule that one sprays for pests only after it has been determined that the pest is going to cause a level of damage above a calculated "economic threshold"—it is assumed that most pest organisms are present at some population level most of the time, and the trick is to identify the time when application is truly needed and worth the economic cost and environmental hazards. No one among his colleagues contradicted these blunders and others. The thought was unavoidable that in the guise of professional help the Mixtecs were going to be led into even deeper and more complex ecological dilemmas than the ones from which they already suffered, with corresponding economic costs that were incalculable.

INIA is not the only federal agency working in the Mixteca. CIDRI, whose study of the Mixtec problems has been cited extensively above, has developed a plan for the recuperation of the Mixteca. This plan, sharing a good deal with similar plans developed by other agencies over the decades, is intended to recognize the complex interrelationships involved in the Mixtec situation. Each part of the solution is meant to depend on the execution of the plan as a whole and thus relies on a heavy commitment of political authority and coordination as well as federal money and the cooperation of state and local government agencies.

Talking to one of the enthusiastic young scholars who designed the CIDRI plan, I learned that a meeting of several hundred Mixtec community leaders had endorsed the plan at a meeting called in Mexico City to discuss it. But the real excitement was over the helicopter trip the Mexican president had made to the Mixteca, during which he had announced to assembled news people his deep concern for the region and his enthusiasm for the CIDRI program. CIDRI personnel saw this as the key factor in determining the success of the plan, but while enthusiasm for the president's words was high, even the young scholar in the flush of optimism in which I found him that day was careful to say, "Of course, the president endorses dozens of plans publicly that he does not carry through with in terms of actual political and budgetary support consistently over time. So, we shall see" (Astorga Lira 1984).

The Mexican government has an unusually powerful chief executive. Mexican presidential candidates are chosen by the leaders of the political party that has been in power for more than half a century. After the leaders have chosen the candidate it is certain that he will be declared the winner in the presidential election, both because PRI (Partido Revolucionario Institucional) commands considerable loyalty among the Mexican populace and because it is not above manipulating the election process and the results when necessary. After the president comes to office, his team selects his own appointees to more than sixty thousand positions in the government and bureaucracy, turning out the old officeholders. Congress has been a rubber stamp for presidential initiatives, and until 1988 the various right- and left-wing political parties had represented no serious challenge to the president's authority. Near the end of his term, the president will be able to play a significant role in selecting his successor. Journalists and scholars speak of modern Mexican history in terms of *sexenios,* the six-year terms of office of the various presidents. Scholars studying the Mexican political system have aptly described the president as "an elected dictator" (Grindle 1977; Alonso 1984; Hamilton 1982).

Many things, however, limit the power of the elected dictator. One is the relatively weak position of the Mexican economy in the larger international economy. Mexico has tried hard to catch up, becoming a significant second-rank industrial power and a predominantly urban society in the process. PRI chose, through every president since 1940, to emphasize economic growth as the engine of progress in Mexico, in the process de-emphasizing the strong sentiments coming out of the Mexican Revolution of 1910 to 1920 and a radical period in the 1930s that favored the restructuring of political economic power and redistribution of land and wealth. PRI points with pride to the fact that its steady postwar record of economic growth is, in percentage terms, as strong and consistent from 1945 to 1979 as any country's, including Japan. Unfortunately, this record of growth is limited in its effects by high rates of population growth, gaping social, economic, and regional inequalities, and a strong dependency on the international economy. This last factor may be the decisive one in determining the immediate future of the Mixteca and other similar regions.

Mexico's growth was financed by heavy international borrowing that has not been paid off as quickly as expected, by real and anticipated petroleum earnings that dropped drastically in the late 1970s, and by government outlays to cushion the worst effects of rapid rural-urban migration and the raw sore of continuing glaring inequalities. By 1987, the situation had produced one of the highest per capita foreign indebtednesses of any economy in the world, amounting in that year to more than U.S. $100 billion. Inflation in the 1980s passed through all the double-digit figures to reach into the hefty triples (up to 120 percent in 1986). International bankers, the International Monetary Fund, and the governments of the major creditor countries pressure the Mexican presidential administration insistently to follow policies intended to reduce indebtedness while increasing the productivity of further private investment. The measures favored by the IMF include reduction in government social expenditures, a reduced role for publicly owned enterprises, reduced government direction and regulation of business activities, and a firm commitment to increasing exports to earn hard currencies for making payments on the debt. Every single one of these objectives goes against the kind of government-directed investment program toward a greater degree of regional self-sufficiency represented by the CIDRI plan for the Mixteca. On the other hand, every element of the plan favors the further expansion of export-oriented agribusiness regions like the Culiacán Valley (Cockcroft 1983: 277–312). (We will take a closer look at why this is so and why the Mexican government will for the time being likely continue to support such trends in a later chapter.)

The future of the Mixtec region now depends mightily on its place in the international economy. The postwar policies of the Mexican government, given new emphasis by the debt crisis, have the effect of treating regions like Culiacán as the sources of growth and progress and of treating regions like the Mixteca as welfare cases. The banking institutions can, at present, cause far more political trouble for the Mexican government than the Mixtecs (Cockcroft 1983: 277–312).

Interestingly enough, the main author of the CIDRI plan to whom I spoke in 1984 became the director of a program rather like the one he had envisioned. But instead of emphasizing his dependency on the Mexican presidency, Astorga Lira has distanced himself rhetorically from PRI as a means of gaining greater support in the Mixteca, pointing out that he is himself a Chilean and that the funding for the program came largely from European government and foundation sources, not from the Mexican government. But critics can point out that the plan, while bringing some benefits to the Mixteca, is not transforming the region in any fundamental way. They also claim that, rhetoric to the contrary, the program is being used systematically to provide patronage to rebuild PRI strength in the region behind a mask of independence. It is certainly too early to judge the effects of the plan as a whole, but it does not for now appear that it will succeed in breaking radically out of the dolorous history of the Mixteca region.

But of course, the life and death of Ramón González serve to remind us of the fact that the Mixteca is not really a welfare case in Mexico's economy or the international economy. Ramón, his father, his uncle, his younger brother, and the family friends, all of whom died early deaths associated with their poverty, were the easily exploited cheap labor upon which agribusiness fortunes and the export economy have been built, in Culiacán, Sonora, Arizona, Texas, and

California. Also, the income earned by migrant labor in Mexico's north and the United States has been more important than social welfare expenditures in maintaining the home communities and the people who provide this expendable labor to power the engines of growth. The Ramón Gonzálezes represent a hidden subsidy to U.S. agribusiness and to the most rapidly growing sectors of the Mexican economy. The size of the subsidy in the form of a stream of ever more desperate laborers increases in proportion to the seriousness of the social and environmental calamity in the Mixteca and similar regions. Such expanding calamities represent a hidden cost of development, a debt that banks and governments ignore. It is Guadalupe González and her still-living children who for now have unwillingly assumed the burden of this debt and who will pay it.

In a larger sense, however, everyone in Mexico, and especially future generations, will pay the cost of building up some regions by allowing others to fall into irremediable decay. Even more serious, if the new, growing regions are dependent on technologies that are environmentally destructive and economically self-defeating, then the entire strategy of economic growth and modernization may fail. Mexico's rapid slide in the 1980s into economic decline, runaway inflation, and severe indebtedness is perhaps a warning signal of development strategies gone wrong.

6
Technology and Conflict

*T*hough no one could have known it at the time, the die was cast for Ramón González, the land of the Mixtecs, and the Culiacán Valley in the years 1938 to 1941. In those years, the first successful synthetic pesticide proved its value; the Mexican government did an about-face in its plans for the development of agriculture; and a powerful group of researchers and politicians in the United States became convinced of the need for a program of agricultural research and education in Mexico and other nonindustrialized countries. All these events would coalesce into a set of technologies and policies that would transform rural society not only in Mexico but in dozens of other countries, large and small, around the world. Looking at the developments of those crucial years in the light of what we have already learned about the Culiacán Valley and the land of the Mixteca gives as clear an answer as can be given to the question, Who killed Ramón González?

The Development of Synthetic Pesticides

In 1939, Paul Hermann Mueller worked for the Swiss chemical company Ciba-Geigy. He was looking for commercial uses for a chemical called p,p'-dichloro-diphenyl-trichloro-ethane, or DDT. Laboratory experiments led him to believe it might be used to kill insects in agricultural crops. Swiss farmers were beginning to feel the pressures of the impending European war and were looking for ways to increase production. The disruption of markets and the political requirements of Swiss neutrality would mean that they would have to produce more of the food their country needed to survive. The Swiss farmers became enthusiastic users of Mr. Mueller's chemical to combat the Colorado potato beetle, the major pest in their fields, after DDT had successfully controlled an outbreak of the beetle in 1939. With the war, refugees from all over Europe began to stream into Switzerland, and many of them were seriously infested with body lice, notorious carriers of the epidemic disease typhus. In 1942, the Swiss army commander had a ton of the new Ciba-Geigy product at his disposal, dusting down the refugees to avert a possible typhus outbreak. In 1943, the American military forces used DDT to control epidemic disease among soldiers living in new and threatening conditions. By 1944, American firms licensed by Ciba-Geigy were manufacturing ten million pounds of DDT per year for use by the military, especially in Asia. At the end of the war, public health workers and farmers, as well as Ciba-Geigy, hailed the substance as a solution to many of the persistent problems of disease control and agricultural pests. In 1945, *National Geographic* featured a photograph of children playing in the fog of a spray truck driving down a New York beach with the caption, "Flying and Biting Bugs of Jones Beach Die in a Cloud of DDT, New Insecticide." The photo illustrated a story called, "Your New World of Tomorrow." Paul Mueller was awarded the Nobel Prize in 1948 for his discovery of DDT's ability to kill insects (Perkins 1983; *National Geographic* 1945; Gips 1983).

Parathion, a chemical that would become a major competitor of DDT in world pesticide markets, came out of World War II research to discover powerful nerve gases for use in warfare and concentration camps—by 1947 chemical companies were selling it as an insecticide. The powerful herbicide 2,4,5-T, later to become infamous as the main ingredient in Agent Orange, used to defoliate vast areas of South Vietnam, was first offered as an herbicide by Dow Chemical Company as a product of the company's war-time research. Farmers had gone into the first years of World War II with only a few pesticides available to them (mainly such arsenicals and metal-based poisons as Paris green and Bordeaux mixture), known to be dangerous to people and to nature. By the end of the war, chemical companies were geared up with the research and production capacity to offer the world's farmers a new array of chemicals to kill crop pests. Because the chemicals were relatively complex and/or new, very few people were prepared to question their safety or appropriateness in the field, and many were prepared to sing the praises of the chemicals because of their proven usefulness during and immediately after the war. Forty years after the end of the war, about one pound of synthetic pesticide for every person on earth is cast into the environment each year—close to five billion pounds of substances designed to kill plants or animals (Gips 1987).

Pesticides would prove to be particularly powerful tools of change in the hot lowland tropics, where pest problems of crops and human diseases carried by insects are especially troublesome. Whole new areas were opened to commercial cash crop production, including the humid Pacific lowlands of Mexico and Central America, much of which had been little suited to large-scale agriculture and dense human settlement until synthetic pesticides became available. The "insecticide revolution," as one scholar calls it, transformed the economies and social life of Central America and of much of Mexico, as we shall see. And, as we shall also discuss below, synthetic pesticides made it possible to change the crops themselves, creating wheat, corn, and vegetable seed varieties dependent on pesticides for their viability in the field (Williams 1986; Chapin and Wasserstrom 1981; Bull 1982).

The Old Ways

Pesticides became one of the tools by which corporations, land-holders, and governments were able to reshape rural society in ways contrary to the initiatives for rural change that had been developing out of the social upheavals and conflicts of the first four decades of the twentieth century in Mexico. During the same years that synthetic pesticides became widely available, the Mexican government was carrying out a program to redirect Mexican politics in general and agricultural policy in particular, a strategy that would quickly find a powerful technological ally in synthetic pesticides. In order to understand the meaning of the new policies that began in 1940, we need to review the history of Mexican agriculture up to that time and some of the conflicts the struggle for control of Mexico's land had produced. The Mexican government policies begun in 1940 would lead to revolutionary change in world agriculture as the approach became embodied in the program of research and promotion called the Green Revolution.

Yet the 1940 policies were not fundamentally new, nor was the reaction they engendered among peasant farmers. A basic tension between the peasants' need for security and the rulers' demands for capturable surpluses underlies centuries of Mexican history. By examining in some detail the historical roots of the Green Revolution, we gain a much deeper understanding of what the Mexican government was setting out to do in 1940. The continuity with the past becomes clear, as does the crucial break with the past embodied in the new technologies that would be created. We also see why the Rockefeller research program was so vital to those purposes and we come to appreciate the crucial role of pesticides. In addition, we can see why the agricultural modernization program promoted in most Third World countries over the last forty years threatens to become a tragically irreversible experiment in technology and culture—it has undermined and displaced sophisticated indigenous agricultural approaches to agricultural research and development.

For centuries, the demands of Mexican elites and their Spanish colonial forerunners (as well as pre-Conquest authoritarian rulers) had been coming into sharp conflict with the needs and values of the majority of Mexican peasants. Most rural Mexicans wanted first and foremost to provide food for their families, to control land to make this possible in a secure and reliable fashion, and to work in the context of a village community devoted to religious and secular values derived from a variety of syntheses of Indian and Catholic belief systems. Where profits could be had by selling farm and garden surplus or in such sometimes spectacularly profitable enterprises as making the popular intoxicant, pulque, Indian peasant farmers were happy to take advantage of such opportunities. But within traditional communities subsistence came first and individual accumulation was restricted by rules meant to protect the interests of the community as a whole.

The peasant desire for self-subsistence and security came into constant conflict with rulers in highly stratified societies. In such authoritarian cultures as the Mixtec and imperial Aztec societies, those who ruled wanted to produce conspicuous wealth in large quantities, far beyond subsistence needs. Thompson's influential theory of the collapse of classical Mayan civilization in the tenth century A.D. proposes that it was growing tensions over noble demands on increasingly stressed peasant producers and the productive capacity of their land that led to one of history's most rapid and spectacular declines of an entire civilization. When the Spanish conquered and replaced native rulers, they wanted to mine silver and gold and produce sugar, silk, indigo, hides, and other exportable cash items to be sold at handsome profits, continuing to sacrifice modest peasant needs to more ambitious goals. Instead of requiring enormous efforts to build pyramids, the Spanish harnessed peasants to the construction of similarly grandiose cathedrals and palaces (Carrasco 1976; Wolf 1959: 69–80, 202–256; Hammond 1982: 188–197; Flannery 1983; Paddock 1967; Stavenhagen 1975: parts I, III; Thompson 1968; Florescano 1971: 12–22).

In order to produce massive accumulations of wealth, the colonial rulers and their heirs after Independence looked for and found ways to tear Indian peasants away from their traditional lands and communities, permanently or temporarily, and set them to work in mines and plantations or in haciendas established to supply the mines and plantations. Forced labor in the forms of *encomienda, repartimiento,* and debt peonage (and black slavery) made it possible to shift land and labor away from basic local food needs into cash crops, mines, and the construction of private mansions, impressive civic buildings, and an amazing number of massive and elaborate churches, which were often quite consciously meant to compete for the Indians' admiration, affection, and loyalty with the monumental civic buildings and temples of the native cultures. In the nineteenth century, the liberal reforms did away with the last vestiges of peasant independence by depriving communities of the traditional rights used to control land, opening the purchase of peasant lands to those wealthy land speculators and ranchers who could always outbid the peasants in a "free market" (Díaz 1976; González 1976; Vanderwood 1981: 39–44).

We know that peasants objected to being torn away from their subsistence activities to be enlisted in mines, plantations, construction projects, and haciendas because there is a continuous history of struggle through the courts, through political movements, through quiet individual and collective subversion, and through armed rebellions. Indians and poor mestizos fought for the right to produce their own food, for the right to control what and how they would produce, and for the authority over the land that would maintain for them some degree of security.

The Spaniards were able to conquer the highly organized societies of Central Mexico in a matter of years, but with outlying groups, such as the Mayans of Yucatán, the Yaquis of Sonora, and the Huicholes of Jalisco, armed resistance against the acceptance of Spanish colonial and Mexican national authority continued off and on until the early twentieth century. Once conquered not everyone stayed conquered, as with nationalist rebellions by the Pueblos that threw the Spanish out of the Southwest for sixty years during the seventeenth century and the frequent rebellions of the Zapotecs of the Isthmus of Tehuantepec.

Where Spanish political authority was firm, there was nonetheless a continual struggle over the land and how to use it. Charles Gibson has noted that the most frequent kind of lawsuit brought by Indians in Central Mexico was a complaint that Spaniards and creoles were allowing or encouraging their animals to graze in Indian gardens. He also cites a case in which a Spaniard proposed to begin a cattle ranch near Xochimilco; in order to avoid this disastrous event, the Indians erected an entire village with streets, municipal buildings, and huts in a single night to present as evidence of traditional use by Indians to be put against the Spaniards' claim of settlement of unused land. Evidently Spanish judges often failed to recognize Indian agriculture as productive activity—they were more sensitive to land use that resembled the traditional Spanish patterns. Spanish colonials disdained Indian crops—they attempted to force greater production of wheat by making it unlawful to pay tribute in maize, but Indians largely responded by growing only enough wheat to pay tribute and continued to grow maize for their own consumption. The Spanish carried on a continuous effort to gain more land for urban development in the Central Valley of Mexico by draining the waters that supported the amazingly productive *chinampa* form of Indian agriculture, but in spite of a massive commitment to the task of drainage, the Indians managed to maintain some of the best of the *chinampa* land through the colonial period, if losing much of the less desirable *chinampa* land in more salty areas of the valley (Gibson 1964: 255–299, 320–321; Gómez Pompa 1978; Rojas Rabiela 1983).

Enrique Florescano tells the tale of another way in which Indians resisted dependence on the Spanish economy and tried to maintain food self-sufficiency against tremendous odds. The eighteenth century was marked by years of great grain scarcity, but Florescano points out that some apparently contradictory data require a closer analysis of the agricultural economy of the time. For the first three-quarters of the century, when grain was abundant, prices naturally fell but, paradoxically, the amount of grain sold declined dramatically. In years of grain scarcity and high prices, three times as much grain was purchased in Mexico City as during times of abundance. Florescano explains the paradox by pointing out that at this point in Mexican urban history the Indian labor force of cities maintained much of their rural ways. Everywhere possible, urban Indians planted a garden in land surrounding their houses or in wasteland at the edge of cities. When large landowners had good crops, due to weather or pest conditions, so did urban Indians. They were not interested in buying grain in such times in spite of low prices because they had their own. During scarce grain years, the crop of the urban labor force was as meager as that of the large landowners, and all the available grain found a market, even at high prices. Large landowners resented the competition from urban laborers producing their own food, and the colonial administration considered the phenomenon an aberration that increased the uncertainties of commercial farming and tax collection—their response was to pass laws making it illegal for urban people to grow their own food (Florescano 1971: 85–87).

In spite of the myriad ways that Indians tried to maintain control of their own land and labor, for the most part they had to submit to the forced removals, seizure of land by legal and illegal means, forced labor, drainage of productive *chinampa* land, progressive erosion under pressure from livestock and plows, and the many changes introduced by a cash economy and a European world view. The demographic catastrophe that killed 80 to 90 percent of the natives of Mesoamerica meant that to the extent that Spanish law recognized the rights of private and community ownership of land by Indians, the titles passed into Spanish and Creole hands anyway because so many deaths meant that in each region thousands of titles were vacated without an Indian claimant. Throughout Mesoamerica, the Spanish colonial years produced similar destruction of the landscape through erosion, deforestation, and willful ignorance of what protective measures had been built into native technology and culture. We have already seen what these changes meant to the Mixteca region of northern Oaxaca (Florescano 1971: 48–70; Gibson 1964: 255–299).

Ecologically speaking, a profoundly disturbing set of forces had been set in motion by the European colonization of the Americas. As historian Alfred Crosby puts it in *The Columbian Exchange* (1973), the eons of almost complete separation between the Western and Eastern Hemispheres meant that the organisms of the two constituted two distinct biological worlds. Crosby writes, "Not for half a billion years, at least, and probably for long before that, has an extreme or permanent physical change affected the whole earth. The single exception to this generality may be European man and his technologies, agricultural and industrial. . . . His effect is comparable to an increase in the influx of cosmic rays or the raising of whole new chains of Andes and Himalayas" (1973: 218–219).

For the people of the Americas, these changes brought the European diseases they could not resist. Unfamiliar plants and animals made their way into wild and cultivated landscapes, changing them in ways that were sometimes obvious and sometimes exceedingly subtle. The changes also brought new crops, new pests, and new ways of growing things. Together, all of these changes created a vast ecological disturbance that we hardly even know how to measure and document. As all the difficult social circumstances of conquest pressed in on the Mexican peasant, at the same time peasant families were awash in the heavy seas of tremendous ecological forces unimaginably new and complex. The native Americans were robbed of much of nature as they had known it just as they were robbed of the autonomy and character of their own societies.

The movement for Mexican Independence that began with the revolt led by Father Hidalgo indicated some measure of the accumulated anger and bitterness shared by Mexico's peasant farmers and rural laborers. Given the chance to revolt, the apparently submissive and passive people exploded in a riot of pillage and the murder of overlords. Unfortunately, Mexican Independence only gave vent to peasant frustrations; it did not solve them. In the Independence Wars, it was the wealthy, conservative landowners and merchants who eventually triumphed over Spanish authority and peasant rebelliousness simultaneously. When the conservatives later lost out to the liberals, the liberal slogans of equality tended all too often to mask unprecedented land grabs and wholesale destruction of Indian communities.

The famous Reforma of 1857, the liberal constitution enacted under the leadership of Benito Juárez after his triumph over the conservatives, was a disaster for peasants. What Spanish Conquest and colonialism and independent Mexican conservatism were unable to do to destroy traditional agriculture and small producers, the wars and liberal reforms of the nineteenth century did much to finish. The open grabs for land and labor by the traditional landholding families were replaced by libertarian and egalitarian slogans capable of disguising further land grabs. Liberal thinkers and politicians like Benito Juárez were convinced that part of the problem in Mexico was corporate landownership as practiced by the Roman Catholic Church and its lay orders of nuns and monks and by Indian communities. The idea was that these corporate institutions stood in the way of entrepreneurship and capitalist enterprise, holding Mexico in the grip of feudal backwardness while Europe and the United States forged ahead, fueled by the energy of individual enterprise. The assault on corporate landholding very specifically targeted a range of agricultural practices thought to be regressive: self-sufficiency, inadequate responsiveness to commercial opportunities, resistance to technological innovation (by which was meant European and North American innovations), landholding as a form of prestige or as a bond for the support of religious or village communities, and the resistance to urban development inherent in protective rural communities based on paternalism or mutual support.

The liberal view of agriculture in the nineteenth century was, in summary, that agricultural production was too traditional, too stagnant, too regulated by the requirements of local communities resistant to technological change. Agriculture needed to change faster and in response to the commercial and technological developments of a wider world. Historians with strong sympathies for such heroic, well-intentioned, liberal reformers as Benito Juárez have largely been those to tell and interpret the tale of the nineteenth century, and in their opinion it was not that this liberal view of agriculture was wrong but that it was corrupted. More recent historians have begun to sketch out another interpretation (Hale 1968).

Historians have found many examples in the colonial and nineteenth-century history of Mexico of haciendas, plantations, and religious estates run with technological savvy and entrepreneurial skill. Indeed, the estates of the religious orders were often resented precisely because of the evident wealth they produced—the ability to farm and ranch productively was one of the charges levied *against* the Jesuits before their expulsion from the Americas in the 1760s. Such efficient farmers as the Jesuits represented unfair competition to many colonial landholders. Florescano, Hernández Xolocotzl, Gibson, Wilken, Gómez Pompa, Gliessman, and other researchers have shown that Indians farmed with unusual skill and efficiency. However, the well-run religious estate, the privately owned hacienda, and the Indian community very reasonably shared a powerful concern with physical safety and survival that came before and limited commercial interests and growth. The population collapse of the sixteenth century led to a severe depression that encouraged both private banditry and the rapaciousness of public officials. The great French historian François Chevalier demonstrated that the formation of the institution we know as the Mexican hacienda was a response to the physical insecurity of the seventeenth century. In the fortified hacienda, peasants, laborers, and the landowner himself could seek safety from freebooting bandits and desperate and cynical officials. The rationale of the religious estate was similar—to provide safety to both the members of the order and their Indian charges in a strikingly dangerous world. The restrictions of the Indian community on the activities of its members were designed to provide a unified front to the corrupting and physically threatening individuals and institutions of the outside world.

During the more prosperous eighteenth century and especially under the Bourbon reforms of the last half of the century, some of the religious institutions were broken up (the Jesuit missions and estates) and haciendas and communities were given some incentives for a greater expansiveness. The contradictory and often irrational measures of reform, however, gave ample reason for landowners, estate managers, and peasant farmers to remain cautious (Chevalier 1972; Florescano 1971; Lockhart 1969; Taylor 1972).

Speaking very generally, it appears that nineteenth-century agriculture, like the colonial estates and Indian communities that came before, was dominated by one major concern: security. Understandably so. Rural institutions had been formed as a response to perhaps the most severe population collapse in human history in the sixteenth century and to a century of chronic depression in the seventeenth century. In the eighteenth and early nineteenth centuries, agricultural producers were battered about by contradictions among old colonial regulations and by often ill-designed attempts at reform of regulation. Large-scale producers felt hamstrung by the peculiarities of the colonial marketplace and responded by attempting to join merchants in forming monopolies and controlling granaries to exclude small-scale producers and manipulate the market. Small-scale producers had to try to survive these attacks on their interests while trying to deal with many of the same uncertainties as the larger operators.

At the end of the colonial period, the new rulers of independent Mexico had failed to develop any consistent agricultural policies. Conservatives, liberals, and regional cliques moved in and out of power. Added to the confusion and in no small part a result of it was the scourge of more than half a century of war. War destroyed crops, orchards, irrigation systems, fences, and buildings. Even worse, it meant that there was little way to judge what the size of the market for a crop might be and whether it would be possible to get it to market and earn a worthwhile return. In many regions, the great walls that surrounded the colonial hacienda building complex were built higher and strengthened throughout the nineteenth century. Indians on their own had to learn how to evade forced recruitment, how to hide animals and store grain, how to appear even more miserable and impoverished than they really were. The kind of enterprise that flourishes in such an atmosphere in an agricultural society is one based on speculative manipulation of uncertainty and poor communications, insider's knowledge, and access to military and political power. Peasants, farmers, and ranchers who did not enjoy access to the tools of political and economic manipulation could only hope to survive such times by learning to evade tax collectors, speculators, bandits, police, and military forces. For a secure food supply, they hedged against disaster by practicing an agriculture based on diversity, intimate knowledge of local circumstances, and conservative use of seeds, soil, and water. They continued to rely on a diet that could be sustained by diverse production with limited resources, as we shall see below (Vanderwood 1981; Díaz 1976; González 1976). Nineteenth-century peasants, ranchers, and farmers were not unresponsive to conditions—they were quite responsive. But conditions argued in every way for caution and insularity.

Porfirio Díaz, Benito Juárez's old general, remained dictator of Mexico for nearly forty years by offering a program that seemed to satisfy the liberals' demand for rapid economic and technological development and entrepreneurial opportunity while at the same time creating a secure rural order. Díaz saw that to consolidate his own power he must resolve the problem of security—he had seen this as administrator of a town in Oaxaca early in his career when he had turned the town toughs and criminals into a militia under his personal control, and his solution for the nation was similar when he came to power in 1876. From the beginning, for this descendant of Mixtec peasants the key problem was security. But under the umbrella of security, dispensing with liberal concerns for political rights and democratic institutions, he did the rest of what liberals had been trying to do throughout the century. His advisors, called *científicos*, or scientists (because they believed in the philosophy of French sociologist Auguste Comte that made political administrators the scientists who understood and manipulated the social order), did everything they could to promote European-style technological innovation. With impressive speed they encouraged the private foreign capital investment that provided the rudiments of communication, road, and railroad networks characteristic of then-modern nations. The Díaz *científicos* brought in technical missions, encouraged new enterprise and investment, and courted foreign investors in every way they knew how. They oversaw the final dismemberment of Indian communities and surviving indigenous political and landholding institutions. While all of this was accomplished in a very antiliberal alliance with the Church and many conservative landholders, the *científicos* believed that the political compromises achieved the tangible technological changes that in the long run would lead the way to a modern liberal nation. Liberal historians have always maintained that the problem with Díaz was that he paid attention to only one side of the Comptian positivist slogan so popular in nineteenth-century Latin America, "Order and Progress." They say that Díaz sacrificed progress on the altar of order. But it would be more accurate to say that he achieved a kind of rough order and proceeded with much of what liberals had before seen as progress. That Díaz's liberal opponents did not like his kind of progress when they saw it was not caused simply by the compromise with order—it was also that they had incorrectly judged what the effects of capitalist, modernizing progress would be on the Mexican nation. The rapid creation of wealth through successful entrepreneurial capitalism did not seem to be automatically tied to more democratic institutions and shared prosperity, as they had imagined. Díaz's policies were an understandable but disastrous compromise between the need for security and liberal idealism (Vanderwood 1981).

Unfortunately, the order achieved by this compromise was a special kind of order. Díaz put an end for nearly four decades to the wars and revolutions contesting for power over the nation. He reduced the banditry and brigandage that had terrified rural people and travelers. He made it possible to plan, build, and run railroads, factories, and commercial farms and ranches without running major risks of losing work and investment to regiments, rebels, or outlaws. But this was order for some at the expense of greater insecurity for others. His order made it even easier to throw peasants off their land, exploit rural laborers, and speculate in land and crops. Díaz's order satisfied the need for economic growth, but it did so at the expense of peasant agriculture and left most rural people even hungrier for an order of their own.

In the Mexican Revolution of 1910 to 1920, peasant armies fought under slogans that were essentially conservative from the peasant point of view. Emiliano Zapata stood for the principle that the "land belongs to those who work it." In participating in the Constitutional Convention of 1917, peasant representatives invented two forms of community control of land, calling them *ejidos* and *comunes*, after the generalized terms that had been used by the Spanish to refer to a variety of indigenous, community-based landholding forms. The *ejido* (the name came from medieval Spain, where a somewhat similar village landholding form had existed) meant that most peasants worked the land with permission from the village, which held title to the land. Village authorities could revoke the permission in the case of behavior threatening to the interests or stability of the community. Because title was held by the village, the land could not be inherited, although descendants could apply for and receive the same permission held by their parents. Both ancient and modern *ejidos* sometimes involved cooperative labor on all or part of the village land but frequently did not. Most of all, village ownership protected both the family and the village from the pressures of the market—the idea was to frustrate land speculators and rich private landlords through legal restrictions that made it difficult for *ejidos* and communes to sell village land. Colonial laws had eroded community landholding and liberal reforms had eliminated it.

In 1917, peasant revolutionaries envisioned the resurrected *ejido* as the principal beneficiary of the agrarian reform also mandated by the new constitution. The *ejido* represented the peasant hope that the agrarian reform would be made meaningful by protecting peasant landholdings through an ancient institution that predated market society and protected the interests of the community from the misfortunes and temptations of individuals subject to the market. While the state would have to enact the agrarian reform and distribute land to the *ejidos*, it was mandated to do so by the constitution and by the strength of peasant militancy as represented by Zapata and his allies. Peasants saw the traditional small community form as primary, as the first line of defense against both the state and the market. Peasants were rejecting the logic of the modern capitalist economy and the logic of modern socialist ideas based on the primacy of the state—they were trying to reinstitute and protect a world they believed they had lost to centuries of exploitation. Among modern ideologies, anarchism was their closest affinity, as the participation of anarchists of various kinds in the camps of Flores Magón, Zapata, and Pancho Villa attest. Unlike many of the urban-based revolutionaries, they were not interested in creating a modern nation-state, capitalist or socialist, so much as they were interested in securing self-determining village communities (Womack 1968; Krauze 1987b; Warman 1978: 117–119).

Breathing life into the *ejido* was one way to try to achieve greater security in peasant life, and we shall shortly see to what extent it succeeded in doing so. Gaining peasant control over landholding and integrating the control over land with peasant-controlled rural communities were fundamental goals of organized peasant movements. As Gibson has documented so thoroughly, peasants learned early in the colonial period that however weak they might be, they nonetheless must try to defend themselves through the law and through institutions. The peasant-inspired articles of the 1917 Constitution are one of the more dramatic and important consequences of the lesson they had learned.

Unfortunately, while insisting on the re-establishment of the peasant landholding community through the *ejido*, peasant leaders forced to compromise with politically and militarily more powerful forces also allowed in the 1917 Constitution for the legal redistribution of land through various "small-holding" forms that would soon permit the land reform process to be turned definitively against the *ejido* in favor of individual, entrepreneurial agribusiness holdings. The small-holdings could be larger per person than *ejido* redistributions, and subsequent distortions of the small-holding laws permitted the creation of large commercial estates and agribusiness empires. (Gutelman 1971: 75–85 shows the tragic traps and contradictions of the 1917 Constitution from the point of view of the Indian community.)

The Technologies of Security

Peasants had never relied on laws and institutions alone to guarantee a secure livelihood. They had also developed a technology of security, a complex technology whose principles and benefits we are only now beginning to understand. Examining it alongside peasant social goals in the early twentieth century helps us to understand how, in 1940 and subsequent decades, the attack on peasant land rights, the *ejido,* and peasant communities went hand in hand with an attack on traditional agricultural technology. The creation of a new, chemical-dependent technology that was to a considerable degree incompatible with peasant livelihood and interests required the simultaneous undermining of peasant legal rights and communities and the destruction of the peasant technology for food security. Recent research on traditional food security technology allows us to understand the connection between ancient institutions and venerable technologies on the one hand and, on the other, between the logic of the modern Mexican development model and the creation of agricultural technology based on synthetic chemicals.

Mexican peasant families left to their own devices relied on a great diversity of domesticated crops and, often, on a wide range of products gathered from wild plants growing both inside and outside the fields. Within each crop, seeds were specifically selected in ways that guaranteed wide genetic diversity within each crop. Each crop was planted in association with other crops in carefully planned schedules and physical patterns to make the most of limited space and available moisture, control pest outbreaks, encourage beneficial insects, fungi, and weeds, and maintain soil fertility. The combination of plants cultivated provided for a well-balanced diet.

One of the best-known and most carefully studied elements of the Mexican agricultural security technology is what is sometimes called the "beans-squash-maize complex." Before World War II, Mexican peasant farmers seldom grew corn alone; rather, they planted maize along with beans and squash. This planting strategy has an amazing number of significant advantages, some of which were well described by Harvard Professor Paul Manglesdorf, whose participation in the planning of the Green Revolution makes him one of the chief people responsible for the decline of the practice he so admiringly describes:

> One of the most remarkable aspects of this complex is the fact that the combination of corn, beans, and squashes is now recognized as furnishing an adequate, indeed an excellent diet. Corn supplies carbohydrates, small amounts of protein, and fat: the beans represent the principal source of protein, but more important still, they contain adequate amounts of some of the "dietary essential" amino acids—the building blocks of proteins—in which corn is deficient, especially tryptophane and lysine. Beans can also remedy corn's notorious deficiency in two vitamins, riboflavin and nicotinic acid. Squashes are valuable in supplying additional calories as well as Vitamin A, and their seeds furnish an increment of wholesome fat in which a diet of corn and beans is barely adequate.
>
> How the American Indians with no knowledge of chemistry discovered that corn, beans, and squashes together provide an adequate diet is still a mystery, but somehow they did and they did this thousands of years ago. (1974: 1–2)

Manglesdorf also praised the way the complex "made efficient use of the land, especially for a people who lacked draft animals to perform the various tillage operations," because the "beans climbed and twined on the stalks of corn, exposing their leaves to the sun without drastically shading the leaves of the corn plant, and the squash vines spread out over the ground between the hills of corn, choking out the weeds."

Manglesdorf, a botanist and plant breeder for whom corn was a kind of obsession as well as a profession, was to come into sharp conflict with another scholar, Professor Carl Sauer, a geographer from the University of California at Berkeley. Although they were to quarrel bitterly over the advisability of the Rockefeller research program begun in Mexico in the early 1940s that was to become known as the Green Revolution, they were both admirers of the beans-corn-squash complex. Sauer cited all the advantages pointed out by Manglesdorf but added some crucial observations beyond those of his plant breeder colleague.

> [Corns, beans, and squash] . . . formed a symbiotic complex, without an equal elsewhere. The corn plants grow tall and have first claim on sunlight and moisture. The beans climb up the corn stalks for their share of light; their roots

support colonies of nitrogen-fixing bacteria. The squashes grow mainly on the ground and complete the ground cover. In lands of short growing season, the corn was planted first, the other two later in the hills of corn. . . . By long cultivation varieties of all were selected, able to grow to the farthest climatic limits of Indian agriculture. Civilized man has not extended the limits of any of them [this was true when Sauer wrote in the early 1950s but has now changed somewhat] and has introduced only a few crops that succeed under more extreme climates. Maize, beans, and squash were grown on the lower St. Lawrence and by the Mandans on the upper Missouri. They are grown in milpas on the Mexican volcanoes to the highest patches of available soil. Forms were successfully selected for growing on the margins of the deserts of Sonora and Arizona where there is only an occasional summer thunderstorm. The Hopi, living in a land of little and late rain, of short summers and cold nights, depend on them and by them have maintained themselves and their fine and gentle culture, our civilization lacking the skills to match theirs for this harsh environment. (1969: 64)

Sauer strongly emphasized the key role of the genetic diversity of each of the three crops that allowed the complex to be successful in such a variety of ways and places. "Our main interest in corn is as feed for livestock; native attention has been given to it as a staple of human food. The corn on Mexican or Guatemalan hillsides that may seem a sorry plant to the visitor from the United States, is likely to be very properly suited to the native diet and the local soils and weather." Sauer takes up in turn each of the three plants of the complex, showing the diversity of seed varieties within each of the three, diversity that had advantages for diet, soil maintenance, storage and preservation requirements, complementarity with other crops and wild plants outside the complex, and staggering of harvest that contributed to stability of food supply and specific dietary requirements. He also discusses the incorporation of features of the complex into religious and political practices. "The experiences of very many generations of corn growers are not to be set aside lightly by the simple and short-range interests of commerce. It is not accidental that a single native village may maintain more kinds of maize than the Corn Belt ever heard of, each having a special and proper place in the household and the field economy" (1969: 64–70).

More recent investigators have found yet other advantages in the corn, beans, and squash combination. Dr. Javier Trujillo de Arriaga studied what he called the triculture in the high plateau of the state of Tlaxcala, where the modernization of agriculture along Green Revolution lines has not yet changed traditional practices in a dramatic way. He concluded that the triculture was strikingly successful at reducing pest damage in each of the three crops compared to damage levels where they are grown alone, even where pesticides are applied. He also compared the overall productivity of the fields to fields planted to single crops. He found that corn produced with traditional techniques in many instances outproduced modern Green Revolution varieties grown with the package of fertilizers and pesticides necessary to the modern hybrids. Much of the literature comparing the productivity of modern techniques to traditional ones *assumes* monoculture, so that researchers compared corn productivity in a triculture field to corn productivity in a monoculture field, without giving any value to the beans and squash produced by the triculture. When the beans and squash are counted, Trujillo demonstrates, the triculture even more clearly outproduces the chemical-dependent monoculture in dry, highland fields of the type that historically supported a large part of the population of Mexico. Trujillo and his colleagues also showed that olive, fruit, and other trees planted at field borders play an important role in retaining soil moisture, that maguey and cactus plants provide important dietary supplements and, along with the trees, sometimes provide habitat for beneficial insects. Beehives aid in crop productivity and provide honey. The selection of seed varieties in the crops maintains a high genetic diversity and a strong element of drought and pest resistance within the crops. In sum, the triculture system provides a high level of productivity, an excellent diet, and considerable insurance against natural misfortune. As Trujillo concludes,

> the triculture should be more widely used in arid regions of Mexico where corn is produced, since it contains inherent mechanisms to face the diverse agricultural constraints. . . . The yield advantage of the tricultures over the corn monocrop shows that the monocrops endorsed by the official credit, insurance and extension agencies, along with an associated technological package (including synthetic fertilizers and pesticides) are not superior to the traditional cropping systems practiced by Tlaxcalan farmers under the constraining climatic conditions that prevail there. (1987: 126)

One of the joys of traveling in Mexico is its great cultural, topographical, and climatic diversity. It is because of this great diversity that Trujillo limits his comments to the kind of climatic conditions that prevail in the region he studied. But what Trujillo found in Tlaxcala, other researchers have found, in varying forms, in regions and cultures very distinct from the plateaus of Tlaxcala. Gary Nabhan of the University of Arizona studies the agriculture of the Papago Indians in the northern Mexican state of Sonora and southern Arizona. He has found that these people make amazingly effective use of scarce desert resources—they have used for centuries some of the techniques of water spreading by which water is encouraged to run off impermeable or semipermeable surfaces into prepared fields, techniques more recently adopted by Israeli agronomists.

In addition to the use of water-conserving strategies, the Papago use their astoundingly intimate knowledge of plants and animals in hunting and gathering to supplement the agricultural diet. They also have turned their attention to the lack of organic material and nutrients in the desert soils of their region and have come up with a variety of effective solutions. The Papagos gather bat guano from churches and mountain caves to spread on their fields. They also gather debris left by floods, ash from fires, crop waste, and the heavy litter left by leguminous trees. "Mesquite and ironwood are key in the nutrient balance of desert washes, since they may both pump and fix nitrogen that can then become available to aid in the growth of other plants. Papago families still seek out places where moist, rich litter has accumulated under these woody legumes, sensing what scientists have only recently confirmed. They then dig up the top two or three feet of organic matter around the trees, and take it back to their plants to enrich them." By locating their fields where mountain washes begin to spread out onto the desert floor and by constructing low, water-spreading fences of woven brush, they help floodwater to dump its load of debris within the fields. "Rummaging through a foot-high drift of [such material called] *wako'la* that had been deposited on a field . . . I was surprised at the organic richness of its contents: rodent dung, mesquite leaves, mulch developed under the trees, and water-smoothed twigs. The farmers . . . take this flotsam, spread it, and plow it into their soil. Enough of this humus comes into their fields to add an inch of organic matter to the cultivated surface each growing season, reducing soil alkalinity and increasing moisture-holding capacity." By such techniques, Papago farmers are able to quintuple the organic content of the desert soils (Nabhan 1982: 123–130).

Nabhan confirms his own investigation of the effectiveness of Papago farming techniques by looking to a study done half a century earlier by an irrigation expert, H. V. Clotts. After visiting hundreds of Papago fields, Clotts dismissed the idea of large irrigation projects in Papago country, saying,

> the Papago Indians, by several hundred years of desert experi-
> ence, are thoroughly conversant with the conditions in their
> country, and with consummate judgement have so located
> their charcos and fields as to secure maximum results from
> the limited rainfall available. We cannot go into their country
> with the idea of teaching them farming or irrigation under
> conditions as we find them in other parts of the country. Any
> attempt to introduce modern farming methods, as we under-
> stand them elsewhere, would result in disaster. The most we
> should do for these people is first to protect them in the pos-
> session of the land which they have beneficially used for hun-
> dreds of years. They should also have educational facilities
> and such modern agricultural machinery as would be adapt-
> able, with some assistance in improving the stock industry. It
> is quite probable that they could be aided in the conservation
> of their very limited water. . . . If thus protected and left in
> peace they are fully capable of working out their own salva-
> tion, indeed they are particularly anxious to do so. (Quoted in
> Nabhan 1982: 128)

Clotts wrote these words in 1917, the same year the peasant dele-
gates to the Mexican Constitutional Convention were agitating for
their rights to land and to be left in peace, believing themselves fully
capable of working out their own salvation and particularly anxious
to do so.

At the other extreme from the desert conditions of the Papago and
the semiarid conditions of the Tlaxcalan plateau, Mexican native
people worked out similarly ingenious solutions to the problem of a
secure food supply in humid, flooded lowlands and in the midst of
highland lakes. The most famous of the techniques to turn the potential problem of a flooded landscape into the benefit of highly productive agriculture involved the construction of what are known as
chinampas.

Chinampa agriculture provided most of the food, flowers, and fi-
bers to the Aztec capital of Tenochtitlán. When Cortés first laid eyes
on the Aztec city, there were perhaps ten to fifteen thousand hectares
(twenty-five to thirty-eight thousand acres) of *chinampas* linked to
the urban markets by a dense network of canals. When the German
geographer Elisabeth Shilling studied the *chinampas* in 1938, she
found that Mexico City's one million people still received most of

their vegetables and flowers from *chinampas*, even though the Spanish and modern rulers of Mexico had drained away or diverted much of the water the *chinampas* required and otherwise undermined the productivity of the *chinampa* fields. Shilling pointed out that until motorized trucks came into common use in the 1920s, a canal went from the *chinampas* in Xochimilco directly to a spot near the great downtown market of Merced. Harvesters picked vegetables, fruits, and flowers in the late afternoon and loaded the produce on barges next to the fields. The barges, propelled by men with poles, arrived near Merced before dawn. The earliest shoppers in Merced would have their pick of the freshest possible food and flowers (Hassig 1985: 47–53; Shilling 1938; Rojas Rabiela 1983; Gómez Pompa 1978; Altieri 1987: 87).

Chinampas are raised beds surrounded by canals such that irrigation comes largely through osmotic pressure, bringing water into the root zone of the plant. *Chinamperos* dredge silt from the canal bottoms or from nearby lakes and marshes and spread it as fertilizer across the fields. When the bed is raised too high for subirrigation through osmosis, the farmers use a scoop on a long pole while standing in a canoe to splash very precise amounts of water on the growing plants. The field borders are originally stabilized with poles driven into the lake bottom, to be replaced in time by border plantings of willows, whose young branches provide the raw material for mats. *Chinamperos* fill the mat grids with semiliquid silt and plant seeds. Each square then provides transplants for the field when the mat is broken apart. Water plants in the canals provide mulches and rich fertilizers.

> Supplemented with relatively small amounts of animal manure, the chinampas can be made essentially self-sustaining. The animals, such as pigs, chickens, or ducks, are kept in small corrals and fed the excess or waste products from the chinampas. Their manure is incorporated back into the platforms. On the chinampas, farmers concentrate the production of their basic food crops as well as vegetables. This includes the traditional corn/bean/squash polyculture, cassava/corn/bean/peppers/amaranth and fruit trees associated with various cover crops, shrubs or vines. Farmers also encourage the growth of fish in the water courses. (Altieri 1987: 87–88, based partly on Gliessman et al. 1981)

Chinampas can produce four and in some cases more crops per year. *Chinampa* agriculture is considered by some to be the most productive form of agriculture known per unit of land, in the same range as the most productive Asian raised-field agriculture, to which it is conceptually very closely related.

Scholars have recently discovered the remains of raised bed agriculture similar to *chinampas* near the ruins of classical Mayan sites and flourishing modern *chinampas* along the Gulf Coast of Mexico. These discoveries lead to speculation that *chinampas*, or variations on similar techniques, may have played a major role in the rise of many of the great civilizations of Mesoamerica. Their prominence in classical Mayan sites, confirmed over large areas by the use of aerial images using side-looking radar capable of penetrating the forest cover, also raises the question of their possible role in the collapse as well as the fall of Mayan civilization (Hammond 1982: 160–164, accompanied by an excellent bibliographic discussion, 312–314).

Chinampas still remain in the suburbs of Mexico City in an area large enough that one *chinampero* complained to me (with scarcely hidden pride) that although he had worked most of his forty-some years in the *chinampas* he still from time to time became lost in them while duck hunting. The Mexican government has recently attempted to preserve some of the remaining *chinampas* in an "ecological reserve" to be protected from the numerous forces of land speculation, pollution, second-home development, and water diversion that have ruined most of the ancient *chinampas*.

Edgar Anderson was one of the botanists whose early work on the classification of corn varieties greatly facilitated the work of twentieth-century plant breeders, including those who created the particular varieties associated with the Green Revolution. Anderson wrote, "When I first went to live in San Pedro Tlaquepaque, a small pottery-making town in western Mexico, I was under the mistaken impression that my Mexican neighbors had nothing but dump heaps and a few trees in the yards behind their homes. As I lived there longer and came to know more about the life of the village, I realized that many of these dump heaps were carefully managed gardens and orchards" (Anderson 1978: 136). Anderson slowly came to recognize that the backyard clutter of many Mexican farm and village homes was not, as he first thought, the product of neglect but represented instead productive gardens based on a sophisticated knowledge of nature and plants. The "dump heap" gardens were difficult to recognize as agriculture because they so closely mimicked the natural vegetation associations of the regions in which they occurred. The household gardener planted fruit and nut trees, useful shrubs, and vegetable plants in patterns that provided shade and sun where they were needed and that through companion planting gave the pest-protection benefits of some plants to other plant species that did not by themselves enjoy such protection. The seemingly careless but in fact very calculated planting practices allowed plants with different root depths to grow side by side to take advantage of water and nutrients from all levels of the soil. Household refuse and animal dung served to enrich the gardens, and some free-ranging animals, especially poultry, spread their rich waste around the garden and gobbled up insects. Anderson's admiration for the sophistication of the dump heap garden and his eloquent description of it in a collection of his classic articles called *Plants, Man, and Life* has inspired other agricultural researchers, such as Wes Jackson of the Land Institute, to begin to think of designing agriculture using the model of natural systems rather than simply extending human artifice to greater and greater removes from the structural characteristics of natural ecosystems.

The agricultural geographer Gene Wilken published in 1987 the most complete account of traditional agricultural techniques in Mesoamerica in a book called *Good Farmers.* Wilken shows that in every aspect of farm technology, the traditional farmers of Mexico and Central America operate with a finely tuned appreciation for what can be done with seeds, water, weather, and soils. Farmers have intimate knowledge of soils, surface water, and underground aquifers. They use fertilizers ranging from cattle dung to the rich refuse thrown out of leaf-cutter ant nests. They organize water cooperatives and even water markets to make thorough and rational use of irrigation water. They delicately balance a crop's need for water with the necessity for good drainage. Farmers shape the soil surface with a variety of terracing techniques adapted to angle of slope, the erosion and water-holding potential of soils, and the stability of underlying rock formations. Traditional farmers plant nitrogen-fixing leguminous crops and trees to help maintain soil fertility. They plant a variety of plants together to control shade and sun, make the best use of water, and minimize pest damage.

> Any lingering images of the lazy, dull, noneconomic peasant farmer must surely have vanished. . . . Given any sort of reasonable reward, traditional farmers respond much like their industrialized counterparts: they identify goals, minimize costs (most of which are labor), and maximize gains by careful use of available resources. Their management methods are also not "traditional" in the sense of being stubbornly preserved anachronisms of dubious or even negative value. In-

stead, they are agronomically and hydrologically sound, based on a thorough understanding of local conditions and plant requirements, and deserve close examination by the scientific community. (1987: 263)

Wilken justifiably speaks of the farmers of Mesoamerica as practicing agricultural science, and he shows that the classification for such essential resources as soil are as sensitive and, from the farmer's point of view, as useful as the taxonomies used by modern agronomists. In many areas, the traditional systems are more refined. "In most cases . . . traditional and scientific [soil] appraisals correspond rather closely, as would be expected. What distinguishes local classification from regional reconnaissance is degree of detail; farmers recognize subtle variations and combinations of soils overlooked by most surveys. Even tiny plots may have several distinct types of soils that farmers identify and for which they adjust cropping and management practices" (1987: 25).

The techniques Wilken describes occur in areas that have for a long time supported dense human populations. Wilken emphasizes that many scholars and popular writers have erred in thinking such traditional agricultural technologies are suitable to low population densities. "It is also not surprising that most of these farming systems support dense rural populations; the association between population and resource management has been noted for many areas of the world. Dense rural populations supply both the need for high productivity and the labor with which to achieve it" (1987: 8).

Wilken does not claim a universal virtue or superiority of traditional techniques over modern ones. In a later chapter, we will discuss some of his cautions about using traditional methods as models for agricultural development. But he does show that in using every resource at their disposal—land, weather, soil, seeds, water, flat and sloping land, soil amendments, animals, and social organization—the farmers have adapted their practice intelligently to their primary goals.

But these are the tools and structures with which traditional farmers control flows of water and energy in their small plots. Do they work? Of course they do! A long record of successful farming, often under difficult conditions, attests to their effectiveness. Are they efficient? As we have seen, that is a more difficult question; the answer depends on one's objectives. Is production paramount? Income? Employment? Ecological stability? Security? If the last two are indices of overall success, then traditional systems should receive high marks. (1987: 9)

The basic objectives of traditional techniques, according to Wilkins, are ecological stability and security. He discusses farming strategies largely as adaptations to limited available physical resources, but his case could be made more comprehensible by seeing the need for stability and security within the social and historical context discussed above in this chapter. During the pre-Columbian period, peasant farmers of the New World often had to deal with the rigid social structures and exploitation by small elites, as in the Mixteca of Ramón González's ancient ancestors. In the Mixteca, this exploitation was aimed very specifically at monopolizing the best soils for the rulers. In Mayan and Aztec societies, the exploitation of peasant labor and land went to building enormous monuments, supporting a leisure class of priests and warriors, and carrying on warfare against other groups. In the post-Conquest period, peasants had to protect themselves against policies deliberately aimed at weakening the indigenous peasant farm economy, muddled policies filled with contradictions and doing unintended damage, outright exploitation through taxes and labor control systems, land speculation, elite control over markets and credit, factional and regional conflict, uncontrolled banditry, and war. In colonial New Spain and independent Mexico, peasant men and women were left to walk a narrow, twisting path where the slightest miscalculation or stroke of bad luck could bring a straight fall into misery.

It is clear that the technologies of security and stability developed by Mexican farmers were not necessarily those they would have developed had they controlled their own social environment. These are to a considerable degree technologies developed to provide security and stability on a slippery slope, whether that slope is literally an eroding hillside or, metaphorically, the exploitive and changeable ways of ruling elites. Ramón González's pre-Columbian ancestors, for example, would surely have preferred that the nobility not monopolize the best bottomlands, but, given that problem, such adaptive strategies as shifting cultivation and the beans-maize-squash complex provided a measure of security in a world not of their own choosing. The incredibly high productivity and stability of *chinampa* agriculture in highland Mexico may have been partially the result of one of the most brutal systems of noble rule over peasants ever devised. The continuing persistence of *chinampas* and the beans-maize-squash complex is partly at least an ongoing response to more modern exploitation of peasants.

Many social scientists have commented on the cultural conservatism and fatalism that came out of such circumstances. The essentially conservative character of peasant communities in Mesoamerica is best represented by the much-studied *cargo* system. In this system, each adult male member of the community has certain obligations that must be met. These vary from culture to culture and village to village, but they usually involve either an obligation to hold offices a certain number of times and help support the activities of the office, an obligation to finance village festivals on a rotating schedule, or a combination of both obligations. Anthropologists have pointed out that this system leads to the necessity of each man working hard to have the resources to carry out his obligation when his turn comes and to the exhaustion of savings at the end of each cycle. The village gains in two ways: essential community functions are provided for, and no man and his household can ever rise far above the others and become exploitive, because household savings that could be invested in individual gain are periodically surrendered to the community (Wolf 1959: 215; Tax 1965; Kunkel 1965).

Anthropologists have also noted that the *cargo* system is linked to a wider world view that governs it—the "image of the limited good." The idea of the limited good is that there is a fixed amount of those things that provide satisfaction in the world, and that if one person has a great deal of those things, others will have less. Life is a zero-sum game. As expressed in the blues lyrics that came out of the peasant culture of the American South, "Don't spend your time trying to be a big winner—If you get fat, somebody else gets thinner," or expressed in the language of ecologist Barry Commoner, "There's no such thing as a free lunch." Given such a view, leveling measures such as the *cargo* system are the only ways to avoid the continuing happiness of one being purchased at the expense of the continuing pain of another. Of course, neither peasant communities nor ecologists have been able to enforce such a view on expansionist European or modern society, where the view of the limited good is rejected in favor of the idea that all can be happier at once by virtue of a continual growth in those things that provide satisfaction, making a bigger and bigger pie. The clash of these two world views is represented in the conflict between conservative peasant technologies that take the natural world as a given and modern technologies that assume nature can be dominated and made to yield more.

Peasant survival requires ingenuity and craftiness, qualities long recognized in literature and folktales where the peasant is portrayed as wearing a mask of stupidity to disguise the sly trickster who lives behind the mask. This craftiness is manifested not just in the peasant who devises ways to cheat his master, lure him into the brier patch, or steal his daughter. The qualities of character and culture that allow for survival under unrelenting pressure are those that have given rise to a very specific kind of agricultural technology deriving from a world that is experienced as sharply circumscribed by Mother Nature and human masters—a technology that could work only through an unparalleled knowledge of local conditions and a complexity of crop choices and a management technique that often comes close to matching the complexity of nature, certainly much closer than anything deriving from the industrial technologies and monocrop strategies characteristic of what we have come to call modern agriculture. Unfortunately, the stability and security of peasant life have seldom been compatible with the purposes of modern nation-states and those who rule them. The desire for growth and modernization along European lines has usually meant an attack on peasant technology and on the cultural and political perspective associated with it.

There is a marvelous photograph that crystallizes the brief moment when the Mexican revolutionary leaders Pancho Villa and Emiliano Zapata held Mexico City against the counterrevolutionary forces. At a dinner in the National Palace, Villa is obviously glorying in his triumph, enjoying the food, the company, the surroundings. But the mustachioed Zapata, who represented the traditional peasantry of central and southern Mexico now caught up in insurgency, cast a sidelong suspicious glare at the happy assemblage of would-be rulers and their sycophants. And well he should have. The new rulers of Mexico in the 1920s would arrange for the assassination of Zapata in 1919 and of Villa in 1924. The sweep of events would in a few years push aside traditional peasant interests in security and stability for another vision of rapid economic growth based on industrialism and an industrial model for agriculture. Only for a few years in the late 1930s would there be any attempt to seriously honor the wishes of the peasantry as expressed in the 1917 Constitution.

Anthropologist Eric Wolf emphasizes that "rebellious campesinos are natural anarchists. . . . The Utopia of the campesinos is the free village. . . . for the campesino the State is something negative, an evil that must be replaced as soon as possible by his own social order of a 'domestic character'" (quoted in Krauze 1987b: 77). In a few sentences, Wolf captures the essence of the problem as it arose for the peasant at the end of the revolution. The revolutionary state would be an even stronger state than the one that came before, with little regard for the free village community. Furthermore, as it happened, the peasant's own social order, that of the village and the household, would be undermined once again by the political and economic goals of the new Mexican government and, even more profoundly, by the destruction of peasant technologies of security and stability upon which the peasant domestic order was founded.

Agriculture and the Revolutionaries

The Mexican Revolution ended with the consolidation of a new government in 1920 based in theory on the 1917 Constitution. The new rulers of Mexico, led by Alvaro Obregón and Plutarco Elías Calles, are known as the Sonoran dynasty. The Sonoran dynasty represented not just a political or regional clique but also a particular approach to Mexican development. The northern state of Sonora, bordering on Arizona and stretching along the coast of the upper Gulf of California (which the Mexicans know as the Sea of Cortés), was in the 1920s a bridgehead for economic interests and technological development from the United States. Calles and Obregón were partners with North American investors in the rapid development of agribusiness and associated industrial enterprises in northwestern Mexico. Obregón was popularly known as the "garbanzo king" because of his monopolization of the trade in the popular Mexican bean through his purchase of a North American trading company. Calles, originally a schoolteacher, had not done well before the Revolution as manager of his father's large estate and of his own rural mercantile firm serving commercial farmers, but he became wealthy through partnership in a variety of business ventures through the 1920s and early 1930s. Although in the guise of a radical, anti-clerical, anti-imperialist government, much of the Sonoran dynasty program consisted of the promotion of North American agribusiness with participation by "revolutionary generals," allied Mexican politicians, and private investors. Although the Sonoran dynasty was eager to eliminate "inefficient" holders of large estates, "the low level of the country's economic and financial resources led the government leaders to give priority to raising levels of production, including agricultural production, and made them reluctant to break up agricultural estates, especially the relatively efficient holdings oriented to commercial production" (Hamilton 1982: 96–97). The political activism of peasant groups as a result of the Revolution was useful against various conservative opponents of the regime, and the prestige of the government depended on making some genuine social gains consistent with revolutionary visions and organizations. By the end of the 1920s, however, even the commitment to the creation of a class of commercially oriented small holders as an alternative to the dominance of large estates as before the Revolution or the predominance of peasant-controlled cooperative *ejidos* as envisioned in the constitution had collapsed. "By the end of the decade, the early orientation to small holdings gave way to the assumption that in the agrarian sector social and economic goals were incompatible; neither the small holding nor the ejido was economically efficient" (Hamilton 1982: 97). Agricultural education, a prime interest of the ex-schoolteacher and failed farmer, would, Calles declared to a journalist from the United States, "constitute . . . the frontal attack in my war against the wooden plow and everything it represents" (Krauze 1987a: 58–59). For the Sonoran dynasty the predominating conditions of agriculture in most of Mexico were just what they were for nineteenth-century liberals: obstacles to progress. Calles and Obregón wanted to rapidly transform the part of Mexican agriculture they believed held strong commercial potential while eliminating or economically marginalizing the rest. Financial and commercial interests in the United States for the most part came to see through the radical revolutionary rhetoric, in spite of a number of standoffs between the Mexican government on one side and the U.S. government and business interests on the other. By the mid-1920s, many American businessmen and policy makers correctly perceived that the Sonorans were men they could work with for mutual gain (Hamilton 1982: 67–68).

During the time of the Sonoran dynasty, Obregón, Calles, and Calles's henchmen who followed him in office expropriated about 10 million acres of privately held land for redistribution, out of a total of about 300 million. But even much of this 3 to 4 percent of agricultural land expropriated ended up in the hands of members of the dynasty or yielded substantial gains for them in the form of bribes, kickbacks, and contracts. Most of the land expropriated did not go to *ejidos* but rather to individual holders of comparatively small parcels (compared to the holdings of the estates that were expropriated and divided). By 1930, Calles stated boldly the position his policies had represented in the twenties. Returning from a trip to Europe in 1930, Calles "announced that the agrarian program was a mistake, that the peons did not know how to use their land, and that production of foodstuffs was steadily declining" (Herring 1962: 374). Calles agreed with the new ambassador from the United States, Dwight Morrow, when he asserted "that Mexico had already taken more land than was needed for peons available to go upon it and could now stop taking new lands and devote its energy to improving the land already taken" (Nicolson 1935: 335).

Calles had every intention of continuing the programs of the Sonoran dynasty that had prevailed since 1920 when he designated his successor for the Mexican presidency to take office in 1934. Although Gen. Lázaro Cárdenas was from the west-central state of Michoacán and lacked the agribusiness connections of Obregón, Calles, and Calles's loyal successors from 1928 to 1934, he was assumed to be a Calles follower. Cárdenas, however, showed a disconcerting independence. From 1934 to 1936 he systematically cut his ties to the Calles machine, finally arranging for Calles's exile to San Diego in 1936. By then, Cárdenas was ready to play on his strong personal popularity and the still-powerful party mechanism to launch his own agrarian reform program in concert with the Constitution of 1917. Altogether, he expropriated about 49 million acres of land and created thousands of *ejidos* to receive and distribute the land among members. Fully half the Mexican population became members of *ejidos*. Cárdenas, under the influence of left-wing members of his government, experimented with collectivizing some expropriated large estates, most notably the cotton farms of La Laguna in the north. He also encouraged a variety of cooperative forms of organization for other *ejidos* (Hamilton 1982).

Encouraged by Cárdenas's agrarian radicalism and in the tradition begun by Calles's minister of education in the 1920s, a new wave of admiration for indigenous culture arose among artists, writers, teachers, and many members of the small Mexican middle class during the late 1930s. The cultural *indigenismo* and the need to show that peasants could provide for national food needs encouraged some interest in the intricacies of peasant agricultural techniques, including investigations of *chinampa* agriculture. Some of this interest was sober and hard-headed, some of it romantic, uncritical, and easily ridiculed when the political climate changed. It played an important role at the time in giving reassurance to a worried urban public that Mexico's peasant farmers were capable, ingenious, and productive, able eventually to provide Mexico all the food it needed (Monsivais 1976).

Cárdenas has often been described as the first, as well as the most, truly popular Mexican president. In running for office he was not content to simply accept the workings of the party to assure him the prize, which it surely would have done. He actively campaigned for office, touring throughout urban and rural Mexico, making himself personally known to thousands. Cárdenas halved his salary on election and insisted on living in his own modest home rather than in the presidential palace. He spent countless hours listening to delegations of peasants and workers. Anita Brenner tells a story popular during his administration of the president's reaction to a list of urgent matters prepared by his secretary, along with a single telegram.

> The list said: Bank reserves dangerously low. "Tell the Treasurer," said Cárdenas. Agricultural production falling. "Tell the Minister of Agriculture." Railroads bankrupt. "Tell the Minister of Communications." Serious message from Washington. "Tell Foreign Affairs." Then he opened the telegram which read: My corn dried, my burro died, my sow was stolen, my baby is sick. Signed, Pedro Juan, village of Huitzlipituzco. "Order the presidential train at once," said Cárdenas. "I am leaving for Huitzlipituzco." (In Meyer and Sherman 1983: 598)

But Cárdenas's power as president was based on far more than his ability to persuade people of his sincerity and dedication. Far more significant was his successful construction of a powerful political party that would remain without serious challenge for at least half a century after he left office. Cárdenas made certain that the gains made by peasants and workers during his administration were repaid by the incorporation of rural and urban organizations into the party and by their complete dependence on it. By continuing also to expand the economic role of the Mexican state far beyond the initiatives of the Sonoran dynasty, Cárdenas allowed the party to erect a vast network of patronage, and potential penalties, that would prove amazingly durable instruments for central government control. While enormously expanding the gains of rural people and urban workers against their domestic masters and by nationalistic measures against foreign governments and investors, Cárdenas also centralized power in the ruling party and, by extension, in the Mexican state. This power, it developed, could be used as well to reduce the influence of workers and peasants and to please foreign interests (Hamilton 1982).

The wave of agrarian reform and rural change naturally disturbed many rich and powerful people in Mexico and a few abroad, but the greatest pressures against the Cárdenas reform program came from his expropriation of foreign oil wells and production facilities in 1938. The expropriation was the culmination of a long-standing dispute between Mexico and foreign oil interests. British and North American firms, among them the Rockefeller-controlled Standard Oil Company, had begun drilling wells during the Díaz period, and in many years Mexico was the most important source of oil imports into the United States. Mexican petroleum was nearly as important to the United States in the thirties as Arab oil fields were in the 1970s.

The 1917 Constitution explicitly revived the ancient Spanish law by which all subsoil rights were the property of the king or the modern Mexican nation. Throughout the 1920s, oil companies in the United States and their allies, such as Albert Fall of Teapot Dome fame, secretary of the interior under Harding, kept up a steady barrage of hysterically worded propaganda against the Mexican government, accusing it of Bolshevism and theft. In spite of reassurances from Obregón and then Calles that the constitution would not be applied in this sense to holdings that predated 1917, the companies maintained their campaign of hostility to Mexico from within the U.S. government as well as among the North American public. A simmering labor dispute during the Cárdenas years brought the matter to a head when the Mexican Supreme Court found in favor of the oil workers against the companies and the companies openly refused to obey the decision of the court. In an atmosphere increasingly heated by threat and counterthreat, Cárdenas decided to expropriate. The decision was crucial to the history of agriculture because while it mobilized nationalistic sentiment for Cárdenas as nothing had before, it also galvanized the hostility of U.S. business interests and their Mexican allies to Cárdenas, leading within two years to the rejection of the whole conception of the Cárdenas reform program. (This story and its consequences are brilliantly told in Hamilton 1982: 216–240.)

Investors urged the Roosevelt government to insist on the return of the oil properties under threat of invasion by Mexico of U.S. troops, the seizure of the Mexican government, and the return of the assets to their former owners. They succeeded in obtaining a declaration of a partial economic boycott of Mexico, but cooler heads prevailed on the question of invasion. This was partly due to the relative moderation of Ambassador Daniels, President Roosevelt, and the U.S. Department of State. But the most persuasive argument against brutal measures against Mexico was that they would undermine the strategy to win the allegiance of Latin America to the American side in the ever more clearly anticipated European war.

The political consequences of the oil expropriation were, in the years 1938 to 1940, very complex and confusing in Mexico. The more conservative forces in Mexico, with strong allies among foreign investors and governments, looked at the presidential elections of 1940 as a crucial test of their power to survive. If the logic of the Cárdenas program were carried further, private capital might have little to say about the future of the country and very limited opportunities for profitable expansion. The newly shaped ruling party could not be opposed, but if it could be won over internally, there would still be a chance to use the new power of the party to achieve a level of national coordination that would make Mexico into a much more efficient and dynamic machine of growth than had seemed possible before Cárdenas's political consolidation. The conservative view did, in fact, prevail, both in the selection of Manuel Avila Camacho as president and in the political program that he carried forward and left to his successors through the 1980s.

The Counterrevolution and the Green Revolution

The new president inaugurated in 1940 clearly represented two things: a close political alliance with the U.S. government and a rejection of the radicalism of Cárdenas, including the agrarian reform. Petroleum production would remain in the hands of the Mexican government, but under negotiated terms acceptable to the U.S. government and the financial interests involved.

In his inaugural speech, Avila Camacho embraced the "vital energy of private initiative" and promised that he would "increase the protection given to private agricultural properties, not only defending those that exist, but also forming new properties in the vast regions not now cultivated." Agriculture would play a new role, not as a base for rural change, but rather as the foundation of "industrial greatness" (Hewitt de Alcántara 1976: 21–22).

The meaning of this inaugural address was soon to be clear as it was put into effect, immediately and over the next forty years. New distributions of land would be largely limited to land not previously thought desirable for cultivation. Very poor quality land would be distributed in the northern deserts to poorly financed *ejidos* who tended initially to think they were being given something of real value. The purposes were to relieve political pressures for continuing land reform and to reduce somewhat the uncontrolled and difficult wave of migration of rural people into the cities. *Ejiditarios* in one case expressed the fate of such farms by erecting a sign along busy Highway 15 with the *ejido*'s name—Ejido Desengaño, *desengaño* meaning deception or disappointment. Other more encouraging redistributions would take place in newly opened irrigation districts financed by the Mexican government and the World Bank. These new farms went mostly to individual commercially oriented landholders, or, as in much of the Culiacán Valley, the original *ejido* owners were displaced through the systematic corruption of the land reform process. (Detailed documentation of the results of these and related policies may be found in CEPAL 1982 and, for Sinaloa and the key link to water rights, in Mares 1987.)

In both types of redistribution, the Mexican government gave little inconvenience to the large landholders who had survived the Cárdenas era, and in some cases such landowners received massive new subsidies from the new irrigation projects, especially in the Northwest. Emphasis would be placed on production gains, not on social change or equity. Agriculture would be seen not as a legitimate way of life for the Mexican majority but primarily as an instrument of industrialization. Agriculture requiring industrial inputs, such as machinery, fertilizers, and pesticides, and farms producing taxable and/or exportable income would be favored over subsistence and small-scale, local market agriculture. President Camacho was careful not to criticize the agrarian reform program under Cárdenas—it was too politically dangerous to do so given its popularity and the popularity of Cárdenas himself—but rather strongly emphasized that the way to improve Mexican agriculture was to make farms more productive.

Just as important to Mexico as Avila Camacho's inaugural speech, and far more important for the future of the rest of the world, was the man who was standing beside Avila at the inauguration ceremonies while the new president outlined his policies. He was the newly elected vice-president of the United States, Henry Wallace. Wallace had been Roosevelt's secretary of agriculture. Before that, he had gained prominence as an active partner in his father's seed firm, Pioneer Hi-Bred. Henry Wallace and his father had developed some of the most commercially successful hybrid corn seeds in the first decades of the century. Wallace liked to think of himself as the "father of industrialized agriculture." He also had a messianic faith in the American small farmer in general and in the techniques of American plant breeders in particular. He was convinced that the New World could teach the Old a lesson, as he had said before the Commonwealth Club of San Francisco in 1939: "Our strength today is not equal to the task of composing the differences which exist in Europe and Asia. Our task, in cooperation with the twenty Latin American republics, is to do a first-class job of laying a foundation for democracy on this hemisphere—for the kind of democracy that will conserve our soil and people for thousands of years to come" (Schapsmeier and Schapsmeier 1970: 40–41).

Wallace's role in Mexican affairs is richly ironic. Seen as a left-wing figure in the Roosevelt administration and by many as a Communist fellow-traveler in his later run for president on the Progressive ticket in 1948, Wallace played a part in burying left-wing initiatives in Mexico. By the same actions that buried the Mexican left, Wallace, considered a strong conservationist at the time, helped develop the Green Revolution strategy that would become a major target of conservationist criticism a few decades later.

Wallace was to boast in his diary that his presence in Mexico avoided a revolution there that would have been disastrous to U.S. interests. One of the two major candidates who had opposed Avila Camacho disputed the election results that made Avila the winner—the candidate had been supported by a crazy-quilt alliance of left wingers who worried about the domestic meaning of Avila Camacho's candidacy and right wingers who wanted to preserve the possibility of an alliance of Mexico with Germany. The Roosevelt administration worried that Avila Camacho's failure to take office would mean a continuation of the irritating Mexican nationalism that had led to the oil expropriation as well as to a possible German dominance over Mexico. It was essential to the Roosevelt administration that the election of Avila Camacho be consummated whether or not it was honest, which Wallace and some others doubted. Certainly Wallace's style, his command of enough Spanish to charm the Mexican press and those he met personally, his willingness to stay in Mexico for several weeks after just having been elected to the vice-presidency of the United States, and his skill with the rhetoric of both populism and Pan-American harmony were influences in overwhelming domestic opposition to Avila Camacho and his policies (Schapsmeier and Schapsmeier 1970: 40–42).

But Wallace was not content to limit his role in Mexico to public relations and diplomacy. He was quick to conclude that North American agricultural expertise, especially plant breeding, could be the salvation of Mexico's poor. Every plant breeder who worked with maize, or corn, tended to be fascinated with Mexico because the hundreds of varieties of corn that plant breeders used to produce new hybrids came mostly from Mesoamerica. On his return to the United States, Wallace invited Raymond Fosdick, president of the Rockefeller Foundation, and Dr. J. A. Ferrell, a Rockefeller Foundation officer, for a discussion in the Senate Office Building in February 1941. Ferrell had earlier pushed the idea of a Rockefeller agricultural educational and research program in Mexico in 1936, encouraged to do so by a former Mexican minister of agriculture under Calles. The idea was dropped during Cárdenas's agrarian reform, but Wallace's invitation represented a chance to revive it (Stakman, Bradfield, and Manglesdorf 1967: 20–22).

Wallace opened the meeting by saying that one of the important

problems facing Mexico was the need for greater agricultural production. He suggested agricultural demonstration programs located
in Mexico's densely populated central plateau.

Wallace further stated that while the Rockefeller name was
generally associated with philanthropy, "it might also be
linked with the oil industry, expropriated property, and the
attendant controversies." When questioned on this point,
Wallace responded that the Rockefeller Foundation was regarded as apolitical and that it had advantages over alternative
programs, such as between the United States and Mexico.
    Wallace also commented that his concern over low produc-
tivity was the result of the "possible danger" that this made
in the face of a high birth rate. Wallace stated that he and the
U.S. ambassador to Mexico, Mr. Daniels, felt that President
Camacho "wants peace with the United States and that a
basis for the settlement of controversies about U.S. property is
in prospect." The Vice President concluded the meeting with
an emphasis on the importance of Mexico in the national defense of the United States. (Ferrell memo, February 10, 1941,
in Jennings 1988: 48)

On the basis of this meeting, subsequent discussions led to the
appointment, two weeks after the initial conversations with Wallace, of an administrative commission within the Rockefeller Foundation and of a survey team to go to Mexico as quickly as possible to
make recommendations about the proposed program.
    Those selected for the survey team were Dr. E. C. Stakman, pro-
fessor of plant protection at the University of Montana, Dr. Richard
Bradfield, professor of soils and agronomy at Cornell University, and
Dr. Paul Manglesdorf, professor of plant genetics and breeding at
Harvard. The foundation decided on the composition of the team because, as the members of the team wrote a quarter-century later,
"'Experience has shown that the greatest practical contributions to
agriculture come through the fields of genetics and plant breeding,
plant protection, soil science, livestock management, and general
farm management.' The committee therefore decided to select men
from the first three fields to constitute the Survey Commission"
(Stakman, Bradfield, and Manglesdorf 1967: 22). Having recalled
that decision, Stakman, Bradfield, and Manglesdorf would also say
in the same chapter, and with no apparent irony intended, that they
undertook their study "without preconceptions." "The mandate was
simple: 'Go to Mexico and find out whether you think the Founda-

tion could make a substantial contribution to the improvement of agriculture, and if so, how?'" (Stakman, Bradfield, and Manglesdorf 1967: 25).

In their book written in 1967, *Campaigns against Hunger,* the members of the study team seem to be somewhat confused about exactly what created the need for the improvement of agriculture, which is not surprising given the circumstances surrounding their mission. They recall that they were worried that agricultural production was not as high as it needed to be. They blamed this partly on the redistribution of land into plots they believed were too small for efficient production. They also took no account whatsoever in their recollections of the inevitably difficult conditions of a country coming out of ten years of internal warfare nor of the difficulties caused by the Great Depression, a Depression that had followed on years of severe economic decline in North American agriculture in the 1920s, leading to the catastrophic North American Dust Bowl of the 1930s. Their confidence in North American ways and their disparagement of Mexican ones, given that historical context, are remarkable. (On the failings of American agriculture in this era, see Worster 1979.)

They cited the high birth rate in Mexico but made no attempt to correlate it with social conditions of the Mexican countryside and ventured no questions about how the program of rural change they were recommending would affect the high birth rates. (As their work radically increased levels of insecurity for Mexico's poor majority, demographic theory would suggest that their program of agricultural modernization actually helped to keep birth rates high.) They talked about food shortages but advanced no analysis whatsoever about what could have led to such shortages other than their preconceived notion that poor technology was at the root of the problem. Economic and social factors were seen throughout their report and throughout their evaluation of their work a quarter of a century later as nothing more than obstacles to technological development. But is this surprising, considering that the team consisted of natural scientists and agricultural technologists alone, men who had been given the task of proposing solutions to a vast and complicated social problem? These men were selected to provide the technical underpinnings for a major social and economic about-face. Governments and investors, not researchers, had decided on the policy changes and the kind of research they wanted. It is only remarkable that in their writings the researchers have never revealed an understanding for the fact that they were pawns in a larger game.

The study team spent five months in Mexico beginning in July

1941. They were accompanied by Richard Schultes, who had just completed his doctorate in botany at Harvard and who had "first-hand knowledge of Mexican flora, Mexican botanists, and the Spanish language." They were helped and advised by Rockefeller Foundation personnel in Mexico and by Mexican officials of the new Camacho government. They traveled five thousand miles in about half the states of Mexico. The task was not always easy, as they would recall.

> The Commission tried to talk with an adequate sample of these back-country people but found that discourse without communication was of limited profit. Apparently there was considerable truth in the official estimate that two or three million Mexicans habitually spoke one or more of 54 Indian dialects, and, although a million of them reputedly spoke some Spanish also, the Commission often consoled themselves with the statements of more experienced observers that "some Spanish" usually was "very bad Spanish."
> Despite this handicap of mutually bad Spanish, the Commission learned to appreciate some of the problems and hopes of the humbler peoples. (Stakman, Bradfield, and Manglesdorf 1967: 27)

In December, the study commission of Stakman, Bradfield, and Manglesdorf were ready to make their recommendations, in spite of the difficulties. They proposed a program of "technical assistance," asserting that "the most acute and immediate problems, in approximate order of importance, seem to be the improvement of soil management and tillage practices; the introduction, selection, or breeding of better-adapted, higher-quality crop varieties; more rational and effective control of plant diseases and insect pests; and the introduction or development of better breeds of domestic animals and poultry, as well as better feeding methods and disease control" (Stakman, Bradfield, and Manglesdorf 1967: 32–33). The study team recommended that these goals be carried out by a research program in Mexico, directed initially by North American researchers, by a fellowship program that would send Mexican students to the United States for advanced training in agronomic sciences, and by an effort to guide and improve agricultural colleges and institutes in Mexico.

> The plan presented assumes that most rapid progress can be made by starting at the top and expanding downward. The alternative would be to start at the bottom and work toward the

top. A program of improving the vocational schools of agriculture and of extension work directed toward the farmers themselves might be undertaken. But the schools can hardly be improved until the teachers are improved; extension work cannot be improved until extension men are improved; and investigational work cannot be made more productive until investigators acquire greater competence. (Stakman, Bradfield, and Manglesdorf 1967: 33)

The Rockefeller Foundation moved quickly to institute, with the cooperation of the Mexican government, the recommendations of the team. The program in each of its phases and taken as a whole came to almost completely dominate Mexican agricultural research and education until it was turned over to the Mexican government in 1961, creating the National Institute for Agricultural Research (INIA). As shown by the research of Bruce Jennings, by that time the Rockefeller program of student grants and fellowships assured that those who directed INIA as well as the National Agricultural University at Chapingo, most of the other major agricultural colleges, and most of the branches of the Ministry of Agriculture for another two decades would be Mexicans trained in ways consistent with the goals and philosophy of the Rockefeller program. The Rockefeller program, along the lines originally recommended by the study team in 1941, would be exported in whole or in part to dozens of countries under joint sponsorship of the Rockefeller Foundation and the U.S. government and would come to be called the Green Revolution (Jennings 1988).

Recent research in the Rockefeller archives by Jennings reveals that from the very beginning scholars with intimate knowledge and long experience of Mexico offered the foundation well-reasoned critiques of the program. A bit nervous about the implications of the recommendations of the study team, one foundation officer encouraged solicitation of scholarly review of the study team document. The criticisms came from natural scientists and social scientists with far more experience in Mexico than the members of the team who designed it. One of these was geographer Carl Sauer of the University of California at Berkeley, who commented on the program in 1941 and in continuing correspondence over several years. Sauer had studied Mexican agriculture since the 1920s and had years of field experience in Mexico. He was very disturbed by the program proposed. Sauer maintained that the problems of Mexican agriculture had far more to do with economic exploitation than with Mexican agricultural practices, which he largely admired. In what seems a thinly veiled reference to the involvement of Henry Wallace, who was first and foremost a plant breeder and commercial seed sales- man from Iowa, Sauer wrote a scathing criticism.

> A good aggressive bunch of American agronomists and plant breeders could ruin the native resources for good and all by pushing their American commercial stocks. . . . And Mexi- can agriculture cannot be pointed toward standardization on a few commercial types without upsetting native economy and culture hopelessly. The example of Iowa is about the most dangerous of all for Mexico. Unless the Americans understand that, they'd better keep out of Mexico entirely. (Sauer in Jen- nings 1988: 51)

Sauer proposed a more anthropologically based approach, recom- mending that the Rockefeller program should study in detail the problems of the peasant household, patiently solving problems as identified by the peasants themselves.

Although the foundation ignored Sauer's views, Sauer continued to take an interest in the program, meeting in 1945 with the direc- tors in Mexico and corresponding with the foundation. He was favor- ably impressed with some of the views of North American research- ers in the program and assured the foundation that he saw value in agronomic research directed at improving native Mexican seed va- rieties. But he continued to have profound doubts about the direc- tion of the program and of Mexican agricultural policy in general.

> I am not interested in the Indians as museum pieces and I am also interested in non-Indian populations that have cultural values of their own as apart from the standardizing tendencies which are flowing out from the urban centers to strip the country of its goods and ablest men and pauperize it cultur- ally as well as often economically. . . . The agricultural situa- tion in Mexico is an exaggeration of that which we have in our experimental work in California; the interest is directed away from subsistence or village agriculture to the needs of city and factory with the attendant emphasis on standardiza- tion of product and yield, also on tariff protection, and on the commodities which the privileged faction of the population can absorb. (Carl O. Sauer, letter to Joseph Willits, Rockefeller Foundation Archives, February 12, 1945, 1.2/323, p. 1, quoted in Jennings 1988: 52–53)

In spite of a good deal of confirmation of Sauer's views produced during the early forties by the foundation's own International Health Division concluding that "attempts to change dietary habits in this region would be a mistake until economic and social conditions can be improved" (quoted in Jennings 1988: 54), the policy of the foundation's Mexican Agricultural Program would continue to focus narrowly on production gains alone. Paul Manglesdorf wrote in 1949, responding to the opposition by Sauer and Edgar Anderson to the directions the program had taken. Manglesdorf maintained that their position was that

> if the program does not succeed, it will not only have represented a colossal waste of money, but will probably have done the Mexicans more harm than good. If it does "succeed," it will mean the disappearance of many ancient Mexican varieties of corn and other crops and perhaps the destruction of many picturesque folk ways, which are of great interest to the anthropologist. In other words, to both Anderson and Sauer, Mexico is a kind of glorified ant hill which they are in the process of studying. They resent any effort to "improve" the ants. They much prefer to study them as they now are. (In Jennings 1988: 54–55)

Jennings points out that this interpretation "did fundamental violence" to the arguments of Sauer and Anderson and demonstrated an inability of scientists such as Manglesdorf to appreciate the social meaning of scientific research (Jennings 1988: 54–55).

The results of the foundation's research program would have confirmed Sauer's worst suspicions of it, though Sauer did not live long enough to submit his own final judgment. Perhaps the harshest view of the results of the Green Revolution program has come from a scientist trained at Harvard by Manglesdorf, Garrison Wilkes. Wilkes was to become deeply disturbed by the way the promotion of a few commercial varieties of the major grains developed in Green Revolution research had displaced the planting of the much wider diversity of grain crops that had prevailed in the world before the successes of the Mexican Agricultural Program and its diffusion around the world. He pointed out in an article in *Science* magazine in 1972 that this greatly reduced crop diversity left humanity at the mercy of uncontrollable evolutionary forces that were sure to produce new plant diseases and pests. With only a few commonly planted crop varieties, it was likely that at some point these new enemies of human crops would be able to penetrate the defenses of human ingenuity as represented by pesticides and cultivation practices, and plant breeders would look, as they had before, for traditional crop varieties that had inbred resistance. But what Sauer had called the "standardization on a few commercial types" would mean that plant breeders would be unable to find the necessary resistant varieties, varieties eliminated through global standardization. The few "backward" farmers who still maintained fields of traditional grains were being pushed out of agriculture, and in their last desperate years of farming were likely to be forced to eat rather than replant the seeds that carried the genetic diversity produced by humanity over millennia in the interaction of human culture and food requirements with the evolving natural environment. Wilkes summarized the issue succinctly, writing that, "quite literally, the genetic heritage of a millennium in a particular region can disappear in a single bowl of porridge" (Wilkes 1972: 1077). Another experienced plant breeder, Walter Galinat, warns that without careful efforts to preserve genetic variability, "we and our mutually symbiotic food plants may vanish like the dodo" (quoted in Kahn 1984: 81).

Another way of saying much the same thing would be to point out that the effect of the Rockefeller program was to eliminate the system of security and stability built into agriculture by thousands of years of peasant technology practiced under widely varying circumstances and adapted to those differing conditions. Narrowing the genetic base, however, was only one of the ways in which the peasant technologies of security and stability were eliminated.

The issue of genetic diversity has raised a particularly lively debate in Mexico. One of the means by which some genetic diversity may be maintained for world agriculture is through the gene banks that store seeds for possible future needs. Such banks cannot truly ensure future diversity because the seeds must be planted from time to time to maintain their viability and consequently lose genetic material with each planting. The seeds produced in each planting within a given environment do not reflect the wider genetic diversity present in the parents that derive from plants grown in a wide variety of ecosystems over long periods of time. But gene banks are a partial replacement for the former diversity provided by traditional agriculture around the world. The United States and some other industrialized nations have such gene banks, sharing material through an international program overseen by the UN Food and Agricultural Organization. For decades, Mexico has argued that since a very large portion of the seeds stored in these banks come directly or indirectly from Mexico, one of the two or three oldest and most diverse centers for the domestication of crops, Mexico should have its own gene bank. Mexicans argue that such a facility would partially redress an old injustice by which the United States and other rich nations have grown prosperous on the genetic research of poorer nations. More important, they argue that the essential difference in interests between rich and poor countries requires that there be a gene bank outside the physical control of the rich nations (Barkin and Suárez 1983).

The dangers arising from the loss of genetic diversity are closely related to the fact that the new agriculture created in the postwar period was designed to be heavily dependent on synthetic fertilizers and pesticides. The men who worked as Rockefeller Green Revolution scientists in Mexico have written accounts of their work that make it clear why this is so. Dr. Norman Borlaug, who won the Nobel Prize for his plant-breeding work in the Rockefeller Mexican Agricultural Program, emphasizes in his writing that the Green Revolution consists not just of improved seed varieties, the Green Revolution consists of a package that must include fertilizers, pesticides, and improved water delivery to make the miracle seeds viable. Without all the elements in the package, the seeds alone represent no improvement and may be worse than traditional techniques (Bourjilay 1973; Borlaug 1986).

A brief history of how the "miracle seeds" were first developed makes it clear why this is so. Plant breeders had noted a distinctive curve related to the production potential of the grain seed varieties they worked on. They had reached a point in seed development where the main limiting factor on the yield of grain was the amount of nitrogen the plant could absorb and use. Yields would go up in direct relation to the amount of nitrogen applied to the plant, until the plant reached a distinct limit to the amount of nitrogen it could utilize. When nitrogen was applied approaching this limit, yield increases would be minimal in response to large increases of nitrogen. When the limit of nitrogen utilization was exceeded, the plant would suffer from additional nitrogen application and yields would fall with the application of more nutrient. Breeders like Borlaug defined the plant's capacity to absorb more nitrogen as the "limiting factor" in increasing yields.

The solution Borlaug and his colleagues sought in Mexico was to breed grains with the ability to absorb more nitrogen and convert the increased nutrients available into more seed while retaining resistance to the almost universally troublesome wheat stem rust diseases. Their first experiments were successful in producing wheat capable of turning more nitrogen into more grain, but the plants produced were too tall. As a result, they "lodged" easily; that is, they tended to fall down because of the long stem and heavy head, making harvest difficult or impossible.

The solution then was to look for dwarf seed varieties that could be crossed with the heavy nitrogen-absorbing varieties. Borlaug was successful in finding suitable dwarfs and he now had a wheat that would absorb large quantities of nitrogen, resulting in dense stands of wheat stocky enough to stand up under the weight of the large head. However, the plant could not absorb the nitrogen unless the nutrient was carried in a medium in concentrations that would not be toxic to the plant, meaning that heavier nitrogen applications would be effective only if there was more water available to carry the nutrient to the plant cells. Such very high quantities of nitrogen could realistically be delivered only by synthetic chemical fertilizers in most circumstances, reducing or eliminating traditional fertilization with manure or other organic material, organic material that otherwise would help absorb and hold water and improve the texture of the soil in ways that would help penetration of nutrients to the roots of the grain plant. Gradual loss of organic matter then meant, in turn, the need for more water and more chemical fertilizers to actually get the appropriate quantities to the plant.

The densely packed short wheat, well-watered and heavily fertilized, made for humid soil and wet grain plants and seeds, with a minimum of aeration by wind, and the result was ideal conditions for the growth of fungal diseases, insects, and weeds. The Green Revolution solution was to apply chemical pesticides to control these new field conditions. The need for pesticides was increased by the loss of organic materials such as manures that tended to provide habitat for beneficial insects and other organisms, maintaining some balance between crop pests and predators on pests. In short, the new seed varieties typically meant the redesign of the field environment and a complete change in agricultural technique, changes that required the use of more water as well as synthetic fertilizers and pesticides (Stakman, Bradfield, and Manglesdorf 1967: 80–93).

While the story of Borlaug's dwarf wheat varieties portrays something of the meaning of the Green Revolution research in Mexico, it alone cannot capture all the complexities and implications of the program. In a few cases, the early researchers actually developed ways to reduce pesticide use in some specialty crops, notably potatoes, although further agricultural modernization led to new potato varieties and techniques that required more pesticides. In recent years, research effort has been directed at building a degree of pest and drought resistance into Green Revolution seed varieties, restoring, to a degree, qualities bred out in the early research, while maintaining some of the high response to chemical fertilizer. There has been some effort to tailor seeds more closely to specific regional ecological conditions, rather than the vast standardization that swept Mexico and the world in the wake of the Green Revolution research, again merely restoring a portion of the adaptive qualities lost during the first decades of the program. Defenders of the Green Revolution often emphasize these new research directions by talking about a second, third, or even fourth generation of Green Revolution development (Borlaug 1986).

However, certain features of the Mexican Agricultural Program of Green Revolution research are clear in spite of the complexities and qualifications. These derive from the nature of the research in a technical sense, but it is the social and ecological results of the research that are really important.

The dwarf wheat story makes it clear that farming with the new seeds would mean greater out-of-pocket expenditures for the farmer. The chemical fertilizers and pesticides would have to be purchased. In many cases hybrid seeds meant that instead of saving a portion of the harvest for replanting in the next season, the seeds themselves would have to be purchased, too, because the hybrids did not yield viable seed in the second generation. Somehow, more reliable water delivery had to be arranged, and even where government-financed irrigation projects absorbed most of the costs, there were almost always additional costs to traditional farmers involved in improving irrigation.

All these increased costs meant that to adopt the new technologies, farmers must have access to large amounts of credit at reasonable rates. Few matters have stirred more controversy than the government's policies on rural credit, but it can be said with certainty that larger, more commercially oriented farms could find credit far more easily than their smaller, more traditional competitors. This was true not only because of the deliberate bias of government credit programs intended to provision cities, encourage the purchase of industrial inputs, and earn hard currency through export, but even more because through most of the post–World War II period the bulk of agricultural credit came from private firms and banks. Private agricultural credit came mostly from marketing companies and banks with little interest in traditional operations oriented to subsistence and local markets and producing comparatively small cash earnings (Hewitt de Alcántara 1976).

Like most governments, the government of Mexico was not content to allow market forces to determine whether the traditional technologies or Green Revolution technologies were more viable.

Since World War II, the Mexican government has offered subsidies to agriculture in many forms: government-financed dams, canals and irrigation schemes, road construction and subsidized national railways to ship agricultural produce, rural electrification and communication networks, fertilizers and pesticides manufactured by nationalized firms and sold at bargain-basement prices, government-financed market development, public agricultural research, education, and extension work, and direct crop subsidies. At the same time, the government has penalized agriculture for the sake of urban and industrial growth, maintaining wholesale and/or retail price controls to keep food cheap for city people and rewarding industrial investments with tariffs, import restrictions, tax incentives, infrastructure investments, below-cost petroleum products, and a whole variety of other mechanisms.

As a whole, these policies in agriculture and industry are the economic development policies of the nation. It can be assumed that while they rewarded some, they punished others. The pattern of these awards and penalties is extremely complicated, the subject of a large literature full of bitter disputes. It is also certainly the case that Mexico far more than most other Latin American governments and many in Asia and Africa did exert some considerable effort toward cushioning the blow of modernization policies on small producers and rural laborers. There has been an ongoing, if very partial, program of agrarian reform and, compared to many other nations, a strong program of education and social services in the country. The Mexican government has from time to time sponsored programs like the Sistema Alimentario Mexicano specifically meant to assist small producers in rainfed regions overwhelmed by the power of larger farmers in irrigated regions. But in the long run, none of these programs has gone far enough or been consistent enough to solve the basic problems of Mexico's rural poor. Steven Sanderson, in his excellent analysis of Mexican agricultural policy, has in fact concluded that Mexico's agricultural policies as a whole have clearly favored the growth of international agribusiness at the expense of the peasant and have, "in a word . . . left the national state without competent mechanisms to govern rural growth and development" (Sanderson 1986: 275).

The evidence strongly indicates that the policies favored large farmers over small, monoculture over diversity, and chemical farming over traditional methods. In addition, they changed the food patterns of the nation, with wheat favored over corn, the North American researchers going a long way to accomplish what the wheat-loving Spanish had attempted to enforce on Mexican farmers for three hundred years. The policies of the last decade also rewarded grains of high yield, but those high yields were mostly of moist, mold-vulnerable grains with lower protein quality than the older varieties. Few dispute that the policies have been designed so that, taken as a whole, they favor industrial growth over agriculture, agriculture that supports industrial growth over agriculture that does not, export and luxury food production over the basic nutritional needs of the poor majority, and industrial cities over rural regions (Yates 1981; Barkin and Suárez 1978, 1985; Feder 1976; Hewitt de Alcántara 1976; CEPAL 1982; Mares 1987; Sanderson 1981, 1986).

The agribusiness valley of Culiacán grows rich using the labor of the collapsing Mixteca, and industrial Mexico City and Monterrey dominate both. One result is that migration rates from the countryside to the city have been very high, at about 7 percent per year in the post–World War II era, twice the rate of overall population growth. As the communities of rural regions disintegrate, the tidal wave of migration makes the cities unmanageable. Benefits and costs of this sort are nearly universal in poorer countries and many rich ones in the modern world, so much so that their implementation and results have come to be called "modernization."

The Green Revolution strategy of crop development relied on synthetic pesticides that had just become available during and after World War II. The new policy direction of the Mexican government begun in the 1940s under Manuel Avila Camacho, emphasizing production gains on highly commercial operations in order to serve goals defined and pursued by elites rather than social change of the kind defined and pursued by peasants, dovetailed with the technical innovations available at the time. In earlier periods of Mexican history, ruling elites had resorted to different means toward the similar ends: specifying that taxes could be paid only in grains preferred by Europeans, legally defining land as unused and available for European settlement if it was not being used in ways congenial to colonial rulers, forbidding urban laborers from feeding themselves, forming legally protected grain monopolies, throwing peasants off their traditional lands, destroying peasant communities, forced labor recruitment, debt peonage, and war. These measures would ensure that Mexican agriculture would yield export crops, provision mines and related industrial activities, and produce readily captured income and taxes. In the post–World War II period, the availability of Green Revolution seed varieties based on expensive chemicals and irrigation projects made it possible to accomplish the similar results through technological innovation and a program of persuasion that led many people to believe the results were desirable for all and perhaps the inevitable consequence of the "march of science."

The term "modernization" itself implies inevitability. "You can't stop progress, can you?" used to be the crude way of ending any and all proposals for business expansion or new public projects, and the term "modernization" invented by scholars proved just as useful or more so, perhaps because its meaning is even vaguer and more easily mystified. What should be clear from the history of the Green Revolution in Mexico is that it was a very specific political project designed in a very particular political context and carried out by technicians whose self-justifications had only the slightest relevance to the real purposes of the project. It was carried out in order to defeat a popular political movement with a different conception of the improvement of human life, and it so far has succeeded in that task.

That the Green Revolution proved highly exportable was due to the fact that in country after country, elites searched for ways to have economic growth without bending to the demands of rural majorities, majorities with different conceptions of culture, technology, economics, and the improvement of human life than those held by the elites. It is this that explains why the Green Revolution in the mold of the Mexican Agricultural Program was adopted by government after government at precisely the times that severe confrontations were shaping up over agrarian reform in those nations. This was the case, for example, with the timing of the first major promotional efforts of the Green Revolution in Indonesia, India, Pakistan, and the Philippines, which together account for more than a quarter of humanity (King 1977).

The Green Revolution is one way of changing rural life and agricultural technology, a way compatible with the desires and plans of relatively wealthy landowners and urban elites, a way incompatible with the demands and cultures of traditional peoples focused on security and long-term ecological stability. Peasants dream of relatively autonomous rural communities within what is assumed to be a closed natural system. Elites dream of greater consumption, growth, and monumental undertakings.

All of this conflict of world views and political purpose remains unusually interesting in Mexico today because while the Avila Camacho policies put into technical shape by the Green Revolution and supported by the governments of the United States and Mexico have prevailed, they have had to contend with a large peasantry still somewhat resistant to such policies. Cárdenas's land redistributions allowed a large portion of the Mexican peasantry to sink their roots deeper into the countryside, making it more difficult than in many other countries to dislodge them. Mexican peasants fill their villages with statues of Lázaro Cárdenas, they belong to a large national organization that still must pay lip service to the agrarian reform ideas of the 1917 Constitution and the Cárdenas government, and they continually mount protests and new organizations to resist the demise of the peasant way of life. Periodic economic crises frequently lead, as in the 1980s, to actual resurgences of peasant agriculture in some regions as people try every available strategy to feed themselves in a predominantly urban economy with very high unemployment.

The Crisis

In 1988, the continued economic crisis led to a split in the ruling party, with Cuauhtémoc Cárdenas, son of Lázaro Cárdenas, running on an independent ticket with a strong propeasant, pro–land reform plank. Many insist that he gained the largest number of votes, although PRI has proclaimed its own candidate the winner.

Mexico has the world's largest city, but it also has a rural population that is today as large as the whole Mexican nation in 1950. Thirty million rural people cannot be ignored even in a rapidly industrializing nation such as Mexico was until the late 1970s. When economic crisis grips such a nation for more than a decade, food prices rise sharply, real income drops by half, and the nation is crippled by foreign debt, then rural and urban people both may be prepared to question modernization. At least, they wonder whether there might be better, more stable ways to achieve some of the benefits of economic development.

As we look at the debate the present situation has engendered over future agricultural technologies, it should not be surprising that the debate is not simply or even primarily a technical one. Those who identify with Ramón González, who died in the fields of Culiacán, are attracted to technological solutions different from those that lure the millionaire farmers and their North American creditors who control the Culiacán Valley. It is almost a truism that new technologies tend to create conflicts—we should now understand that new technologies also often arise out of conflict.

7
Consumers, Workers, Growers, and Experts

*I*n February 1987, I accompanied a reporter for the *Wall Street Journal* to the Valley of Culiacán. He was easily able to arrange contacts with the businessmen and growers of the region. I had volunteered to go along to ensure that in the brief time available to him he would see the realities in the fields that his other contacts would not likely show him.

One of the reporter's contacts was then the head of the National Union of Vegetable Growers (Unión Nacional de Productores de Hortalizas). This man was smart and politic in what he said, avoiding dramatic or excessively defensive statements. He admitted, for example, that farm workers were often exposed to very dangerous pesticides and said that outsiders were justified in keeping the pressure on growers for improvement in pesticide safety and the living conditions of farm workers more generally. He treated us to a fine dinner in a private men's club on the third floor of a building housing offices and a parking structure. Behind the locked doors of the club, his friends and business acquaintances were eating and getting drunk.

Soon, the friends had joined our table, apparently attracted by the presence of the inquiring foreigners. They first attacked our host for talking to us at all, arguing that "we never should have elected you in the first place, because you don't understand about these bastards." Then they turned their anger on us directly. Since some of these men spoke English, the reporter spoke no Spanish, and I spoke both, the conversation became very complicated at times by the crosscurrents of the two languages. Some of the men were embarrassed now and again by the openly displayed hostility emanating from their friends and tried to defuse it by changing the subject or suggesting platitudinous resolutions to the issues under discussion. "We all work to make things better when we can, for the workers and the consumers, don't we?" At times, the angry men would focus on the reporter, asking me such questions in Spanish as "Where did you get this Jewish prick?" Early on one of them said, "Aren't you afraid to come down here where we kill gringos who get in our way?" but if one can retract a rhetorical question, he did so by saying soon after, "No, don't worry, we're not going to kill you."

At another point, one of the men spoke in a low voice and confidentially to me in Spanish, saying, "You obviously know Mexico. You know Spanish. You understand. And then you bring this man down here." He paused and nearly spat as he said, "You are a traitor! That's what you are. A traitor. How can you do this?"

The conversation was occasionally interrupted by a strolling guitar player who would begin playing the melody to a romantic ballad, and the angriest of the men would stop in mid-sentence to break into song in a fine, controlled, almost operatic voice, "When I met you in the Moorish garden . . ." to suddenly break it off with another challenge or obscenity thrown at us.

Some of these men were growers, but most were businessmen tied into the agricultural economy of the valley in one way or another. Two sold insurance to business firms. One owned and managed a factory making packing boxes for vegetables out of sugarcane waste. Another sold agricultural machinery.

One of the most revealing moments of the evening came as we were discussing the dangers to farm workers from pesticide use.

The man given to intermittent serenading cut into the conversation, asking, "But wait a minute—do your readers, the readers of the *Wall Street Journal*, give a damn about what happens to Mexican farm workers? Well, do they?"

The reporter admitted that his readers' interest in the health of Mexican farm workers was probably minimal.

"Of course they don't give a damn! And we don't care if they do! We have that situation under control, and I don't care what they think about it. But you're here to hurt us . . ."

The reporter interjected, "You can't say that. I'm here to get a story. Whether it hurts you depends on what I find."

"Don't give me that shit. No reporter who has ever showed up here from the States has ever been here to do us any good. OK, you're here to hurt us. We understand that. And the only way you can hurt us is to damage our market. That means consumers. The only thing you can do is to frighten consumers into thinking they will be poisoned from eating our vegetables. Or frighten investors who might worry about the consumer reaction to our products. That's the way you could hurt us, right?" The reporter conceded that the man might be correct, although the story he published in his newspaper did give considerable weight to the pesticide dangers as they affected farm workers (*Wall Street Journal*, March 26, 1987, pp. 1ff.).

In my experience, many other reporters had assumed from the beginning of their investigations that the angry man from Culiacán was right. Over a period of three years, I received calls from newspaper, radio, and television reporters asking for information about the problems of Mexican pesticides. They would soon center their questions on the problem of dangers to the consumer from pesticide residues and on their belief that pesticides widely used on vegetables and fruits in Mexico were chemicals that had been banned for use in the United States.

I would interrupt them at this point to say something like the following, "Look, in my opinion, the danger to consumers from pesticide residues is an important public health issue that deserves attention. However, I am not completely convinced that pesticide residues from produce shipped from Mexico or from within the United States pose a really serious danger to consumers. Furthermore, the pesticides I observed in common use in Mexico are ones that are now legal in the United States and that are widely used in the United States, but that are highly hazardous in spite of their being legal. The powerful issue here as far as I am concerned is the blatantly abusive use of pesticides that does serious damage to farm workers and rural residents and to the environment and to the productivity of Mexican agricultural land into the future—these are major issues of human rights, of the preservation of nature's diversity, and of the long-term productivity of good farmland."

In most but not all cases, the reporter would then say that the story as I had framed it was certainly important, but that the audience for the story would not be interested in anything that did not influence its health or well-being directly. Often, he would say something like, "Unfortunately, nobody here in the United States is going to get excited about the health of Mexican farm workers, or a bunch of migratory birds killed by pesticides in Mexico. You have to hit people where it hurts them directly." The reporter would often then ask for the names of the people who were mainly concerned about the pesticide-residue-on-foods issue and hang up after getting that information.

In writing popular and scholarly articles, cooperating with journalists on a variety of other stories, and speaking on the issue of pesticide abuse on dozens of occasions in Mexico and the United States, I have given a great deal of thought to the viewpoint that people only care about what affects their health or their pocketbooks directly. If the viewpoint is correct, then one necessarily rather narrow set of political and technical alternatives presents itself as the best way to achieve a solution to the problem. If the viewpoint is incorrect, or only partly correct, then a much wider and more ambitious set of strategies becomes attractive.

It is also possible that these two ways of looking at the matter establish a false dichotomy or opposition between two perspectives. It might be that the question of immediate self-interest is tied into larger and more generous concerns in complicated ways, as complicated as many other human emotions, intellectual concerns, and political viewpoints. While the vast philosophical, sociological, and political literature on the question of self-interest in formulating social concerns is much too complex to examine here, it is useful, I think, to imagine what strategies might derive from using different assumptions about what motivates people to solve problems. To begin, we will look at this in terms strictly limited to the question of pesticide abuse and then move to the larger questions about social equality and environmental protection raised in previous chapters.

What if people only care about what affects them immediately and directly? If that is the case, then there are certainly some groups of people who might be assumed to have a strong interest in eliminating the abusive use of pesticides. The first and most obvious group of self-interested people are all of those who are repeatedly exposed to dangerous levels of pesticides. Farm workers, rural residents, people who work in manufacturing, packaging, transporting, and formulating pesticide mixtures have a clear reason to want to see pesticides handled responsibly.

The conditions described in earlier chapters make it clear that there is little or nothing an individual worker can do to reduce his or her exposure other than refuse work, work which in the case of migrant farm workers especially is vital to the survival of the laborer and the family. But in many countries, organized labor and industrial workers have played a leading role in reducing workplace dangers of all sorts, including those in agriculture. Labor unions have written and lobbied for protective legislation and maintained a constant critique of regulatory practices. This is true at the general level of establishing basic legislative principles and such agencies as the Occupational Health and Safety Agency in the United States and the federal government public and occupational health programs in Mexico. Revolutionary labor organizations in Mexico forced the adoption of an ambitious labor protection code within the Mexican Constitution of 1917, including health and safety protections. At the international level, the International Labor Organization of the

United Nations plays an important role in initiating investigations of worker safety in the field and putting pressure on other international organizations such as the Food and Agriculture Organization and World Health Organization to pay attention to field worker safety issues. But no one believes that such broad regulatory principles or the establishment of agencies in themselves protect workers. The generalized measures must be complemented by specific implementation and enforcement actions that ensure that laws and bureaucrats do what they are supposed to do (Bull 1982; Boardman 1986: 47–51).

Organized labor has always been crucial at this more specific level as well. For example, in Lathrop, California in 1978, pesticide manufacturing workers began to suspect that they were suffering from sterility as a result of exposure to the chemicals they worked with. In casual conversation, the men at the plant had found that the wives of the ones who worked in a particular pesticide unit had been frustrated in their attempts to conceive children. A union officer arranged for sterility tests to be done on the men, tests they had to be cajoled and pressured into taking, confirming that the men had very reduced sperm counts. These tests and the pressure of the union and publicity the union helped to arrange forced the company to admit that it had data, sixteen years old, that indicated the pesticide caused sterility and that it probably caused cancer. The pesticide, DBCP, is used against nematodes, a kind of roundworm that is a serious pest of vegetable and fruit crops. Subsequent investigation showed that DBCP was persistent in soil and could be expected to contaminate groundwater. These findings played a major role in establishing a test program for rural wells in California's Central Valley that showed disturbingly high levels of groundwater contamination, spurring additional legislation in California and other states. The California Department of Food and Agriculture reluctantly agreed to heavily restrict DBCP use, and a national ban came two years later when it was shown that DBCP also constituted a threat to consumers. Many other nations followed suit, although for a time, DBCP manufactured in Mexicali found its way into California fields. The accumulation of DBCP in soils means that wells all over the world will continue to be contaminated for a long time to come, but at least with further applications banned there is an upward limit on how much contamination will occur from this pesticide. Without the active role of the union, the health of the manufacturing workers, farm workers, and the public might have been much more seriously damaged over a much longer period of time, and there is nothing to ensure that the problem would ever have been identified.

The United Farm Workers Union has lobbied for many years for improvement in pesticide regulations and enforcement in the state of California, and the UFW's relative strength in California is one of the major factors that accounts for California's more rigorous pesticide regulatory schemes. The UFW has been important in the California legislature, but it has also been crucial in helping workers identify and react to specific abuses. Partly as a result of union lobbying, California has a fairly strict set of rules about how long workers must be excluded from fields that have been sprayed with acutely toxic pesticides. These rules are sometimes abused, often resulting in poisoning of whole crews of workers. Where the union has contracts with growers, it plays an active role in insisting that workers not be forced into recently sprayed fields and, where abuses have occurred, at pursuing legal remedies. A farm worker who complained about pesticide safety violations in California fields was told by his foreman, "There's no union here. Shut up and get to work." Not only the union but the threat of unionization can certainly play an important role in pesticide safety.

In the 1970s the UFW was in the ascendancy in California. By 1973, the UFW had signed contracts for 40,000 workers on 180 farms. These contracts were mostly for workers in lettuce and grapes, where acutely toxic pesticides are used heavily under conditions that tend to lead to direct worker exposure. In the wake of this successful organizing drive, the liberal Democratic legislature and governor had established the Agricultural Labor Relations Board (ALRB), with a number of board appointees strongly sympathetic to labor. During this period, improvements in farm labor housing, wages, fringe benefits, and field safety occurred regularly even where workers were not organized but where growers feared that they might be. With the election of a conservative Republican governor allied strongly with agribusiness and serious internal problems in the UFW, there has been a steady decline in the real wages and working conditions of farm workers. By 1986, the UFW had only 115 contracts covering 21,000 jobs, about half the number of jobs covered in 1973 before the creation of the ALRB (Wells and West 1988).

In the area of pesticide safety, the recent record is somewhat mixed because of the activity of environmental organizations and regulatory agencies at the state and federal levels, resulting in new regulations and compensating somewhat for the reduced pressure from the union. However, the number of pesticide poisoning incidents has climbed steadily during most years of the last decade (Barnett 1988; Wasserstrom and Wiles 1985). At present, the UFW has focused its energies on a campaign against the use of hazardous pesticides. The union has organized a national boycott of grapes to bring attention to the problem of farm worker pesticide problems, gaining the support of almost all national labor unions and circulating an unprecedented amount of information on the issue to a public that otherwise would hear little or nothing about the problem.

How much can such efforts accomplish, and can they be successfully duplicated in Mexico and elsewhere? In the United States, the question has yet to be clearly answered and will depend strongly, as discussed below, on the degree of general public support given to the union's efforts. In Mexico, as in many other poorer countries, the basic problem with union militancy on such issues is that there is a superabundance of cheap labor available. Unemployment and underemployment in the countryside have been running at 50 to 60 percent for more than a decade. We have seen earlier that the crisis of one region alone, the Oaxacan Mixteca, can provide the vegetable growers with most of the labor they need, not counting the many other regions of Mexico with similar labor surpluses that can be tapped at any time. And gains in one country may contribute to harm in others—one thousand banana workers have suffered from sterility as a result of DBCP poisoning in Costa Rica after chemical companies lost their market for the pesticide in the United States (Thrupp 1988).

In addition, in Mexico as in many countries, union organizations are seriously compromised by their dependence on ruling political parties and the state. In the Culiacán Valley, for example, all farm --workers belong to a union that is affiliated with a national labor organization that forms part of the ruling national political party, PRI. For the most part, the policies of the national unions and their affiliates are determined by labor bureaucrats in negotiations with PRI, and local organizations who rebel against these policies are severely chastised, suspended, put in receivership, or, in some cases, attacked by paid thugs. In Culiacán, the result is that few workers are even aware that they are members of a union, their dues being deducted from their paycheck without their realizing the purpose of the deduction. The union is important in such areas as insuring that the government-set minimum wage is paid and in negotiating periodic increases in the minimum wage (which seldom keeps up with the rate of inflation), but there is almost no organizing effort in the fields and no significant activity on such issues as pesticide field conditions. The union is part of the apparatus for enforcing the will of the ruling party and the state, not an independent force militating for workers' rights. Since the policy of the state in this case is focused on the drive to increase agroexports, demands of the union seen as a serious obstacle to promotion of export crops will be largely ignored (Posadas 1982; Mares 1987).

As a result of this situation, some independent organizers have tried to establish militant unions in the Mexican fields. One active organizer in the vegetable fields of Culiacán, a young Mixtec man, was imprisoned in 1978 during a strike of twenty-five to forty thousand farm workers and severely beaten each day of his two-week imprisonment. The strike succeeded in gaining somewhat improved living conditions in the farm worker camps (see Chapter 2) but did not address the problem of pesticide safety (Posadas 1982). In 1980 and 1984, the same leader confessed both that the conditions were not ripe for a repeat strike and that the union remained uncertain about how to formulate demands with regard to pesticide safety, due both to the complexity of the problem and the union's lack of technical information about it. In 1987, during an organizing campaign in the vegetable fields of San Quintin, Baja California, Sr. García ran on the ticket of a left-wing party for federal deputy. After losing to the official PRI candidate, the union organizer was imprisoned again. Internal problems within his organization have led to additional difficulties in organizing workers.

The problems of such independent unions are immense because organizers must try to organize desperately poor migrant workers against the sometimes violent opposition of the well-financed growers. There are very few migrant agricultural workers' unions in the world who have ever been able to do this successfully for any period of time. But in addition to these obstacles, the independent unions in Mexico must also survive the disruptive and sometimes violent activities of the official PRI union, the ruling party, and the Mexican state. As David Mares concluded from his study of labor organizing on the Culiacán Valley vegetable farms, "Grower control over the production process and state determination of the basis for organization and mobilization of farmworkers have doomed efforts to wrest significant benefits from industry for the benefit of labor" (Mares 1987: 166). It is not realistic to expect that such odds can be overcome without a major change in the political climate and a shift in political alignments in Mexico. The ruling party and the state would have to be won over to the side of militant organizing among farm workers or at the very least neutralized as a force. There is no prospect of this occurring without the rise of very serious oppositional threats to PRI of a magnitude which it has not experienced in the last fifty years. Later, we will consider whether there is reason to believe that such major political changes might occur in Mexico, but we can safely conclude that without them, labor organizations alone will not be able to markedly decrease pesticide abuses.

One veteran of the battles over pesticide safety observes that labor organizations alone have seldom been able to achieve major advances in reducing farm worker exposure to hazardous chemicals. Ralph Lightstone, of California Rural Legal Assistance, notes that even when DBCP had clearly been shown to be highly dangerous to workers, the federal EPA did not act to ban its use until it was shown that the chemical was contaminating groundwater. "Nobody at the federal level felt they had to move to protect farmworkers—as in so many cases, only when there is a threat to consumers or the general public was there any action taken. That's why it's so essential to link farmworker safety issues to consumer issues" (Pesticide Education and Action Project 1987).

Is it possible that consumers concerned about pesticide residues in their food could force farmers in Mexico and elsewhere to reduce or eliminate pesticide abuse? There are a number of hopeful signs that they could be of real help. Vegetable and fruit exporters in Mexico and other major exporting countries fear that they may lose a share of their market to consumer worries. As we have seen above, such worries were one of the reasons growers in Culiacán shifted away from the use of chemicals that have been heavily restricted or banned in the United States. Growers in Culiacán, with financial help from the Mexican government, established pesticide residue testing labs in Guasave and Culiacán to sample produce before spending the money to ship it north. These labs serve to identify pesticide residues in crops that might cause bad publicity or enforcement actions at the border and also to furnish the growers with independent data which they sometimes use to dispute the results obtained by FDA labs (González 1987; USGAO 1986).

The pesticide testing program established by the Food and Drug Administration of the United States is woefully inadequate, as shown by a General Accounting Office (GAO) study. The GAO study demonstrated that there is not enough testing, that labs do not produce results for weeks, months, and even more than a year in some instances, and that the program does not test for enough pesticides to ensure safety. Mexican fruits and vegetables had roughly twice the levels of pesticide residues as comparable domestically grown produce, but only rarely at levels the EPA considers a threat to health. But test results do not serve to protect consumers at all against a given load of vegetables that might be highly contaminated—the results simply do not come in quickly enough. The delay at the labs ensures that the produce will have been sold by the time a problem might be identified by lab test results. Furthermore, the FDA tests for a very limited number of chemicals, far fewer than those used in the field. The legal standards are based on exposure to single chemicals, while a consumer may be exposed through various foods to many different pesticides that may have additive or even synergistic negative effects on the human body. The FDA limits are based on average consumption figures for the United States, failing to take into account that some individual or regional diets may include many times more tomatoes, eggplants, avocados, artichokes, or strawberries than other diets or cuisines. In any case, the toxicological data on the human health effects of pesticides, particularly at chronic low levels of exposure of the type experienced through residues on foods, are so poorly developed that no one really knows what standards are appropriate (USGAO 1986; National Academy of Sciences 1987).

The results of the FDA testing program only serve to put growers on notice when certain chemicals are showing up above legal limits, with the warning that enforcement action might be taken if the problem continues. The FDA defends its program by saying that the rarity of high levels of illegal residues shows that there is no great need for greater vigilance, and that in any case the federal government is unwilling to put up enough money for a significantly stronger program. The California Department of Food and Agriculture has drawn similar conclusions from its random test programs.

Consumers and consumer organizations have been strongly dissatisfied with the defense of the testing program. They continue to pressure for a stronger residue testing program and for much greater funding for basic toxicological research. They also pressure the EPA and FDA to use more conservative assumptions about health effects. "Chemicals are now considered to be innocent until proven guilty— the burden of proof should be reversed so that companies must show that a chemical is harmless before it is allowed on food crops" (Pesticide Education and Action Project 1987).

The debate over what constitutes a safe exposure level and how it might be determined and enforced has now been waged intensely for about two decades, and it is clear that if it continues to be waged technical point by technical point and chemical by chemical there will be little reassurance satisfying to consumers.

The Federal Insecticide, Fungicide, and Rodenticide Act passed by the U.S. Congress in 1972 officially recognized that the existing data under which thousands of pesticide products were registered for legal sale in the United States were inadequate. (As pointed out in an earlier chapter, U.S. research and standards strongly influence regulatory decisions in Mexico and elsewhere.) Congress gave the EPA two years to set up a program to produce new and more reliable data and to require that each new pesticide be "reregistered" using the new data. Congress ordered that once established, the program should reregister all pesticides within an additional two years, leading to a completed reregistration of pesticides by 1976. No one seemed at the time to realize the size of the task being proposed by the new law. By 1984, only 26 of 600 pesticide active ingredients had been reregistered. At the present rate, reregistration is projected to go well into the twenty-first century, not taking into account new products that will come onto the market that will have to undergo the same testing procedures (Wasserstrom and Wiles 1985).

As if the delay were not enough, the program was seriously undermined by Industrial Bio-Test Labs, a private testing laboratory convicted of producing a large body of fraudulent data used in the reregistration process (Mott 1986). In addition, toxicological research continues to raise new issues, such as the health dangers of solvents, propellants, dilutants, and other so-called inerts used in pesticides in addition to the active ingredients. More potentially frightening is the evidence that pesticides may cause long-term damage to the nervous system—the organophosphates so commonly used in the Culiacán Valley are prime suspects. Other studies suggest that many pesticides probably cause a suppression of the immune system's ability to ward off disease (Metcalf 1982; Olson 1986).

The continued complexities and disappointments of the regulatory process have led many consumers and consumer representatives to conclude that the only long-term solution is to sharply reduce if not eliminate the use of synthetic pesticides on human food crops, whether imported or domestically grown. In the 1970s the argument for sharp reductions in pesticide use was advanced by many people who knew little about agriculture, and it was largely rejected by the agricultural community. In the late 1980s, the terms of the debate have shifted strongly. As we shall see in greater detail below, chemical agriculture has been put on the defensive. For the moment, let's look at the specific role consumers have played in calling chemical agriculture into question.

A shopper may now go into a chain grocery store in some parts of the United States to find fruits and vegetables labeled with information specifically identifying the farming techniques or risk levels associated with particular bins of produce. At fifty-two Raley's stores in California's Central Valley, some produce is labeled "organically grown," some is labeled "no pesticides used in the growing of this product," and some is labeled "no detectable pesticide residues." Literature available in the produce section, at the check-out stands, and in large newspaper display ads and on television and radio explains that all produce sold at the store is tested not only by government labs but more closely by a privately contracted lab through a company called Nutri-Clean. A pamphlet distributed by the store says that Raley's "lab testing for pesticide residues" is "working with science to bring you wholesome, nutritious, fresh fruits and vegetables." The pamphlet cites congressional staff reports, a General Accounting Office study, and a study done by the Agriculture Board of the National Academy of Sciences that maintain, respectively, in the words of Raley's, that "nine out of ten registered pesticides lacked vital health risk information," that "only 1% of our food is tested each year for pesticide residues . . . [and that] FDA testing does not cover many of the pesticides which are used or which may be present in the food," and that the "NAS in Washington D.C. issued a report saying that nine highly consumed fresh fruits and vegetables may contain levels of pesticide residues that can cause tumors." The pamphlet says that chemical manufacturers and some farm organizations "feel the dietary dangers from pesticides have been exaggerated, while consumer groups and many scientists feel the danger has been understated." Raley's goes on to say, "We do not know who is right or wrong on the issue. It is going to take more time and research to accurately determine what adverse health effects we may suffer from long-term dietary exposure to pesticides. In the meantime, Raley's prefers to err on the side of safety." "Whenever possible, we want to offer you fresh fruits and vegetables grown without the applications of any pesticides. However, we realize that, for the present, pesticides are still necessary to produce many of the foods we eat. We encourage our growers to use pesticides prudently, leaving no residue in our food."

Interviewed privately, Raley's executives say freely that while their innovations in marketing of pesticide-free or relatively pesticide-free produce came partly from a sense of social responsibility, the program is highly desirable to Raley's because "there is money to be made in satisfying the consumer demand for safe food." At least one other medium-sized chain in California and one on the East Coast have followed Raley's lead, and the huge Safeway chain has now begun to offer a modest selection of organically grown produce. As might be expected, chain stores searching for produce grown without pesticides have made many farmers more venturesome about taking the risks to convert to pesticide-free farming (Raley's 1988a, 1988b).

Small health food stores and grocery cooperatives focused on "natural" or "organic" foods have been doing what Raley's is doing, or more, for many years. But these were clearly serving a small and very special public, a public which may now be seen as pioneers for a wider consumer movement involving the larger commercial chains. The pioneers made it possible for groups of farmers to organize to offer "certified organic" foods, grown under legislation in a number of states designed to establish a kind of reassuring brand identity. The legal requirements of the organic certification process are meant to protect genuine organic producers from the unfair competition of those who do apply synthetic pesticides and fertilizers but who might claim that they do not. The certification usually succeeds in establishing a higher market price for the food bearing the organic label. While many organic farmers claim that the requirements of the law are being violated by many dishonest farmers, there is no doubt that the certification laws have helped to build consumer awareness and the larger market that firms like Raley's are now trying to serve.

Rumors abound that fruit and vegetable brokers and large chains plan to make it difficult for Raley's and other renegade chains to buy produce. The idea is that Raley's is undermining consumer confidence and unfairly playing on the public's fears, and that they must be disciplined to avoid the spread of the contagion. But there is no indication that if such a conspiracy exists it is succeeding. The entrance of Safeway into the organic food marketing business would seem to indicate the battle is moving the other way. Raley's competitors have published rejoinders in their ads that maintain there is no reason to be concerned about pesticide residues because farmers and government work effectively to be sure that residues do not pose a public health threat, but it seems highly unlikely that where government and industry officials over many years have failed to calm consumer concerns a grocery store ad is going to succeed.

Strong warnings are in order, however, about the changes that consumer demand may generate in the fields. A good example of the limitations of the Raley's strategy, for example, came during a lecture I had been asked to give Raley's executives regarding conditions in Mexican fields. They showed strong interest in the facts I was presenting. After I showed slides of a field operation where men wearing no protective gear sprayed tomato plants with acutely toxic insecticides while women and children worked in the still visible chemical fog, the vice-president in charge of the meeting turned to the produce buyer. He asked whether Raley's bought any produce from the farm in the slides, and the buyer assured him that these farms were one of Raley's main sources. Although the vice-president showed considerable discomfort over his reply, Raley's has continued marketing the tomatoes from this farm, easily identified by the brand name labels glued to each individual tomato. These tomatoes are marketed by Raley's during the height of the summer season when tomatoes are readily available from farms throughout California's Central Valley where the chain stores are located. While Raley's markets these tomatoes at times with no claims about the lack of pesticide residues and at other times with signs proclaiming that laboratory tests have shown no pesticide residues, shoppers might be easily misled by the overall advertising strategy of the store. In February 1989, the California Department of Food and Agriculture withdrew bell peppers and cucumbers from Raley's after shipments from ABC farms registered well above legal limits and some consumers reported illness apparently caused by eating the vegetables.

But the more serious issue raised by this example is the fact that it is entirely possible that farms in Mexico or anywhere else might be able to severely abuse farm workers and the environment with acutely toxic pesticides while producing a crop that will show no detectable residues at market time. Many of the most severely toxic pesticides degrade fairly rapidly under normal field conditions (see Chapter 2). In many cases, there is a direct relationship between the high level of acute danger to humans and animals directly exposed to the chemical and the hyperreactivity of the chemical that allows it to be broken down rapidly. As discussed in Chapter 2, the chemicals banned in the United States have most often been those whose special dangers to the environment are associated with their persistence. The chemicals used in their place are often nonpersistent but much more dangerous on immediate exposure. The general trends in pesticide use tend to favor nonpersistent chemicals that under proper conditions leave relatively small quantities of residues but that are terribly dangerous to people, animals, and in some cases plants that are exposed to them at or near the time of application. The very hazardous insecticides parathion, methamidophos, chloredimiform, and aldicarb and the herbicide paraquat are clear examples of rapidly degrading substances that may pose very serious field use dangers but if used cleverly may leave few residues. Chemicals restricted for use to protect one set of interests may increase the level of dangers of another kind, as has been the case for farm workers exposed to more dangerous chemicals as a result of environmentalist and consumer concerns over the implications of persistent pesticides. The political consequence is to weaken the link between consumer interests and the interests of those exposed in or around the fields. As might be expected, farm worker organizations have downplayed this factor in order to encourage consumer attention to farm worker concerns, and chemical companies and growers tend to overemphasize or oversimplify the rapid degradation of pesticides to reassure consumers. The precise degree of linkage between consumer interests, environmental interests, and the interests of farm workers and rural residents is not a given but rather changes as the technology of pesticides changes, severely complicating judgments about what constitutes safe pesticides or safe pesticide use. To some who are generally skeptical about the safety of pesticides, this is simply one more argument in favor of sharply reducing pesticide use overall.

In recent years, the consumer and farm workers groups concerned about pesticide use have picked up some strange bedfellows in the United States. Some growers in the United States who must compete with imported produce look at the pesticide abuses in other countries as an unfair competitive advantage. These growers are increasingly convinced that the ability to use pesticides without serious regulatory oversight represents reduced costs and/or greater management flexibility that those operating in most states in the United States do not enjoy. Combined with the cheaper land and labor costs and government subsidies to pesticide and fertilizer use that prevail in Latin America, the comparative advantage to foreign growers looks very substantial. The dissatisfaction with this situation has risen dramatically in recent years because of the rapid rise in the quantities of fruit and vegetable imports overall and most particularly because of the increasing quantities of produce imported during summer months. When imports came only during the winter, the only U.S. growers who were concerned about them were those in Florida and a few other small areas in the southern United States who could also offer winter crops. But with the increasing imports from areas like Baja California that export a large summer crop, growers in the Midwest and western United States have become alarmed. In addition, Mexican growers have been moving more and more into crops such as broccoli, cauliflower, and asparagus, which can be readily grown in various parts of the United States, especially California and Florida, during the winter. It all adds up to serious, year-round competition (Moulton et al. 1988).

A substantial body of U.S. growers are now arguing before Congress, the California legislature, and the International Trade Commission for some kind of regulation of vegetable imports based partly on the argument that unregulated pesticide use represents unfair competition. They are also lobbying for more stringent pesticide residue testing and enforcement programs at the U.S. border. These groups have been collecting evidence on pesticide abuse in Mexico and Chile especially to support their argument. Congressman Leon Panneta, Senator Pete Wilson, and other California representatives have been under pressure from California growers to do something to reduce foreign competition and have solicited information and testimony from those investigators in the United States who are familiar with foreign pesticide use patterns. The General Accounting Office of the U.S. Congress in 1988 carried out its own research into the issue.

There are, however, serious limitations to the effect the pressures of some U.S. growers and growers' associations may have on the imports of foreign produce or on the conditions used to grow fruits and vegetables abroad. The first is that many growers in Florida and California are themselves large firms with financial investments in fields and farms south of the border. Generally speaking, the agribusiness firms and growers with strong interests in Mexico, Chile, and the Caribbean are the largest and politically most powerful members of the traditional agribusiness lobby. Such firms as Castle and Cook, which sells large quantities of Mexican produce under the Dole label, Campbell's Soup, Del Monte, and General Foods are among the giants involved in financing foreign production in Mexico and a variety of other countries where pesticide use conditions are abusive. Frank Maggio, whose cellophane-packed carrots are familiar to millions of American grocery shoppers, finances farms in Culiacán and Baja California and has been doing so since his father became involved in the Mexican business during World War II. Maggio runs a large corporate spread in the Imperial Valley, where he engaged in a long labor dispute with the United Farm Workers leading to the death of two men connected to the union and a U.S. Supreme Court judgment in Maggio's favor. The companies and people in the United States who provide the bulk of financing for foreign imports are political heavyweights who might be expected to prevail over the much less influential individual U.S. growers (Maggio 1987).

The transnational agribusiness firms enjoy the support of strong allies in the arguments with domestic producers. The International Monetary Fund, in its negotiations with the debt-ridden Latin American governments, insists on the importance of promoting agricultural exports as a way of attracting new investment and earning hard currency toward debt repayment. The World Bank and the Inter-American Development Bank, while nominally supporting a greater role for agrarian reform and improved domestic food production,

continue to finance large-scale irrigation and agribusiness development projects strongly reliant on agroexports to pay back the loans
contracted to build the schemes. The U.S. government Agency for
International Development is committed through Reagan's Caribbean Initiative as well as through other long-standing programs to
the promotion of agroexports from Latin American countries (Kissinger et al. 1984).

Logically enough, these banks argue that their investments and
the accompanying investments they call forth from other governmental and private sources will be good money thrown after bad if
the United States imposes tariffs, quotas, or other restrictions that
would reduce the ability of Mexico and similar countries to export
to the United States. They argue that the strategy of agroexport promotion contributes to the resolution of the debt crisis and is therefore essential to saving the international economy from ruin—ruin
that would follow on the massive defaulting of the large debtor nations of the Third World such as Mexico, Brazil, Indonesia, Nigeria,
and the Philippines, all countries where agroexport promotion is defined as essential. Combined with the power of the jovial verdant be-
hemoths of agribusiness, these agencies can fairly reliably block the
efforts of relatively small, independent U.S. farmers.

In addition, it is important to note that the U.S. farmers involved
are not interested in pesticide abuse as a human health or environmental problem. They are simply convinced that a farmer allowed to
use pesticides freely can lower costs unfairly. The strategy of the
U.S. growers is to erect trade barriers that, in one form or another,
would represent additional costs to foreign producers. If their argument for trade barriers is correct, it also might follow that the additional costs imposed on foreign growers might be compensated for
by even more careless use of pesticides, more abusive treatment of
land and labor, or both. There is little question that Mexican growers would argue forcefully to the Mexican government that any new
trade barriers justified some new special favors for agroexporters,
and the International Monetary Fund's constant pressures would
tend to support their demands.

We have seen above that when environmentalists and consumers
tried to improve conditions in the Mexican fields by calling attention to the use there of chemicals banned in the United States, the
Mexican growers switched to other chemicals that under Mexican
conditions posed equally or more serious dangers of a different kind.
Oblique efforts to change the situation in Mexican fields by address-
ing only a part of the problem may not result in the changes consumers, environmentalists, farm workers, or Mexican peasant farm-

ers might desire. The growers of Culiacán have shown their talent for adaptive strategies other than those imagined by their critics.

Growers in the United States have an additional problem in their quarrels with foreign producers. Pesticide abuse is a serious problem in the United States. Epidemiological studies reveal that herbicides that have been in common use on American farms and on suburban yards and golf courses since the late 1940s cause greatly elevated levels of lymphoma and soft-tissue sarcoma among farmers who have applied them regularly. Cancer deaths among farmers in patterns that suggest pesticides as the cause have been found in Nebraska, Kansas, and Iowa. A recent survey of farm workers in the United States shows that most have experienced pesticide poisoning at some time in their lives and that farm workers are generally convinced of the inability to protect themselves against injury by pesticides through existing agencies and organizations. Other studies done over the last two decades in Florida and California have shown similarly routine problems farm workers experience with pesticides. State-documented pesticide poisoning incidents have been on the rise in the relatively well-regulated fields of California, in spite of constantly tightened rules for pesticide use and a set of state standards for dismissing many probable cases of pesticide poisonings from state statistics. Cancer clusters in the California towns of McFarlane and Fowler have not yet been shown to be linked to heavy surrounding agricultural pesticide use, but many of the parents of children who have died in these towns are themselves convinced of the connection. In Florida, the Atlantic seaboard states, and the upper Midwest, various studies show an ongoing, serious problem with farm worker poisonings. As state after state adopts groundwater testing programs, pesticide contamination shows up again and again, from Long Island to California's Salinas and Central valleys. There is no study in the United States that systematically documents routine abuses of the type recounted in this study of Mexican field conditions. But it is also the case that few attempts have been made by independent and qualified observers to actually stay in the fields day after day, week after week, as in the work I did in Mexico (Moses 1988; Wasserstrom and Wiles 1985; Barnett 1988; Clements 1981; California Assembly Office of Research 1985; USEPA 1987b; Blair 1979; Burmeister 1983; Hoar et al. 1986).

One exception is the Owens and Owens study financed by the EPA, in which Dr. Emiel Owens, with twin doctorates in economics and biology, worked as a laborer in farm worker crews throughout an agricultural season on the Atlantic seaboard. In contrast to my Mexican fieldwork, in all cases the cooperation of the growers had been solicited by Dr. Owens and his brother, an agricultural scientist, before the season began, alerting growers to be on good behavior. In spite of the fact that growers knew their pesticide practices were under close scrutiny, however, field workers showed steadily declining levels of acetylcholinesterase, indicating organophosphate poisoning, throughout the season. One worker with a very depressed ACHE level died before lab results reached the field supervisors, and Dr. Owens himself ended the season a sick man who needed three months to recover from a series of mysterious ailments. Scientists reviewing the study have claimed that there were a number of errors made in the design and conduct of the study that throw its conclusions in doubt (Owens and Owens 1982).

Nonetheless, as virtually the only data gathered on a group of farm workers throughout a season, the study can certainly be taken as a caution by growers in the United States tempted to claim that conditions are dramatically and reliably better in their own fields than in Mexico. It has historically been the case that farmers have been reluctant to question the results of pesticide use anywhere in the apparent fear that criticism leveled at others might reduce the availability of chemicals the farmers believe they need or bring down regulatory actions that would fall on themselves. The current wave of criticism of Mexican and other foreign growers is remarkable for its novelty.

Whatever the interests and claims of competitive growers in the United States, it might seem obvious that Mexican farmers who use pesticides would have an interest in reducing their costs by reducing pesticide applications. But it is not so obvious. Farmers buy pesticides because they believe they have advantages over alternatives. Pesticides are an especially persuasive alternative under monocrop conditions in subtropical or tropical environments such as those that occur in export fruit and vegetable or cotton agriculture in Mexico. All the traditional means for controlling pests break down when tens of thousands of contiguous acres are planted to crops that are attacked by many of the same pests, and this is doubly so where there is no frost to annually reduce pest populations. What makes such agriculture attractive is the opportunity to sell the product in much richer economies, selling luxury or semiluxury goods to tens of millions of buyers who can afford to pay prices unthinkable to the workers and their countrymen who produce the crops. Farmers and their creditors are willing to accept the risks and costs of pesticide use to take advantage of this opportunity.

Nonetheless, to the extent that they can continue to sell export crops, the farmers of Culiacán and similar regions would be happy to find ways to cut pesticide use. Pesticide purchase and application costs account for 15 to 35 percent of all production costs. In an earlier chapter, we reviewed some of the reasons why pest control specialists believe that pesticide applications could be cut by as much as 50 percent with no loss in the quality or quantity of crop harvested. We also saw why the monocrop conditions of the agroexport regions like the Culiacán Valley discourage attempts at primarily biological controls on pests, meaning that pesticide use rates may be cut but will almost certainly remain high. Often, field experiments attract growers to innovative solutions, but the inevitably ambiguous results of a limited number of experiments under controlled conditions and the disputes among pest management professionals about the broad implications of these experiments leave growers doubtful and anxious. The easiest solution, season by season and day by day, is to keep using lots of pesticides.

For example, pest management specialists have recently enjoyed strong success in tomato fields near the Sonoran city of Guaymas by what is called "pheromone confusion." Insect sex-attractant chemicals are sprayed on the fields in sufficient quantities to make it very difficult for insects to successfully find mates, because their own emission of pheromones is drowned out by the more massive application. The experiment has given excellent, low-cost results. However, it was conducted on relatively isolated fields with modest established pest populations far from monocrop regions—pest control specialists are skeptical of the possibility of repeating these successes season after season where monocrop conditions prevail over tens of thousands of hectares, with much higher established pest populations.

The prospect for radical reductions in pesticide use in intensive pest-vulnerable monocrops tends to stimulate strong arguments among scientists, agricultural researchers, and growers. The arguments boil down to the different underlying perspectives on the role of human activity with respect to nature held by the partisans of the three major groups in the dispute. One group, often known informally to the others as "nozzleheads," believes that heavy pesticide applications are absolutely necessary to agriculture. Growth in pesticide use throughout the world is taken as a necessity for feeding humanity. Humans can and do control the agricultural field environment through their combined use of cultivation techniques and chemicals, and nature will always yield and accommodate eventually to human cleverness. This group strongly influenced the wording of the UN Food and Agricultural Organization code on pesticide trade and use in 1985, with the preamble to the code stating that *rapidly growing* pesticide use was a human necessity, particularly in developing countries. This generalized necessity would be even more compelling in pesticide-intensive crops such as cotton and vegetables. There is little doubt in the minds of this group that people can learn to design and manage biocides used in agriculture with sufficient skill to avoid unacceptable health or environmental consequences. In the first three decades of Green Revolution research, this view held nearly undisputed sway in Mexican and North American research institutions and extension services, with only a small group of biologists and agricultural researchers opposed to it. In this view, the subsequent historical record documenting problems caused by pesticides only suggests guidelines for improved design and management and emphatically does not call into question the wisdom of wide-scale use of the chemicals. Advice given by professionals working from this viewpoint will always tend to deemphasize the problems of pesticides and emphasize their advantages. Since in Mexico as in the United States most pesticide advice to growers comes from pest management advisers who work for chemical companies, it is this perspective that most strongly influences whether pesticides will be used or not (UNFAO 1985; Perkins 1983; California Department of Food and Agriculture 1978: chap. 4).

John Perkins, in his excellent work on pesticide research and policy, maintains that this perspective is essentially discredited among pest management professionals, but it is important to realize that his conclusion is drawn from the recent literature in the field and not from the actual practices of agronomists and growers nor from the work of agricultural researchers who continue to use a chemical control model but who know they must give lip service to a more ecologically oriented perspective. As seen in previous chapters, the chemicals-first perspective tends to be predominant among working agronomists, growers, and regulatory officials in such areas as Culiacán, with a few dissenting voices and a lot of window dressing to convince outsiders that other methods are under active use or consideration.

A second camp, with a strong and growing influence over universities and research institutes around the world, tends to believe that pesticides are and will remain an important tool in agriculture but that wherever possible safer, more sustainable alternative techniques should be substituted. The perspective of this group is informed by the numerous problems that pesticides have produced around the world—growing pest resistance, secondary pest outbreaks induced by pesticides, pollution of soils, air, and water, human health hazards, and danger to wildlife. Generally using Integrated Pest Management (IPM), the members of this school of thought use the release of sterile males, pheromone traps and confusion, parasites and predators of pests, breeding of resistant crop varieties, careful control of humidity, soil composition, and nutrients to disrupt the life cycle of pests, and other techniques to reduce pest damage. They also emphasize the careful monitoring of pest populations and calculations of the threshold of economic damage to determine when pest control measures are really needed. IPM usually involves some chemical use when safety, cost, or crop protection make chemicals more attractive than other techniques or when use of some chemicals is necessary in combination with other methods (Flint and van den Bosch 1981).

Growers of cotton in northeastern Mexico, sugarcane, wheat, and soybeans in the Culiacán Valley, and citrus in southern California and Central Mexico use IPM routinely. In all these cases, growers still use chemicals, but in reduced quantities compared to previously used methods. A popular technique in cotton used in parts of Mexico and Nicaragua involves some combination of sterile male releases, releases of parasitic wasps that kill cotton pests, trap cropping, and conventional spraying. Trap cropping consists of planting narrow strips of the crop in question earlier than the main crop, encouraging pests to congregate and reproduce there; workers then spray the trap crop heavily, severely disrupting the reproductive potential of the pest within the main crop area. The use of these techniques in Nicaragua, for example, has made it possible to reduce pesticide use by about 60 percent from a very high starting point. Pesticides remain an essential part of production, but at a level that is more subject to some controls, reduces health and environmental hazards, and slows the buildup of pest resistance and secondary pest outbreaks (Hansen 1986; Gips 1987; Dover 1985; Swezey and Daxl 1983).

IPM programs similar in conception to the one in use in Nicaragua have proved their value in dozens of instances around the world, and there is no doubt that some of the techniques will prove useful in Mexican vegetable and cotton fields where they have not yet been applied. This is not to say, however, that the use of some IPM techniques can be expected to necessarily eliminate or even seriously reduce the abusive use of pesticides because, as is generally recognized, the subtropical intensive monocrop conditions of Culiacán-style agriculture defeat any ambitious plans for management through biological means alone.

A second approach to IPM is to engage in massive campaigns to eradicate some pest organisms altogether from large regions or from

whole continents. For decades, the U.S. Department of Agriculture
has conducted eradication programs against the screw-worm fly in
cattle and against the cotton boll weevil, in both cases working
closely with the Mexican government. Eradication campaigns have
seldom if ever been successful for long, and many university agricultural researchers in the United States in particular believe that eradication is a fundamentally wrong-headed approach to pest management because its methods are too expensive and often destructive
and because the chances of eliminating rather than controlling any
pest are so small. The proponents of eradication argue that if greater
resources and cooperation could be marshalled in the campaigns,
the initially high use of pesticides for eradication could make possible decades of very low pesticide use due to the elimination of the
most important crop pests. Whether eradication deserves the name
Integrated Pest Management brings forth heated arguments from researchers similar in passion to arguments among believers in the re-
spective faiths about whether the Anglican or the Roman Church
deserves the name "Catholic." In any case, because of the wide variety of pests in areas like Culiacán and the ability of most of these
pests to maintain populations outside agricultural fields, eradication
is simply not an option for most pests in the main areas of pesticide
abuse in Mexico (Perkins 1983).

More recently, some IPM researchers have turned to genetic engineering to extend their control over pests. By manipulating genetic
material within plants and pests, researchers hope to produce crops
that are more resistant to pests. They also have already experimented with genetically engineered pest parasites. For example, researchers at the University of California developed a parasitic mite
that is resistant to insecticides, allowing fruit growers to use both
the parasitic mite and insecticides simultaneously to control pests.
Researchers in Davis, California funded by Monsanto Corporation
are working to genetically engineer crops to be more resistant to herbicides, allowing increased herbicide use without causing unwanted
crop damage. Chemical companies like Monsanto argue that genetic
engineering in the long run offers an alternative to pesticides, but it
is notable that most of the money being spent for pest control techniques through genetic engineering is actually directed at permit-
ting larger quantities of pesticides to be used while avoiding some of
the negative consequences for crop production. The environmental
and human health effects of the increased pesticide use are not under
active consideration in this research (Doyle 1985).

In spite of the obvious contradictions, eradication and genetic engineering aimed at increasing pesticide use go forward under the

name of Integrated Pest Management, to the great annoyance of many of the pioneers in IPM. The justification for use of the IPM label is that IPM does not necessarily mean less pesticide use but rather a more complex integration of chemical with biological controls, choosing the most "appropriate" mix of techniques for the circumstances (Doyle 1985: esp. 208ff.).

The anger from the IPM researchers and practitioners over what they regard as the cynical misuse of the IPM label for eradication and genetic engineering approaches is only partly justified. From the beginning, the philosophical basis of IPM has been rather poorly defined and therefore open to the kind of exploitation of the concept that is now occurring. The predominant view of human interaction with nature through technology expressed in IPM is that human management of the agricultural fields must take into account the great complexity of nature in order to achieve more successful manipulation of the natural environment. Farmers must manage not only the crop but all the organisms, water, nutrients, and soil to secure high-yielding crops. Human ingenuity has essentially no limits in its ability to control the environment of the agricultural field, but cleverness must be developed far beyond simple-minded chemical solutions alone. In an apparent attempt to avoid being accused of unscientific sentimentality or vagueness, IPM promoters have always been at pains to emphasize the pragmatism and greater scientific sophistication of IPM—in so doing, they have mostly emphasized cleverness and stayed away from searching questions about the overall context of modern agriculture. This has often been so in the literature even when privately IPM proponents express their ideas in terms of extremely passionate devotion to the necessity for ideals of ecological balance or harmony that are essentially nonverifiable by scientific means. A Mexican doctoral candidate studying under professors of this sort complained of the distortion of IPM ideas in Mexico by asserting that the best-known promoters of IPM in Mexico were "applying such classic IPM techniques as sterile-male or parasite releases just as though they were so many chemicals with no more regard for the overall ecological circumstances than that used in exclusively chemical controls" (Trujillo de Arriaga 1983).

One of the disadvantages of IPM from the point of view of large-scale growers like those in the Mexican vegetable industry is that it implies a great deal of constant and careful monitoring of field conditions and the adjustment and readjustment of a large variety of chemical and nonchemical means to keep pest populations at acceptable levels. Such extremely close management, aside from the expense per se, implies a devotion to the task of pest management that most growers, and certainly the majority of growers who are essentially absentees from daily farm management, would prefer to spend tending to the hard-fought politics and economics of agribusiness and to making money by investment of income into non-agricultural businesses. The agronomists whom growers hire to take over full or partial management of the farms usually have only a superficial knowledge of Integrated Pest Management techniques. In addition, the supervising agronomists hired by growers fear that if they rely on a previously untried method that fails to protect their employer's crop, they may be fired, while if standard chemical approaches used for years fail, the agronomist will not likely be blamed. Working under inherently pest-vulnerable conditions in subtropical monocrops, agronomists and growers alike are reluctant to try new methods, as are the banks and marketing firms who finance them. Purchase and application of broad-spectrum pesticides represent an easy way to free their attention for other matters, especially when honest and capable pest management specialists proposing other techniques are not prepared to give simple and straightforward advice in the way pesticide salesmen are. As Ing. Trujillo lamented, "One of the problems of those of us who would like to see an ecological approach to IPM is that we don't have a simple list of easy answers available for each pest problem—and if we did, we would not be following a genuinely ecological approach based on a fresh analysis of circumstances" (Trujillo de Arriaga 1983).

As if the disputes among conventional proponents of growing pesticide use and those promoting Integrated Pest Management and eradication efforts were not confusing enough to growers and practicing agronomists, yet a third and growing school of pest management philosophy now competes vigorously for attention and support. Whether using the name organic, sustainable, regenerative, ecological, or traditional, the essential point of view is the same. The vision is of agriculture as a human activity always subordinate to dominating natural processes. The key to productive, stable agriculture in this view is to recognize that fields of crops and whole agricultural regions are part of a world that is wider, more enduring, and ultimately more powerful than any human management efforts. Humans manage the field and the growth of crops, but if they are wise they recognize natural limits to their ambitions. In this view, the growth of tens of thousands of hectares of contiguous intensive vegetable crops must always be considered a very hazardous activity for which no purely technical solutions can be devised.

From the ecological point of view, the hazards of Culiacán-style agriculture do not necessarily come exclusively or even primarily from the use of pesticides. The apparent need for such massive use of biocides in the fields comes from the radical imbalance created by the drastic simplification of the natural environment involved in such a concentration of a single kind of crop. "Persistent and damaging pest outbreaks are signs of an ecological illness" (Trujillo de Arriaga 1983). In addition, the careful micromanagement techniques used in the technology of stability and security in traditional agriculture discussed in the previous chapter have been abandoned in agribusiness regions in preference for simple-minded chemical techniques adapted to very large scale production of pesticide-vulnerable crops. Wherever farmers replace intimate knowledge and management of the field environment of the kind practiced by skilled traditional farmers with "magic bullets" like pesticides or "miracle seeds," the constant threat of massive pest outbreaks and rapid soil damage can be expected as a matter of course. This may be true as well if Integrated Pest Management techniques such as sterile male or parasite releases are used with little regard to their overall ecological requirements or consequences—biological control methods applied as though they were chemicals out of a barrel are likely to be unreliable and disruptive of any balance between field organisms that might be established with more careful attention to the requirements of healthy "agroecosystems" (Trujillo de Arriaga 1983; Altieri 1987: 161–162).

Growers and agronomists in Culiacán often mention failures of pheromones, sterile males, and parasites to solve the problems of growers who have tried them, applying a standard of the quick fix which is the same standard they have used in evaluating pesticides—a standard which the glib promises of some IPM specialists seem to make appropriate. For example, parasitic wasps or application of the bacteria bacillus thuringiensis may eliminate caterpillars but reduce competition and lead to greater outbreaks of other, nonlepidopterous pests. Or, as the INIA scientist noted in Chapter 2, introduction of parasites or predators of pests may be defeated by heavy spraying on the surrounding fields owned and controlled by other farmers that through drift kills the parasite or predator populations back to ineffective levels. Or, high pest populations on adjoining fields may quickly move into IPM program fields because the overall conditions of monocrop or pesticide use by neighbors make it difficult or impossible to maintain pest parasites, predators, or sterile males in high enough populations within the field being treated. While manuals on IPM warn against such possible outcomes, there may be a crucial failing in the overall IPM philosophy as described by many of its proponents.

This failure in its simplest form is the unwillingness to simply say that some kinds of agriculture should not be practiced at all. A doctor might recommend to an emphysema patient that he move away from polluted city air and maintain a course of medications and physical therapies proven to reduce the ravages of the disease. But if the patient insists on continuing to smoke two packs of cigarettes a day, the recommendations will have little value. If either pesticides or IPM are seen simply as ways to manage and control the individual field environment, with little regard for what happens either in adjacent fields or in the noncultivated areas surrounding the agricultural region, analogous failures are inevitable.

Perhaps the best examples in this regard are the truly astounding and unmanageable outbreaks of locusts, grasshoppers, and boll weevils that sometimes occur in northern Mexico and that are legendary in much of Africa. These outbreaks typically begin in either uncultivated areas or in monocrop agricultural fields, building such high populations that breaking the momentum of the plague becomes virtually impossible. There are solutions other than generalized, hazardous, and very expensive pesticide application to thousands of tens of thousands of square miles of territory of the kind coordinated and carried out under FAO auspices. Alternative solutions involve a significant increase in crop diversity and reduction of monocrop combined with the control, by pesticides if necessary, of the usually well-identified breeding sites that give rise to the plagues. But such an approach is not available to IPM practitioners or pesticide promoters operating at the level of individual fields or even individual regions subject to pest migration from other fields or regions. It necessitates national and international coordination at the stage of planning cropping patterns rather than at the point when the plagues are already devastating huge portions of the continent. In addition, some observers in Africa believe that the enormous national and international bureaucracies and commercial enterprises built up to deal with the periodic plagues mightily resist efforts to resolve the matter through application of ecological analysis precisely because sensible ecological solutions would cost so much less and employ far fewer professional personnel (Ahmed 1988).

In January 1977 an ancient Mexican disease of wheat called *chahuixtle* threatened devastating losses to Mexican national grain production. The progress of the disease was greatly facilitated by the monocrop plantings that offered the fungus tens of thousands of contiguous acres of wheat—not only wheat, but wheat of a very few Green Revolution varieties susceptible to the disease. Eventually, the Mexican Ministry of Agriculture coordinated a campaign that averted disaster, partly by recruiting pilots from the Mexican air force to fly pesticide spray planes. Much of the crop was saved through the massive aerial application using fungicides we now know to be highly carcinogenic. The agronomist Carlos Manuel Castaños wrote movingly of the hurried campaign of coordination necessary to protect Mexico's wheat crop and offered in the end some principles for avoiding future disasters of the same sort. Among his recommendations was a strong argument against excessive monocropping of wheat and excessive reliance on a few varieties of wheat, but his idea of limits on monocropping were that "it should never be permitted to sow more than fifty percent of a twenty thousand hectare area of wheat to the same variety if in previous seasons the variety had shown signs of susceptibility to chahuixtle." After warning of the necessity to pay attention to other factors that might promote the growth of the fungus, Ing. Castaños concluded that it is "better to prevent than to cure." What is obvious from his commentary is that, even after suffering the dramatic epidemic, his overall critique of the conditions leading to it remained completely within the context of a kind of agriculture that cannot allow any genuine diversity—ten thousand hectares of a single area planted to one variety is monocrop and subject to all the ills of monocrop, pious warnings notwithstanding (Castaños 1981: 355).

Part of the problem is that technologies are not simply collections of tools with directions for their use. Technologies address problems and offer partial or total solutions to those problems. The creation of a given technology implies a critical analysis of what the problem is. A building technique perfectly applicable to the construction of skyscrapers may be a serious threat to people and the community if it is used where earthquakes are common or soils are unstable. Good construction techniques must be adapted to the environment in which they are used. The same is true, at a typically much greater level of complexity, with respect to agriculture. Critical analysis requires a willingness to say that some construction techniques are never safe anywhere and that others are safe in some locations but highly dangerous in others. Civil engineers have never felt embarrassed at putting forth such principles. The same kind of analysis can specify that some kinds of agriculture are never stable and that others are safe only under strictly defined conditions or in particular regions. When a bridge or building falls down, the victims of faulty design are obvious, including the taxpayers and those injured or killed in the collapse, and it is frequently possible to hold firms or individuals responsible for fraud or incompetence. When faulty design of agricultural systems results in failures, it is almost always hostile nature or the individual farmer who is held to blame, and the victims are not so clearly identifiable, even though crop failures may leave more dead and injured than in the case of poorly built structures. In what case have agricultural researchers ever been held directly to blame for the systems they have designed?

The immediate interests of Culiacán farmers cannot withstand critical analysis and they are not interested in seeing it applied. They are enthusiastic about biological control techniques only if the methods can be shown to be reliable and effective under conditions that maintain the growers' domination over excellent land and endless supplies of docile, cheap labor in a large area dedicated to essentially monocrop conditions. If the techniques do not meet these conditions over a period of years, growers reject them. The continued breathless naiveté of IPM proponents who do not understand this only serves to offer false hope and prolong the agony of the Culiacán Valley and similar regions.

The ecological viewpoint demands a social as well as a scientific analysis and critique. The desire of a grower for a pest management solution in his field may come from his attempt to bend both nature and society to ambitions that destabilize both nature and society in harmful ways. If this possibility is not admitted, the search for control over the pests in fields where crop choice, field size, ownership, and profit expectations are taken as givens may be further destabilizing, even if the methods use biological means rather than chemical ones. It is the ultimate ecological arrogance to assume that unrestrained human desire is an evolutionary fact to which the rest of nature must be adapted.

The dangers of ecological arrogance may be easily overlooked in the consideration of strategies for change based on the interests of particular groups, without considering the meaning of the larger social and ecological contexts. In politics, such interest group analysis frequently leads simply to the most powerful of the interest groups granting minor concessions to weaker ones while demanding and expecting more resilience from the natural world. The end result fails to adequately protect the weaker parties and largely neglects the larger society's interest in maintaining or enhancing the quality of the environment.

Farm workers, rural residents, consumers, and growers on both sides of the border clearly have some self-interested reasons for working toward reduced pesticide use. Experience in the last few years seems to show that marginal improvements in the situation may from time to time result from pressures from one or another of these groups, sometimes working together. As we have seen, however, each group is limited in its ability to fundamentally change the situation, whether through lack of power or because of the complex and often conflicting nature of their interests. The present situation was built on the fact that hundreds of thousands of farm workers in Mexico desperately need work, that millions of consumers enjoy having the whole array of fruits and vegetables available all year round, and that agribusiness in general and chemical companies and farmers in particular can make high profits from satisfying the desires of consumers. For each group, pesticides used to achieve these things may present problems, but the main focus of each group is centered on the advantages of the present arrangements.

Does this mean that the best that can be hoped for is marginal change, that some significant pesticide abuse will continue indefinitely, and that the social inequities, the long-range destruction of agricultural productivity, and the ongoing assault on the environment will continue essentially as before? It seems that this would be a reasonable conclusion if one holds to the belief that people are motivated by little more than their immediate self-interests. The mean-minded perspective that says that consumers are satisfied if they are not poisoned, that farm workers will endure if they continue to find work, that growers will respond only to threats to their profits, and that no other identifiable body of people or interests in society may be brought to bear on the problem locks us into necessarily mean and partial solutions. It excludes a genuinely ecological perspective or a broad social perspective.

The problem of poor peasants who use pesticides unsafely or irrationally out of ignorance would be one kind of very serious problem if it existed in isolation from other major social issues. Where responsible governments were willing to address problems of the peasantry, it could be attacked largely through education of peasants and extension workers, on the one hand, and restraints on the advertising and sales techniques of chemical companies, on the other. However, the real problem is usually not a failure to realize that it is important to carry on peasant education and the restraint of unethical commercial activity. The problem is that ignorance and unregulated promotion of dangerous technologies are intimately tied to the political relationships and the ideological assumptions that determine how a nation is ruled.

Abusive use of pesticides is usually the result of a whole set of problems that indicate the loyalties and purposes of the people and groups who hold power. The problem of tens of thousands of acres of fine farmland given over to large-scale commercial production of crops through abusive use of pesticides by millionaire farmers is strongly related to the continued powerlessness of peasants and farm workers and their relative ignorance of the safeguards needed in applying modern technologies. This situation is in turn related to much broader political and economic issues involving a nation's relationship with the rest of the world. The fact that 70 percent of pesticides used in the Third World are used on products for export to wealthy nations expresses the strong connection of pesticide abuse to the particular kind of relationship that predominates between rich and poor countries. In addition, the Mixtec farm workers who tie Culiacán to the collapsing traditional regions of Mexico also tie Mexico to California, Texas, and Arizona and force us to ask how the unequal development of Mexico will affect the future of the United States.

In the summer of 1989, I observed in Nicaragua an encouraging example of what can be done when there is a real convergence of interests working toward safer pesticide use. Under the sponsorship of the nonprofit U.S.-based organization CARE, Dr. Douglas Murray had organized an exceptionally thorough pesticide safety program in collaboration with Nicaraguan government agencies. Inspectors systematically cleaned up pesticide aircraft landing strips, retrained pilots, designed container washing and disposal systems, and worked out systems of manifests to keep track of pesticides in trade and use. "Popular educators" visited newly formed cooperatives to hold lectures and discussions with members on pesticide safety and conservative pesticide use. The woman I observed on a cooperative outside León not only managed to elicit the very active interest of the coop members in the topic of safe pesticide use, but at the same time communicated a great deal of related information about nutrition, health, and environmental protection. Serving to strongly reinforce her words and provide baseline data on coop members in case of possible poisoning incidents, government health workers took ACHE samples from each worker during the morning's session. Practical pest management advice aimed to reduce pesticide use buttressed the technical work that had been carried out by the team from León's National University under Sean Swezey and Ranier Daxl's guidance. Unfortunately, however, two factors may prove to have fatally undermined these efforts, both of them underscoring all the larger political and economic problems that set the context for pesticide abuse. The first is the Sandinista government's uncertain dedication to crop diversification under constant pressure for foreign exchange earnings. The second is the defeat of the Sandinista government in elections in February 1990 by a coalition with strong backing from the agribusiness and landholding elite that promoted disastrously hazardous use of pesticides and cotton monocrop in Nicaragua in the past.

Many people have already recognized the way pesticide abuse transcends narrowly technical matters or the interests of isolated groups in society. For example, connections with broader issues account for the fact that organizations that call themselves consumer organizations do not concentrate their attention on consumers. The International Organization of Consumer's Union Organizations, headquartered in Penang, Malaysia, brought together people from every continent in 1983 to form the Pesticide Action Network. Since then, PAN, which unfortunately shares an acronym with Mexico's large right-wing political party, has played an important role in coordinating the efforts of people all over the world to fight pesticide abuse. They have put pressure on the Food and Agriculture Organization of the United Nations to adopt a voluntary code governing manufacture, commerce, and use of pesticides. In November 1987, they succeeded in placing into the code the provision that countries importing pesticides should be notified of the regulatory status of the pesticide in the exporting country, along with documentation on the reasoning for the regulatory decisions made. These accomplishments, while symbolically important, are very limited by the voluntary nature of the code and the complete lack of any effective mechanism for monitoring code compliance by signatory countries.

The more important work of PAN is carried on through its individual members. They include a government pest management specialist from the Sudan, an agronomist and teacher from Colombia, a labor organizer from Kenya, peasant organizers from Indonesia, a university professor of pest management from Senegal, community organizers and environmental scientists from India, community organizers, science reporters, and pest management specialists from the United States and Europe, and agricultural researchers and a pollution chemist from Mexico. PAN serves to link these people together for information sharing and mutual support. These functions are vital, especially to people working in poorer countries for whom obtaining basic toxicological or agronomic information on a farm chemical may prove an incredibly expensive and time-consuming task. In some countries, information on pesticides may come almost exclusively from pesticide sales firms, and information on alternatives may simply not be available.

The reason that the international Consumer's Union works with all these people is that the leadership of the organization is convinced that pesticide abuse is merely symptomatic of the development process gone wrong. "Pesticide problems as we see them are just one part of an unjust system and erroneous ideas of develop-ment," says Anwar Fazal, the original organizer of PAN and a staff member of the Consumer's Union in Penang. The consumer stands to suffer, not necessarily as someone who may eat pesticide resi-dues, but as a person whose life will be more difficult due to the irra-tional patterns of production and resource use characteristic of eco-nomic development. PAN maintains that it is futile to attempt to control pesticide use within the boundaries of regulatory actions—it is instead necessary at the global level to work toward the creation of an alternative "sustainable agriculture [that] is ecologically sound, economically viable, socially just, and humane" (Fazal 1986; Gips 1987: v–vi). At an international conference in Ottawa, Canada in 1986, PAN announced a statement signed by twenty-seven inter-nationally known experts on pesticides and pest management en-dorsing this broad approach to the problem and calling "on members of the worldwide scientific community to join us in developing agri-cultural systems fundamentally different from those promoted over the past fifty years" (Ottawa Declaration 1986).

As consumer organizations and many agricultural scientists have defined the issue as one of urgent interest to society as a whole, so increasingly have farm worker organizations. In the United States, the United Farm Workers has called for a boycott of table grapes as a way of focusing public pressure on the need for more vigorous pesti-cide regulation and promotion of alternatives. The publicity used by the union points out the dangers of pesticides to farm workers, con-sumers, and rural communities. Some technical experts have criti-cized the campaign on the basis that it blurs the distinction between the dangers to farm workers and those to consumers of particular kinds of pesticides targeted by the union, and they are certainly right that each pesticide presents different kinds of dangers to different groups. What these experts fail to understand is that the boycott is based on a generalized lack of confidence in expert opinion by a large segment of the public. People can reasonably be skeptical of expert rationalization of pesticide safety without hoping that they them-selves can make good judgments on each individual chemical, and people can reasonably be persuaded to attack the generalized men-ace of lavish pesticide use without wishing to enter into the myriad facets of expert debate (*Sacramento Bee*, August 24, 1988, pp. 1ff.).

Mexico serves as an excellent example of the problems of regula-tion of pesticide use for countries that have not proceeded so far along the path of economic development and agricultural moderni-zation. The Mexican experience makes it clear that some of the so-lutions proposed to eliminate pesticide abuse simply will not work.

Mexico has tens of thousands of trained agronomists and an extremely large and advanced system of agricultural extension compared to other countries, directed by people with doctorates in agricultural sciences from the most prestigious universities in the world. It has a well-developed body of law and regulations on pesticide use and a relatively large regulatory apparatus. These things, governments and chemical companies say, should lead to effective solutions. Mexico makes it clear to other nations that this is illusory. The problem is not a narrow technical issue but a broad social issue involving basic decisions about how nations should seek to become more prosperous.

8
Theory and Consequences

A grower in Culiacán leaned against his pickup as he and I gazed from a levee top at a camp of two thousand farm workers housed in poultry sheds. He was a big, handsome man with silvergrey hair, of Greek ancestry but with three generations of Mexican citizenship behind him. It was a man like this who had ordered that Ramón González and his friends work in a recently sprayed field and who repeats such orders again and again year after year.

Tossing his toothpick from lunch onto the ground he said, "The poor devils have done nothing to deserve this. But if men like me don't work to develop this country, they will never know anything better."

The growers of Culiacán usually justify their actions by saying that growth in high-value agricultural exports is essential to national development. Since 1940, Mexican presidents have consistently put forward policies based on the same idea. Large private investors, domestic and foreign, mostly agreed and profited accordingly. Since the beginning of serious economic crisis in the early 1970s (interrupted briefly by a petroleum-fueled recovery in the late 1970s), the International Monetary Fund, World Bank, and Inter-American Development Bank have gone from supporting such policies to insisting on them as a precondition of debt refinancing and new development loans. The idea is that successful promotion of agricultural exports earns hard currencies for debt payment, encourages foreign investment, and, because of the kind of technology associated with export agriculture, induces industrial growth (see David 1986 for a review and critique of these ideas generally, and Esteva 1982; Cockcroft 1983; Sanderson 1986; and Mares 1987 for a discussion of their implications in Mexico).

Crops grown for the domestic market, on the other hand, do not earn hard currencies and must be sold to fewer people with less money. Because of the limits of the domestic market opportunities, private investors will not be attracted. Domestic market crops can be much more readily grown with relatively cheap and simple traditional technologies, so agricultural growth will be less successful in stimulating industrial growth. As a result, the argument goes, industrial employment will grow slowly, new goods and services will not be easily forthcoming, economic activity as a whole will be retarded, and employment, tax revenues, and social spending will all remain at low levels. There will be little savings available for new investment because of the generally slow pace of development, and the economy could be expected to fall into permanent stagnation and deeper indebtedness.

The argument is also based on the assumption that the low level of education and skills of the local population makes it difficult for the economy to offer higher-value technological goods to the domestic or international markets. The earnings from the agroexport business will help provide the revenues to improve public education and social welfare, eventually upgrading the skills of the population and allowing the country to compete across the board with more highly industrialized, prosperous nations. It is a catch-up game being played, and because they must catch up to those already highly developed, the present sacrifices of the poor and underdeveloped must be correspondingly great.

This viewpoint has dominated Mexican economic policy in general and agricultural policy in particular since 1940. As discussed in earlier chapters, in Mexico more than in many other countries the state has intervened to soften the blow on the rural poor and has sometimes undertaken major initiatives sustained for a few years at a time to support poorer farmers and landless rural people. The ameliorating, if inconsistent, Mexican state policies have made things far better for the rural majority in Mexico than in many otherwise comparable countries. But as observer after observer has concluded, the large-scale agribusiness and export orientation has always overwhelmed the ameliorating effects of contradictory policies. We now have nearly fifty years of experience on which to judge the Mexican model of rural development. Does it work? Are there any more attractive alternatives?

These are obviously terribly complicated questions that have been argued bitterly in the political arena and unceasingly in scholarly debate. Judging all the details of this controversy is too large a task for this work. What we can do in the next two chapters is to take a look at that part of the controversy that is particularly relevant to the idea of founding the nation's agriculture on a more ecologically sustainable basis while at the same time improving life for the rural majority. Are there other choices, other lives possible for the younger brothers and sisters of Ramón González?

From the end of World War II through the 1960s, Mexico enjoyed one of the highest rates of overall economic growth in the world, averaging about 7 percent per year, never as spectacularly high as a few years of Brazilian or Japanese growth, but more steady. The country industrialized rapidly, and agricultural growth rates were usually high as well. For a brief period in the 1960s the country for the first time in the twentieth century achieved a goal long thought to be essential to Mexico's security and sovereignty, the achievement promotors of the Green Revolution had often promised—the country was self-sufficient in basic foodstuffs. Most of the nation became literate, and much of the population experienced formal education at higher and higher levels. The government offered improved medical care in most cities and greatly extended the availability of primary health care in the countryside.

Tourists from the United States could spend most of their time in urban neighborhoods or especially prosperous towns that gave the impression that there was no great difference between Mexican society and their own. Mexico's steel and chemical plants, automobile factories, and modern buildings led many Mexicans to be extremely uncomfortable with labels applied to their nation, labels like "underdeveloped," "Third World," or even "Latin American."

The ruling elites of Mexico and much of the self-satisfied middle class were then severely shocked by a set of events that would undermine their confidence and threaten their apparently solid positions in an ever-richer nation. Mexico's government had successfully attracted the international Olympic games to Mexico City in 1968. Mexicans could show off their proud accomplishments to the world. But the world, and much of Mexico, was in a self-critical mood in 1968. Many Mexicans were deeply dissatisfied with their country's progress and believed that the Olympics offered a chance to pressure the government for much-needed changes in the direction of national development. The Olympic games and the confrontations they sparked turned out to be a prophetic warning about the dangers ahead as well as a painful reminder of the not-so-hidden costs of the progress of the recent past (Cockcroft 1983: 237–276; Yates 1981; Meyer and Sherman 1983: 668ff.).

The Olympic protest movement was one of those shaky alliances between dissident peasant and labor organizations and leftist students that provoke more contempt than fear among those who rule Mexico. PRI, the ruling party, had decades of experience in intimidating, co-opting, and breaking such movements. To PRI's surprise, things were not so easy this time.

The movement began in an unlikely way. Street fights broke out between students at two rival high schools in Mexico City in late July 1968. Police and military ordered to stop the fighting killed one student and injured many others, including teachers. Protest marches and the celebration of Cuba's Revolution on July 26 led to the death of four more students, with hundreds wounded. Police and military units fired bazookas at student barricades, and students threw Molotov cocktails. Violence continued to escalate in a pattern similar to events in the United States, France, and other European countries in the same year. What was different in Mexico was that the protest against the repressive tactics of the police sparked further and much wider protests by peasant and labor groups as well as by discontented students and lower-middle-class urban organizations. Sharp regional inequalities, high unemployment rates, low wages, miserably inadequate urban housing, transportation, water, and sewage services contrasted all too sharply with the millions of dollars the Mexican government was spending on the Olympics. An amorphous and divided student movement grew quickly into a broad-based protest against all the frustrations of the economic development policies the government considered an unqualified success. Protests brought hundreds of thousands of people into the streets by late August. Police imprisoned thousands, killed between ten and thirty more, injured hundreds, and would not account for many people who "disappeared." The protests shook the government's confidence, especially as the world press wondered whether the Olympic games could actually be carried out safely in Mexico City.

A week before the opening of the games, on October 2, police opened fire on a gathering of hundreds of thousands of protesters in Tlatelolco in the Plaza de Tres Culturas formed by a high-rise apartment development project. No one knows how many died (the police claimed forty-nine), but independent conservative estimates put the numbers at three to five hundred, with hundreds wounded, arrested, and disappeared (Cockcroft 1983: 240–241).

The movement was temporarily broken, but it had succeeded in forcing even the most complacent members of the middle class and the wealthy elites into recognizing that the economic and social progress they proudly displayed to the world was a thin veneer over a set of serious problems that would not be easily solved. For a few brief years in the mid-1970s, confidence rose again with OPEC-driven rises in oil prices making for a rich and easy harvest. But outside of those few years, persistent, periodic economic crises in the 1970s and a seemingly permanent national depression in the 1980s made it impossible to regain for long the self-satisfied glow that had led the Mexican government to dare host the Olympic games.

In 1977, before the crisis struck in the early eighties, economists could paint a picture that included an impressive prosperity for some marred by a deepening inequality that left much of the population in bitter poverty. For better or worse, the nation went from being two-thirds rural to being two-thirds urban—certainly urbanization meant that more people had access to the kinds of goods and services that Europeans and North Americans take as signs of development (although using 2,500 as the definition of an urban population may seriously distort the real meaning of these figures). About 10 percent of the population lacked for little—they received 50 percent of the nation's wealth and enjoyed lives as materially rich as their counterparts in any other nation. "Another 20 percent of the income goes to the 20 percent just below the elite: shopowners, wholesalers, army lieutenants, state doctors, union bureaucrats, federal schoolteachers, civil servants, and skilled industrial operators." These people could call themselves middle class, enjoying decent, durable housing, an array of household appliances including television sets, and a family car or reasonable hopes for getting one soon. About 17 percent of income went to "20 percent who are regularly employed wage earners" who mostly enjoyed fairly decent lives in modest circumstances.

But 50 percent of the population, whose share of a shrinking national income was to decline further in the eighties, received only 13 percent of the nation's income. "In 1958 the top 5 percent of income recipients had 22 times the income of the bottom 10 percent; by 1977 the share of the top layer had jumped to 47 times that of the bottom. Thirteen million Mexicans (in 1977) live in extreme poverty, and have incomes below those in Latin America's poorest country, Haiti. Over half the population suffers from malnutrition, including 90 percent of preschool children, who constitute one-third of the population." A severe housing shortage meant that a million families were homeless and 40 percent lived in houses consisting of one room with primitive roofing—about half of all Mexicans in 1977 lived in housing that lacked "sewage services, toilets, potable drinking water, electricity, running water, floors other than dirt, social security, adequate footwear, or an income of more than $.25 (U.S.) per person per day" (Cockcroft 1983: 2–3).

Re-examination of Mexico's progress revealed that her most basic economic problems had not been resolved but deepened. The Mexican government had understandably chosen to hasten the process development economists in Europe and the United States recommended to poor countries. The government, always speaking the language of revolutionary nationalism coined in the Mexican Revolution, guided economic development with an array of policy tools. Government economic planners had at their disposal all the power of the state as created by Cárdenas through the oil expropriation and popular agrarian reforms and his crafty construction of an unchallengeable ruling political party. They controlled prices, offered subsidies, nationalized firms, and entered into broad economic agreements with domestic investors and transnational banks and corporations.

What the government economic planners did was straightforward enough, much as it has been dismissed as peculiar or corrupt mismanagement by economists and scholars with the advantage of hindsight. They milked the soil and other natural resources of the nation and its poorest laborers by squeezing money out of agriculture, forestry, fishing, and construction, making credit in these sectors expensive and keeping prices low. They did much the same with a few industries over which the state had especially strong control, like transportation, electricity, and services heavily patronized by foreign tourists. They tried to make it possible to continue to offer low urban wages and encourage those industries that classically led the early phases of industrialization by subsidizing food processing and marketing and textile manufacture, providing cheap food and shelter to urban workers. They encouraged industries that could quickly and easily begin to manufacture substitutes for goods traditionally imported. They hoped that these policies would lead, as much of the foreign advice they received said they would, to strong investment in heavy industry, making machinery for agriculture and other industries (Cockcroft 1983: 145–185).

The plan ran into two closely interrelated problems. The first was that by wringing money out of agriculture they made it unattractive to produce food. Peasant landholders chose to continue to farm only at a clear economic sacrifice to themselves. "In 1965 the average rural wage was more than a small-parcel farmer could earn through self-exploitation, and in the next decade more than two million hectares of rain-fed farmland was abandoned, much of which had provided the bulk of the food staples for the provincial centers and the big cities" (Cockcroft 1983: 254). In the following decade, another four to six million hectares would be abandoned, and much that was not abandoned was worked for the highest possible short-term gain to try to stay on the land a few more years. Wealthier farmers shifted when they could from basic food production to export products sold in richer economies and to such crops as sorghum to be used as feed grains for meat production, meat destined to feed urban minorities and for export. By the early 1970s, Mexico again became a major importer of basic foodstuffs while simultaneously the quality of much of Mexico's traditional agricultural land suffered the deterioration inevitable from farmers who could not afford conservation measures. Migration to the cities and the United States escalated. With burgeoning dependent city populations and basic food production collapsing, the state was severely pressed to keep food prices low through price controls and subsidies (Barkin and Suárez 1985; Barkin and Dewalt 1984; Toledo et al. 1985; Yates 1981).

If the state allowed the price of food and basic goods to increase rapidly, however, wages would sooner or later have to follow, and that brought up the second major problem in the way of Mexico's economic progress. Mexico found that international competition in manufactured goods was stiff, as other poorer countries tried to do what Mexico was attempting and as the already-industrialized nations, fully recovered from the devastation and disruptions of World War II, looked aggressively for new markets. The only way Mexico could hold down imports and offer any exports was to use its one great advantage: low wages. (By the 1980s, wages in Mexico were lower than the older low wage champions of Taiwan, South Korea, and Singapore.) But how could wages be kept low if the nation could not feed itself cheaply and had to rely on imports from higher wage economies for such basic items as food and clothing? How could a development strategy requiring increasing inequality be sustained politically in the midst of economic and political crisis?

A recent comparative study of the income statistics of Third World countries provides an international perspective on what Mexico's supposedly revolutionary state had achieved. While Mexico's per capita income of $1,429 was high by Third World standards, 14 percent of its population lived in poverty as defined by standards considered appropriate to the Third World—poverty not as relative deprivation but as the inability to meet basic survival needs. Countries like South Korea and Zambia had a smaller percentage of their populations living in poverty although their per capita income was half Mexico's. Taiwan, with per capita income in 1975 of $1,075 against Mexico's $1,429, had only 5 percent of its population in poverty as against Mexico's 14 percent. Perhaps even more indicative of the severe inequalities of Mexican society is the measure of income received by the lowest 40 percent of the population. In Mexico, the poorest 40 percent received 8 percent of the income, with only Tanzania and Peru showing greater inequality, at 7 percent, of thirty-four nations surveyed. In ten nations of the thirty-four, the poorest 40 percent received twice the proportion of national income that those in Mexico did, with Taiwan at 22 percent, Sri Lanka at 20, and Yugoslavia at 19. Considering that the reduction of inequalities had always been defined by the ruling party as a primary goal, it is clear that Mexico's economic growth policies were completely at odds with the nation's ideals (Ahluwalia, Carter, and Chenery 1979, in David 1986: 125).

Mexico's financial planners decided to go heavily into debt to resolve the problems and contradictions that had become obvious in the patterns of growth revealed by the political crisis in the years after the Tlatelolco massacre in 1968. Any initial worries about the implications of a rising debt seemed trivial when the petroleum boom of the mid-1970s offered what seemed like a permanent way out. So Mexico borrowed more heavily to expand its oil production and refining capacity while using oil income to maintain food subsidies and expand basic social services, building the party's patronage networks and shoring up political support as it did so. But with the drop in oil prices in the late 1970s, the game was over. Mexico's debt skyrocketed, the government was forced by its creditors to cut back on food subsidies and social spending, and overall economic growth declined sharply. The price of basic necessities went through the ceiling along with the debt, and the spreading unemployment and severe competition forced real wages down by more than half in the last decade.

How can Mexico compete now? There are three ways the Mexican government and its international creditors consider viable. One is to lower wages more, along with lowering social expenditures for such things as education, housing, and medical care. The second is to open the country wider to foreign investment by regulating it less and taking a smaller cut for social needs and the requirements of the state. Nationalized firms are to be sold to private investors. The currency is to be devalued even further (it had gone from 12.5 pesos to the dollar in 1973 to 2,300 to the dollar in 1988) to encourage exports. Licensing and patent requirements are to be liberalized to the advantage of foreign investors, and environmental or social legislation to regulate corporate activities is to be de-emphasized. The third is to sell the products of Mexico's natural environment at the cheap prices the international market will bear. The prices of minerals and tropical agricultural products have in general declined for about fifteen years as country after country has experienced the same problems as Mexico and must sell its natural endowment at bargain basement prices for the same reasons Mexico must. So, in the view of the international bankers and Mexico's policy makers, everything must be done by the Mexicans to keep prices low and competitive. Under these circumstances, Mexico can continue to pay off some portion of its debt to creditors in richer nations and the wealthier nations can control any serious competition in industrial commodities from countries like Mexico while enjoying cheap prices for primary commodities needed for manufacturing and tropical luxury goods.

Export vegetables fit this strategy in every respect. They depend on the lowest wage sector in the Mexican economy—migrant workers. The industry costs very little in social expenditures because no one expects to receive less in the way of schools, housing, medical services, and social security than migrant workers, and few groups have so little power to pressure for such services. Foreign investors are enthusiastic because of the cheap land prices, favorable government concessions given to encourage investment, low wages, and the sale of the product in high wage markets that investors know well and can control without Mexican interference. Lack of environmental regulation and reduced penalties for sending earnings out of the country make winter vegetable production in Mexico even more attractive. Frequent, drastic devaluations of the Mexican currency ensure that the vegetables will remain competitive in foreign markets. The obvious inability of the Mexican government and economy to pay off its enormous debt in the foreseeable future virtually guarantee that all these conditions will remain favorable to those who invest in the fields of Culiacán (Moulton et al. 1988).

While the international advisers associated with the banks and governments in creditor countries insist that such a strategy is the only way to keep Mexico and similar countries on the track toward catching up with the wealthier nations, a growing number of analysts maintain the contrary is true. Critics of the present economic growth policies recommended by the IMF and foreign investors say that the path followed by Mexico was certain to produce the debt crisis and certain to further emphasize the gaps between the wealth of countries like the United States and the poverty of countries like Mexico.

Cheryl Payer, who had worked for the International Monetary Fund, published a study in 1974 that prophesied with amazing clarity the events since that time. In *The Debt Trap*, she showed that the policies of the international lending institutions would not lead to an end to indebtedness and the economic and political dependency that went with it. Instead, she maintained, the policies of the international banks would open Third World economies to exploitation of labor and natural resources at cheap prices, strengthen local elites addicted to expensive imports and eager to invest much of their earnings abroad, create internal social tensions to be controlled with large outlays for military and police, weaken the domestic markets for food and manufactured goods by keeping wages low, and deepen the dependency of the weaker economies on the stronger economies of developed, industrialized nations.

Although her analysis was dismissed by most mainstream developmental economists, it is telling that while the mainstream analysts were predicting further growth and greater stability for such economies as the Philippines, Nigeria, India, Brazil, Iran, Indonesia, and Mexico, Payer predicted a disastrous debt crisis that would forestall further growth and create radical instability. It is clear that Payer's view was more accurate, as foreign debt levels in such economies, including petroleum exporters and petroleum importers, have risen by 500 to 1,000 percent since the appearance of her book. The political regimes of the early 1970s have either been replaced by dramatic revolts or are holding precariously to power. Yet the banks maintain policy demands on these nations that are essentially unchanged—skyrocketing Third World debts have made it a more dangerous game, but a game from which the highly industrialized nations have for now emerged economically stronger. Most of the Third World economies in question have sharply increased agricultural exports but have a more difficult time feeding their people and attaining overall economic growth.

The basic strategy of economic growth promoted by the banks and the U.S. Department of State comes from a particular way of understanding the historical development of Europe, the United States, and Japan, nations which, from this perspective, have already achieved what other nations strive for. The Harvard economist Walter W. Rostow and Harvard political scientist Samuel Huntington laid out the foundations for such modernization strategies in the 1950s and 1960s, and while many developmental economists would now claim that they have moved far beyond the theories of those days, the essential ideas have not changed (David 1986).

Rostow analyzed the economic development of England from the breakup of medieval society on. For a variety of reasons, in the late Middle Ages largely self-sufficient manors and villages began to take advantage of new commercial opportunities, especially by producing wool for a growing textile industry. As these opportunities grew, those who controlled land worked to move peasants off their land to make way for more sheep grazing. Dispossessed peasants went to the cities searching for work, where they soon constituted a large and easily exploitable labor supply. The availability of industrial labor encouraged new industrial investment, which raised the demand for raw products to feed the expanding mills, leading to continuing cycles of peasant displacement and urban growth. This process was often violent, as in the long struggles over the Enclosure Acts that made it easier for landholders to dispossess peasants. As it proceeded, the process picked up speed because the new urban labor force required food (cheap food, if wages were to remain low and investment attractive), the growing of which made for new commercial opportunities for landowners, growing food with increasingly large-scale methods requiring more and more industrial inputs, which further stimulated industrial growth. When this whole process had gone far enough, the urban labor force could begin to exert some pressures for its own rights and advantages while guaranteeing a growing market for agriculture and industry alike. In the final stages, the economy achieved "the take-off into sustained growth" that led to the "age of high mass consumption" (Rostow 1959, 1971).

Increasing concentration of landholding among a few, accelerated movement of the poor from the countryside into the city, burgeoning urban slums, and the creation of a class of aggressive entrepreneurs ready to profit from the process were not part of the problem of developing nations—they were in the long run the solution. The seemingly chaotic pattern of economic growth has to be appreciated for the salvation it represents. The only serious problem is for government to manage the situation capably without killing its inherent creative dynamism.

Critics have tended to agree that England and perhaps a few other countries achieved a certain degree of generalized prosperity through a process that included much of what Rostow described. They have pointed out, however, that he minimizes the centuries of violence and pain involved. Rostow also assumes that because this is one way to generalized prosperity, it is likely the only way. In suggesting that the same thing can be repeated in country after country, he had to downplay the fact that countries begin with different resource endowments and different social forms that condition what can come later. He also had to deny that the dominance of England and Europe generally over much of the rest of the world aided substantially in the development process. The availability of cheap, exploitable labor and resources obtained on Europe's terms must be held to be insignificant. The relationship between the extermination of whole peoples in North America, Australia, and parts of Africa and the subsequent availability of their land, forests, and minerals to English development is dismissed as an important factor. The idea that cultures may be quite different in defining the terms of happiness and prosperity cannot be credited. The notion that the timing of one nation's development in relation to opportunities created or denied by contemporary conditions in other nations or by the prevailing state of technology cannot be considered as a major factor. At bottom lies the stubborn conviction that in modern economic growth European-style lies the answer to the basic problem of human happiness— surely everyone agrees and hopes to emulate Europe and the United States. If they do not agree, they will simply be eliminated or pushed aside by those who do.

With Rostow having laid down the economic interpretation, Huntington provided a fuller view by specifying the political development that supposedly must accompany the process, calling the whole thing, economic and political, the process of "modernization." Rostow, Huntington, and a small army of scholars and development theorists have spent the last thirty years developing and elaborating on their ideas. A key point was to insist that while this pattern creating generalized prosperity has many significant variants in the stories of different nations, the outline remains the same everywhere. The task is to interpret the story and its policy implications within the historical context of each nation.

Rostow and Huntington both had the opportunity to do so in advising the Kennedy and Johnson administrations with regard to the war in Vietnam. Huntington emphasized the necessity of political control over the countryside—it could be taken for granted that the cities involved in rapid change would oppose existing governments and so ruling parties would have to be anchored in the more conservative rural areas. "The stability of a government depends upon the support which it can mobilize in the countryside." Huntington believed that what he called the "Green Uprising" that would inevitably break out over "bread and butter issues" in the countryside would determine the future direction of modernizing nations, because rural demands would always be essentially conservative and therefore stabilizing.

Huntington also urged the necessity of "institution building" to carry political conflict beyond the stage of raw grabs for power. While both Rostow and Huntington saw clearly the powerful conflicts inherent in the struggles of Third World countries, their analysis tended to encourage a social engineering approach among U.S. policy makers. Huntington urged a focus on political organization, saying, "He controls the future who organizes its politics." The problem that Rostow and Huntington could not solve was that the economic policies of Green Revolution–style modernization in the countryside ran into direct conflict with the ability of governments to gain support from the "rural masses" for the construction of stabilizing institutions (Huntington 1968: 433, 461).

Rapid technological change in agriculture is deeply destabilizing and often correctly seen by Huntington's "rural masses" as an imposition of urban elites in coalition with the relatively wealthy rural elites who seize control of the new technology for their own benefit. For that reason among others, political factions in Third World countries did not tend to see the world the way Huntington did and did not wish to organize politics along the lines he imagined as most beneficial. The Green Uprising tended to be both more complicated and less conservative than Huntington predicted. It often challenged rather than stabilized the hold of modernizing elites on national power.

U.S. policy makers, attempting to follow modernization theory, often found that they could not enter into the game as defined by the modernization theorists without first achieving physical control of the rural populace through brutal measures. Military and police control, and in the modern context counterinsurgency warfare, were essential to safeguarding the modernization process from the pained protests of those who did not appreciate its long-term benefits. The world view that seemed liberal and comprehensive from Harvard and Washington translated into reactionary defense of admittedly corrupt systems of privilege in the terms of Third World politics.

Modernization theorists have never been able to transcend this contradiction. They have not been able to see the way modern agricultural technologies radically destabilize relationships in the country while deepening old divisions and hatreds. More generally, they have never been willing to admit that the experience of countries pursuing prosperity is fundamentally different at different periods of history and that the internal cultures, politics, and physical environments of different countries present truly diverse problems. A general theory of modernization is thus continually frustrated by the variety of nature and human experience.

An excellent example of Rostow's kind of thinking appeared in the U.S. *Department of State Bulletin* in January 1985. In an article entitled "Central America: Agriculture, Technology, and Unrest," Dennis T. Avery of the Bureau of Intelligence and Research describes the changes in technology and crops that have occurred in Central America since 1945. He shows that the "peasant lifestyle" has been severely constricted by the rapid growth of rural population combined with the collapse of the traditional peasant economy as a result of the increasing dominance of export agriculture—especially cotton and beef production. As more and more peasants must attempt to make a living on less and less land with the technologies of production becoming more and more expensive and less accessible to peasants, the peasants are caught in a "crossfire." "Guerilla insurgents," according to Avery, exploit the situation to recruit peasants, explaining the growth of successful peasant movements. The physical and economic dispossession of peasants and landless laborers associated with the growth of machine and chemical-dependent agriculture explains the "unrest" that preoccupies State Department policy makers in Central America. At first suggesting that only rapid increases in agricultural productivity—more of the same—can resolve this crisis, Avery in his conclusion finds that the only real hope is to wait until industry takes over from agriculture as the chief economic activity of the region. Until then, he says, it will be essential to maintain military and police control over the population as they rebel against the rapid growth of modern agribusiness in Central America.

At greater length and in less analytical detail, the Kissinger report published in January 1984 reached similar conclusions for similar reasons. Commissioned by President Reagan to shore up support for his policies in Central America, Dr. Kissinger's committee called for the growth of the agroexport economy, arguing that it had been the most dynamic source of economic growth. What had created the most rapid economic growth was automatically assumed to be the source of future prosperity and well-being, as Kissinger's old Harvard professors Rostow and Huntington would certainly lead one to conclude.

"But what if this assumption of policy makers is incorrect?" Robert G. Williams asks in his exceptionally thorough historical study, *Export Agriculture and the Crisis in Central America* (1986). "What if the rapid economic growth of the 1960's, instead of creating conditions for social stability, did the opposite? After all, the social unrest of the 1970's followed closely on the most remarkable growth period the region had experienced in this century."

The economy of Mexico, much larger and more diverse than that of all the Central American nations combined, has been able to turn a much greater share of the returns of economic growth based on the agroexport model into real prosperity for a sizable portion of the population. In addition, Mexico repeatedly has turned to the earnings from petroleum exports and tourism from the United States to dig its way out of the economic binds created by its pattern of growth, options largely unavailable in Central America. Nonetheless, as we have seen above, the questions raised by Williams apply all too well to the Mexican economy as well.

Why is it that Mexico has fallen into economic stagnation for the better part of two decades? From World War II until the late 1960s the country enjoyed one of the highest and most consistent economic growth records in the world. Going into the years that produced the shock of rising petroleum prices for most developing countries, Mexico was able to greatly expand its petroleum exports and reap the benefits of cheap domestic petroleum. Tourism and agricultural exports have simply grown and grown, encouraging foreign investment quite successfully.

Advisers to the international public and private banks and the U.S. State Department have a simple answer. Mexico has relied too much on the state, creating inefficient state-run companies and spending too much on social services and education. The state-directed economy with a one-party state created ideal opportunities for corruption. The dynamism of the economy was lost to the inevitable corruptions of a strong state.

Would these advisers have preferred a weak state in the following circumstances: one of the world's highest population growth rates, creating a population in which nearly half the people are under fifteen and a million new job seekers come into the economy every year; one of the great historical epics of movement out of the countryside, doubling the size of Mexico's cities every decade; the rise of the world's largest city, making for staggeringly expensive and delicately balanced efforts to keep 19 million mostly desperately poor people reasonably content with their lot; erosion of the nation's land and pollution of its air and water proceeding rapidly and with no creditable plans for control? Only a strong state could manage such difficult processes. Time and time again, the essential bargaining tool of the Mexican government in dealing with international bankers is to suggest that more pressure for debt repayment may weaken the state and create a social hurricane no one will know how to control. The bankers have always been able to understand the importance of this political fact of life.

Every large growing economy comparable to Mexico's has required a strong state. This is true whether the nation has used the agroexport model discussed here or not. Brazil, Indonesia, India, Nigeria, China, South Korea, Taiwan are good examples. Whether they call themselves capitalist, communist, or socialist, economies undergoing very rapid change must be directed by a strong state. The theorists who fantasize that the military and police control they themselves recommend can be effectively wielded by a state without great economic power and a strong ruling party are the true utopians. They want the old Mexican formula, *pan y palo,* bread and club, to be applied by a state with no direct ability to give out bread. This is seldom, if ever, possible for very long. The strong state is not the problem with the economic growth pattern of nations like Mexico. Without a strong state, such growth cannot be achieved for the simple reason that radically unequal growth deprives too many people of what they desperately desire and require for the stability of their families, villages, and towns: secure access to land, fair terms for participation in the larger economy, employment at decent wages, protection of homelands and traditional communities.

The problem with an economic growth pattern based on massive dispossession of rural people for the sake of modern agribusiness export development is that it creates contradictory trends in the economy and the larger society. These trends eventually undermine further growth. As we have seen in looking at Culiacán and the Mexteca, the process concentrates earnings and investment in a few regions of agribusiness growth at the expense of the neglect, ruin, and abandonment of other regions. Within the agribusiness region, a small minority captures most of the income by ruthless monopolization of land and exploitation of labor. The work force directly employed in agriculture receives little in social benefit—they do not become more educated or more skilled or more secure. Instead, their health is broken and their communities shattered. The industrial technologies used in agriculture—fertilizers, pesticides, petroleum, machinery—do contribute to general industrialization, but at a rate that falls far short of the ability to absorb the overall rate of population increase and the much more rapid rate of urban growth. In addition, all of these manufactured inputs and the expensive irrigation works that make them valuable in the field have been heavily subsidized by the Mexican government and come at the expense of alternative economic investments and the upgrading of the labor force through health care, better housing, and education. Much of the income earned from the subsidized investments flows out of the country to safer places. The grower who took us to his private club, for example, said, "Agriculture has been good to my family, but I'm now putting everything I can into safer, more productive investments. I have an electronics factory in Tucson that is getting most of my attention." Capital flight is an endemic problem in Mexico, and it is a problem that becomes worse as the economy becomes less stable, creating a vicious circle (Cockcroft 1983: 157–158).

The problem of capital flight is one aspect of the broad problem created by economic development based on growing inequalities.

Mexico is one of the wealthiest countries of Latin America. Much of its economic policies have deliberately created a middle class larger than in most other Latin American countries. But Mexico is near the bottom in Latin America in terms of the severe distance between rich and poor. As Mexico turns more and more to the kind of un-equal development preferred by the country's international credi-tors, such as the growth of agroexports, the country confronts a problem it had not experienced in as severe a form as many other nations that have had nothing but unequal development, countries like Brazil. Unequal development leads to a small national market. If earnings are concentrated, most people do not have much to spend, and if they do not have much to spend, those who sell must look elsewhere. The economy becomes, to that extent, more dependent on foreign economies as markets to sell agricultural crops and manu-factured goods, and the nation is therefore less able to control its own affairs.

Accepting these consequences as inevitable if not desirable, eco-nomic advisers to the Mexican government recommend that the country now follow the path taken earlier by export-oriented manu-facturing economies like South Korea and Taiwan. Aside from whether Mexicans want to become a low-budget appendage of the United States, which is what this means in Mexico's case, the ad-visers are ignoring a crucial difference between the two Asian coun-tries and Mexico. In both Taiwan and South Korea, thorough-going land reform policies laid the basis for broader economic growth. In both countries, the land reforms were done not because of the politi-cal strength of the peasantry or because of enlightened national gov-ernments except in the sense of enlightened self-interest. The chal-lenge of radical changes occurring on mainland China and North Korea led policy makers in the United States to prevail upon the leadership of South Korea and Taiwan to undertake serious land re-forms as a means of reducing the possible influence of communist examples and movements. The economic result was to provide the kind of rural security that led farm families to choose smaller fami-lies, as increased labor would be applied to a small but securely held piece of land. Rural security kept rural-urban migration rates to a manageable level. Both countries had ancient, highly evolved, and very productive traditional agricultural methods improved through some modern technological advances, including some of the Green Revolution seeds. Productive agriculture on a dynamic traditional base with reasonably equal land distribution and selected modern technological improvements played its role as a stable economic factor in overall development, providing food, employment, and security to rural people. But the agroexport model of Mexican development followed since 1940 goes in precisely the opposite direction, undermining what tenuous security rural people gained out of the half-completed land reforms of the postrevolutionary period. (See the next chapter for a fuller comparison with the economic policies of Asian nations.)

As the agroexport model gains, domestic food supplies fall. Agricultural resources in Mexico go increasingly to export vegetables and cotton and sorghum and other feed crops for export cattle. Basic grains for human consumption become more expensive and a larger share must be imported. The degree to which the majority of the Mexican people must simply accept these conditions as another reduction in real income, which has fallen by more than half in the last decade, and the degree to which these costs will be controlled by price controls imposed on producers of basic food crops or consumer subsidies become matters for negotiation with the nation's creditors. The security of the Mexican farmer who wants to grow beans, squash, and corn and of the Mexican laborer who must survive on these foods relies only a little on their own activity. For the most part, the livelihood of each Mexican is a negotiating point between national political leaders and domestic and foreign financiers. Feeding the nation and providing its basic necessities are not seen as the main way the nation earns its income while providing for its needs—providing for basic necessities are now seen by the men at the negotiating table as a production cost that must be minimized to increase the attractiveness of export agriculture and industry. This perspective is the one applied to colonies and not to independent nations. A colony provides benefits for a colonizing nation at a calculated cost. The welfare of the people is no longer the object of the nation's labors and policies—it is a cost paid to gain advantages for other people who enjoy a dominating position.

The protests at the Olympic games were not the end but only a milestone along the path of protest of Mexicans against having their nation put in such a subordinate role. Two subsequent decades of protest have included peasant invasions of privately held land, labor strikes, and insistent pressure for the democratization of the political process. In 1988, all the protest movements crystallized in an extraordinary national presidential election that would shake the Mexican political system to the core.

On July 6, 1988, Mexicans went to the polls to cast their votes for federal representatives and a new Mexican president. The official candidate, guaranteed to win by virtue of the overwhelming power of PRI and its willingness to intimidate, buy votes, manipulate public sentiment, and cheat, was Carlos Salinas de Gortari. Salinas holds a Ph.D. in economics at Harvard, where he wrote a doctoral dissertation devoted to the question of how the government might obtain more consistent loyalty from growers and peasants in rural Mexico who received government benefits. The candidate of the right-wing opposition party, PAN (Partido de Acción Nacional), was Manuel Clouthier, a multimillionaire vegetable grower and agribusinessman from Culiacán. One of his main talking points in the election was for the end to any government authority over land distributed in agrarian reform programs to allow dynamic agribusinessmen like himself more room for expansion, increasing the growth rate of agroexports (Salinas 1982). Clouthier died in October 1989 in a head-on collision with a truck south of Culiacán.

The third candidate was Cuauhtémoc Cárdenas, son of Lázaro Cárdenas, the Mexican president who did most to redistribute land and to attempt to gain more independence for Mexico through nationalization of its most valuable resources. Cuauhtémoc Cárdenas had served the ruling party put together by his father for many years, as a federal senator and as the governor of the predominantly rural state of Michoacán in central Mexico. He had attempted to win the presidential nomination for 1988 from his colleagues in PRI but was unsuccessful. Alienated from PRI, he launched his own candidacy, and at the last moment gained the endorsement of another attractive candidate on the left, Herberto Castillo, who dropped out in the interests of a unified front. Shortly before the elections, one of Cárdenas's chief campaign aides and his assistant were assassinated by unknown assailants who stole none of the money or valuables carried by the two men but who made off with their copy of a large notebook detailing the Cárdenas strategy to attempt to blunt the ability of PRI to control the election through fraud. The final election results were unreasonably delayed while many concluded that Cárdenas had won, based on partial returns and polling place surveys of voters. Days after the election, the PRI-controlled election commission announced the victory of Salinas with 50.3 percent of the vote, but granting Cárdenas more than a third of the remaining votes and admitting that Cárdenas had won a majority in Mexico City and various states. By these official statistics, PRI had received its first serious challenge since 1940. Millions of Mexicans assumed that PRI had simply stolen the election from Cárdenas (Barberán et al. 1988). Cuauhtémoc Cárdenas presents the clearest possible opposition to the development policies of the Mexican government that have led to the present debt crisis and to the rise of large-scale, export-based agriculture like that of Mr. Clouthier's Culiacán Valley.

Andrew Reding asked Cárdenas in an interview for the *World Policy Journal*, "After more than half a century of rule by a self-described revolutionary party with a program of economic development and social justice, Mexico today is in the midst of its worst economic crisis since the Great Depression and possesses one of the most inequitable distributions of wealth in the world. What, in your view, has gone wrong?" Cárdenas replied,

> I think the present crisis is due to the overall abandonment of the goals of the Mexican revolution as they were institutionalized in the constitution of 1917. This departure from the revolutionary program did not occur all at once, but began in 1940, when Manuel Avila Camacho succeeded Lázaro Cárdenas to the presidency, and has progressed steadily since then. Successive governments have tied Mexico's development to ever-expanding foreign investment and to ever-increasing foreign indebtedness, while failing to develop the rural economy or provide economic and political support for the ejidos and other peasant organizations.

Cárdenas went on to point out that urban labor organizations "are used by the government to co-opt and control social demands and foreclose opportunities for political participation." He emphasized that "the foreign debt has become the central fact of Mexico's economic existence." The outgoing administration "has made debt servicing a top priority . . . while giving no priority at all to economic development or attending to the social needs of vast sectors of the population."

> These imbalances have been compounded by a neoliberal economic program that seeks to develop an export-oriented economy while rendering us dependent on decisions made by foreign investors.
>
> Mexico has experienced repeated attempts at export-oriented policies since the beginning of the century, and these have never resolved the country's problems. We of the Democratic Current believe that the surplus necessary to promote development should be generated primarily by strengthening internal markets. This cannot be done without a substantial increase in the purchasing power of salaries, which has declined by more than 50 percent in the past five years.

After a discussion of the need for political democratization of the country and a strategy for achieving it, Cárdenas described his view of a mixed economy and of methods for handling the debt crisis. To increase the rate of economic growth, he recommends development of domestic industry and job creation through the use of Mexico's advantages in petroleum and mineral ores. "We also have great potential for meeting job demand in the area of natural resource development. Soil enrichment programs, fisheries, and reforestation projects all offer enormous possibilities. *We must adopt an ecological approach, one that stresses prudent management of our soil, water, and plant cover and move wherever possible toward multiple use of resources"* (emphasis added). Later, Cárdenas discussed the revitalization of Mexican agriculture and again stressed that all agricultural development "must of course be supplemented by an integrated policy of resource management in the rural sector. Special attention must be devoted to renewable resources such as soil and water, which are fundamental to our economic well-being but nevertheless have been grossly neglected," adding that "this has always been the case, not just under the present government."

After further emphasis on the necessity of good resource management and the protection of water resources from exhaustion and contamination, Cárdenas emphasized that "we cannot afford to be careless about managing our natural resources at a juncture in our country's history when we face such powerful population pressures on those resources" (Reding 1988).

Cuauhtémoc Cárdenas's view of Mexico's troubles and possible solutions is by no means new. This view has been repeated in and out of government for at least two decades and has sometimes even appeared as the rhetoric of PRI representatives. Luis Echeverría, who as minister of the interior in the government of Díaz Ordaz had ordered the police repression in the Tlatelolco massacre before the Olympics and who enjoyed the advantage of high petroleum prices during much of his own presidential administration (1970 to 1976), put forward some policies and an overall perspective that shared much with Cárdenas's ideas. Echeverría, however, was constantly stymied in his attempts to carry forward his ideas by resistance within the ruling party and government, and many questioned his real devotion to his program of redistribution and nationalism. Many people are more prepared to believe that Cárdenas's break with the party makes him more believable and the long-run chances of instituting his policies in a government headed by him much more encouraging.

It can hardly be overemphasized, by way of caution about Cárdenas's candidacy and rhetoric, that Mexico's agricultural ministry and other ministries and bureaus with responsibility for the nation's resources and environment will find nothing disturbing in Cárdenas's words expressed in such general terms. They have long-standing programs supposedly devoted to rational resource use and environmental protection, but as we have seen in the cases of pesticide regulation and soil restoration in the Mixteca, these programs are seldom very meaningful except as political window dressing. The political question for the future is whether better resource and environmental policies will be given a chance to become really significant by being placed within a political strategy and an overall program of national economic development consistent with environmental protection.

Cárdenas's political challenge has shaken U.S. State Department officials deeply, as they tend to see it as the beginning of the fall of the largest domino in a series that began in Nicaragua. As in Central America, they cannot appreciate the destabilizing effects of the development model that has been followed for so long, emphasizing foreign investment and export expansion. Their view of the world limits them to seeing subversion in a man like Cárdenas, rather than seeing the inevitable reaction to development policies that have raised and then undermined the hopes of whole peoples. The future of the two countries will be strongly influenced by how far the U.S. government can progress toward understanding the real causes of dissatisfaction in Mexico.

The rise of Cárdenas and the popularity of his views raises another crucial question. Does Cárdenas have a plausible vision of how to combine economic development with environmental and resource protection? Limiting the question to agriculture, what would be the concrete changes that would lead to protection of Mexico's public health and its soils, air, water, wildlife, and fisheries? To answer that question, in the next chapter we summarize the mistakes made in the past in the form of basic assumptions about agriculture and development, suggesting the progress that could be made if Mexico were to work from different assumptions.

9
The Modern Agricultural Dilemma

As it happened, in the same summer of 1988 when the Mexican government was so shaken by its first serious electoral challenge in decades, the worst fears of many scientists for the health of the earth seemed to have assumed a terrible reality. Newly released scientific studies of the greenhouse effect and the depletion of the ozone layer threatened humanity with the prospect that our careless pursuit of technology's cornucopia may have already irreparably damaged the ability of the earth's atmosphere to sustain us safely into the future. Smaller things, like the epidemic disease that killed thousands of seals in the North Sea, raised disturbing questions. Were the seals made more vulnerable to disease by the massive and continued poisoning of the oceans that goes on year after year, or perhaps more particularly by the massive release of synthetic pesticides into the Rhine River as a result of a fire at the Swiss Sandoz chemical plant and a release of other pesticides at the same time at the nearby Ciba Geigy plant? The Union Carbide Company seemed to have succeeded in 1988 in reducing its liability for the death of thousands at Bhopal through the release of a gas used to manufacture pesticides. An Indian court assessed the company for damages in an amount that was very close to the company's insurance coverage—the company would escape with minor financial damage from one of history's most terrifying industrial accidents. Anyone but Union Carbide stockholders who noticed this could hardly have been comforted. Summer bathers in 1988 found beaches around New York City closed by health authorities due to the huge quantities of medical wastes, including thousands of hypodermic needles, that insistently washed up on shore. Among other things, people worried, with hysteria that expressed a deeply anxious mood better than it did the actual dangers, that a needle encountered on a beach might carry AIDS, the deadly virus that forces us all to realize that humans are far from triumphing over disease or even understanding what generates it. As the Mexican ruling party found reasons to worry about the newly powerful voices of dissent, all of humanity found profoundly disturbing reasons to question the assumptions that make up the modernizing faith.

In this book, we have been dealing with some of the troublesome aspects of the modernization of agriculture in Mexico, a program of modernization that has been imitated in dozens of other countries. The technological and much of the political content of the program was designed by people in the United States who came to have great influence over Mexico. The ideas of these powerful people came primarily from their own experience in the United States—they sought to share their successes with others, motivated by many things fair and foul. While we cannot analyze here all the assumptions that went into the building of the modern world—the background to the destruction of the Rhine, Bhopal, the greenhouse effect, and the depletion of the ozone layer—we can look at some of these ideas with a little new understanding, particularly those regarding the modernization of agriculture. In doing so, we sum up what we learned on the road to Culiacán and gain some wisdom about which paths to choose toward a healthier, more secure planet.

The modern agricultural dilemma arises out of a tension found in almost all modern nations. The highly localized adaptations needed for ecologically healthy agriculture and healthy, stable rural communities are often in conflict with the apparent requirements of rapidly industrializing nations and an expanding international economy.

The growth of industrially based agriculture in the Culiacán Valley and of industrial centers like Mexico City contributes mightily to the environmental and social ruin of regions like the Mixteca. As we have shown, many booming agricultural regions like the Culiacán Valley actually depend on the cheap labor produced by ongoing social and environmental disaster in other regions—and worse, the technologies being used to produce present growth make it likely that the new regions, the Culiacán Valleys, will follow the old into social and environmental decline.

The larger economic strategy of industrialization and export promotion followed by nations like Mexico locks societies into patterns of debt and dependency that make it very difficult to change course in spite of the disastrously mounting social and environmental costs. The modern agricultural dilemma is part of an even larger quandary faced by modern society. Can the pursuit of economic prosperity be made consistent with ecological health and some reasonable degree of human equality?

Agriculture researchers and policy makers have proceeded on the basis of a set of assumptions—some of them explicit and some of them implicit and unexamined—that have guided modernization efforts. They assumed, first and foremost, that the basic problem of agriculture is the need for greater yields per unit of land area. Technologies, they believed, should address the need for higher productivity before all other needs.

Second, they assumed that techniques used by farmers in traditional cultures are largely obstacles to improved agricultural practices, part of the background of mass ignorance to which the scientist and technician were to bring enlightenment. Deep-rooted ethnocentrism precluded serious questioning of the assumption that good farming and agricultural science are mostly European in origin. Systematic investigation of the cultural and ecological value of traditional techniques was not necessary or even desirable.

Development economists and policy makers associated with the project of agricultural modernization quickly built into the research programs a third assumption, that agriculture should be designed such that its development would provide the basis for rapid industrialization of poor countries. Agricultural researchers designing seeds dependent on industrial chemical inputs and improved irrigation consciously or unconsciously bent agriculture to satisfy the requirements of the industrializing assumption.

Early successes with systematic plant breeding, chemical fertilizers, and synthetic chemical pesticides quickly led to a fourth assumption. Researchers began to believe that agriculture had no crucial link to wild nature. They were sure they could control the field environment without needing to worry much about the influence of the broader natural environment or events in surrounding agricultural fields. Since they thought they could control the field environment, they did not worry much about the particular ecological and cultural conditions of the tropics, which varied so greatly from the researchers' previous experiences in temperate climates.

When the program based on these four assumptions began to cause serious social and environmental problems, researchers retreated into a fifth, comforting belief. They argued that technologies are socially and ecologically neutral—nothing inherent in technologies themselves shapes their social or ecological results. What problems may have arisen from modern agricultural technologies arose from faulty implementation of the technologies by governments and firms. Socially neutral tools had simply been badly used.

Let's look at each of these assumptions, summarizing their origins and their consequences, and moving on to suggest the advantages of new ideas to guide policy and research. At the same time, we can review the work of some of the people and organizations using alternative assumptions in the task of building a more socially and environmentally desirable kind of agriculture.

Is Productivity the Problem?

The assumption that long prevailed among agricultural scientists that productivity per unit of land area is the basic problem of agriculture has long been questioned by social scientists and has been especially strongly discredited by researchers in the last twenty years. The process of technological modernization begun in Mexico in the 1940s focused on raising yields, with the idea that raising yields would boost farm income, feed urban people, and provide the incentives and savings for industrialization. As we have seen, this program ran into bottlenecks it has not been able to resolve, with millions of dispossessed farmers, skyrocketing food prices, and massive foreign indebtedness draining off earnings.

As early critics of agricultural modernization suggested, low yields were not the basic cause of rural poverty or economic stagnation in the 1940s. Raising productivity did have some positive economic results, especially for the urban industrial economy and for the richest rural people, but it did not solve the problems of the rural poor or create a permanent base for generalized prosperity, as it was argued it would do.

The planners of the modernization program in Mexico believed that yields were typically low in Mexico, but they possessed little data to sustain their belief and undertook no serious analysis to discover the causes of low yields. They did not account for the specific tumultuous events of the decades leading up to their program, nor did they account for the usual consequence of the early stages of agrarian reform—that in addition to the necessarily disruptive events of a major reform, less food enters the market because farm families are able to afford to eat more of the food they produce rather than sending it all to market. The technicians who planned the program, with little advice from economists or social scientists, operated on what to them was the common-sense presumption that if Mexican farmers were poor it was because they lacked the knowledge to produce more—a conclusion that people like Carl Sauer challenged at the time.

Research in recent years has produced the evidence that Sauer needed at the time to bolster his own notion of the problems of Mexican agriculture. Researchers have shown consistently that low agricultural productivity is more often the consequence of economic and political barriers than the result of technological failings. The most common barrier is poor financial returns to farmers associated with the lack of strong effective demand for growing food production. The low wage economies of poor countries do not allow the existence of a strong, profitable market for food. If farmers do not receive a fair price for their labor, they have no reason to produce. If, over time, they cannot sell their crops profitably they will have to farm for simple subsistence or give up farming, because very few farmers in the world can live without some cash income. As poorer farmers go out of business, larger farmers with more and better land and better financing take over agriculture (Griffin 1987).

Unfortunately, to stay in business, these larger farmers using labor-saving technologies and the best land bought with large cash loans must also earn cash, and they find that poor people in their own countries cannot purchase the crops at a price that will bring a profit to the producer. So, they must sell to the small domestic markets of middle- and upper-class people or to people in rich nations. In some instances, they sell to poor people in their own country by using their government to channel subsidies to the large farms (in both the United States and Mexico, a small minority of large farms earns the overwhelming share of subsidy payments). With the subsidy payments and with grains obtained in foreign food aid programs, the domestic market for the small farmer's crops is even further undermined by government-financed competition.

The peasant farmer then cannot sell his crop and often cannot keep his land. The family become landless rural laborers, sharecroppers, or migrants. Research in country after country shows consistently that it is among the landless rural people and the poorest of the urban slums that hunger prevails, with the hunger sometimes compounded by wars and natural disasters usually strongly related to the process of rural displacement described here. "The fundamental cause of hunger, then, is the poverty of specific groups of people, not a general shortage of food. . . . The persistence of inequality, poverty, and hunger is due not so much to the absence of growth as to the characteristics of the growth that has occurred" (Griffin 1987: 5–8).

The hungry, landless poor are the victims of inequalities of power and wealth in their own nations and of the overall problems of low-wage economies in an international system. They are not suffering primarily from the technological failings of their nations' farmers. In the long run, the governments of poor countries cannot sustain the burden of the subsidies given to the commercial agricultural sector. For this, among other reasons, they are undermined by indebtedness and the kind of financial collapse that threatens Latin America and that has become common in much of Africa, as both conservative and radical economists have pointed out. The heaviest burden of debts that cannot be repaid falls on the landless poor, who cannot feed themselves and who cannot find wage labor or government-supported social services in a bankrupt economy.

As we saw in the previous chapter, this necessarily highly generalized description applies fairly well to Mexico, especially when the drive for industrialization based on exploiting agriculture is added to the picture.

The assumption that productivity was the problem was not really the result of a thoughtful but faulty analysis on the part of the plant breeders, plant pathologists, and crop protection specialists who designed the research for agricultural modernization. It was a preconception on the part of these researchers who had spent their entire lives dedicated to the single-minded task of raising productivity. The economic and political analysis had been carried out by the policy makers whose primary purpose, as we have seen, was the protection of economic privilege and the promotion of a political alliance between conservative factions in Mexico and the financial community and government of the United States. For them, the drive for higher productivity made perfect sense.

The drive for higher productivity could accomplish a number of important goals of ruling elites in Mexico and the United States. By promising abundance for all, it blunted the political momentum for redistribution. Using the kind of technological expertise in agriculture for which the United States was noted, it created a gift that Mexico could receive from the colossus to the north with the idea that the gift would strengthen Mexico. Gratitude promoted alliance. Based on changing the agricultural economy from reliance on native rural resources to reliance on industrial products, the program created opportunities for both Mexican and foreign investors that had not previously existed. The successful promotion of industrial growth based on agriculture would presumably strengthen Mexico as an independent nation with its own industrial base, although it quietly tied Mexico to an eventually crippling dependence on foreign investors and markets, as manifested in the current debt crisis. The program for rising productivity flattered and strengthened the professional and middle classes newly important to the Mexican regime, as it relied on expertise, education, and technical advancement rather than on economic restructuring and redistribution.

In sum, the notion that production was the problem served to promote the growth of markets and investment opportunities and to tie the Mexican economy more closely into the international economy and to a closer relationship with the United States in particular. It succeeded in building a much stronger industrial economy in Mexico. Mexico could produce far more goods, fulfilling many consumer needs and creating new consumer desires to be fulfilled by more growth. New taxable revenues meant the expansion of social services and education for more than two decades and the creation of a small but politically vital middle class of professionals and technicians. The nation became predominantly urban.

Working on the production problem, however, deepened social inequality and destroyed much of the peasant economy. It forced very radical social change under the peculiar modern conditions of falling death rates due to public health improvements in the midst of poverty and insecurity, thus fueling high population growth rates that cancelled out much of the production gains. The technologies of increased agricultural production were severely damaging to the environment and public health and nearly eliminated the older, more complex systems of crop security created by Mexican farmers over the centuries, as described in a previous chapter. Most obvious to Mexicans today, in chronic economic depression and debt, it was based on a heightening or intensification of the basic problem of Mexico—a country endowed with great riches and an industrious, inventive people, yet locked into a highly inequitable political and economic system that frustrates the creation of a strong, stable local market. Even those who dominate Mexico, internally and externally, are faced with the fact that a country with highly unequal distribution of land and low wages has limited opportunities for long-range growth—it submits to plunder rather than becoming stronger and more independent.

There is a crucial distinction that must be made here. I am not saying that Mexico should not have worked to produce more over the last decades. I am saying, along with many others who have studied the problem, that Mexico would have produced more while resolving some of its inequalities and economic bottlenecks if the problem of agriculture had been seen as a problem of incentives to producers, the question of incentives tied strongly to the issue of distribution of land and wages. Stronger domestic mass markets for food and other goods would have rewarded producers, and fewer peasants would have had to leave the land, selling off to export-oriented commercial farmers or simply abandoning good agricultural land. With strong incentives and a growing internal demand, everyone, from peasant farmers to government policy makers, would have had every reason to encourage research on improved agricultural techniques consistent with a broadly shared prosperity in the countryside.

It has been pointed out by an agricultural economist who is enthusiastic about the agricultural modernization that took place in Mexico that there was a tremendous confusion about what the technological innovations called "miracle seeds"—the basis of the Green Revolution—actually achieved. He cites a study by Mexican researchers showing that if Mexico had carried out the same programs of dam building, irrigation, transportation, communication, and credit without the Green Revolution miracle seeds, Mexico would have achieved two-thirds to five-sixths of the production gains it reached by using the new seeds (Yates 1981: 11).

This is a very important conclusion, although a very believable one. Since all the improvements but the seeds and their specific requirements were well known to produce increases in productivity, it would be odd to believe that their contribution was in this one case insignificant compared to the importance of the new seeds themselves. Considering all the sacrifices made for that 16 to 33 percent of the gain over previous production levels, perhaps Mexico would have done as well without the new seeds.

In addition, the study made no projections about what could have been achieved with improvements in infrastructure combined with a program of agricultural research different from the one that produced the miracle seeds. Is it possible that some combination of continued serious agrarian reform, the anthropologically based research approach of Carl Sauer, and the investigation of what was best in traditional techniques, along with all or part of the infrastructure improvements, could have achieved something better at a lower social and environmental cost?

With respect to the development of particular techniques, millions of Mexican peasants, heirs to a tradition of agricultural innovation going back thousands of years, given the proper incentives could have developed the regionally adapted technologies they possessed to meet new production needs. Demographic research tells us that secure landownership and a stable agricultural economy would have dampened population growth rates, reducing the need for dramatic jumps in production. The results would have been more comparable to the rapidly industrialized economies of East Asia (such as Taiwan), which were built on the foundation of ambitious agrarian reform policies and were the first among developing countries to control rapid population growth, except that Mexico would have had the advantage of starting from a much more prosperous base than most of the destitute, war-ravaged East Asian nations.

This view of the problem, focusing on internal demand and effective incentives for agricultural growth rather than on a fixation with technological solutions per se, has been put forward by Mexican opposition figures for decades and was the view held by President Lázaro Cárdenas. It is now the perspective, as we saw in the previous chapter, put forward by Cárdenas's son, Cuauhtémoc, in the first serious challenge to PRI since 1940. The right-wing opposition represented in the 1988 elections by the Culiacán agribusinessman and grower Manuel Clouthier also focuses on the role of incentives but seeks these incentives in additional export growth, based on keeping domestic social spending and wages low, continuing to concentrate domestic income and land into fewer hands—essentially the program of the International Monetary Fund. Agricultural issues continue to be strongly related to the structure of the overall economy. Mexico has never had its choices more clearly defined.

While a program of agricultural development based on improved internal markets as an incentive to growth and development would not guarantee that Mexican farmers would shun pesticides or avoid pesticide abuse, it would certainly reduce the kinds of abuse now prevalent. Overall, pesticide use would decline because smaller producers growing basic food products rather than monocropped export products like cotton and winter vegetables would have less need for pesticides and less incentive to use them to replace their own labor and management skills. Even where pesticide use continued, there would be fewer farms with large absentee owners utterly dominating powerless thousands of desperate, landless people. The smaller class of landless laborers would have greater bargaining power. As shown above, most pesticides used in Mexico, and in the Third World, are used in luxury export crops and cotton, not on the basic food crops that would be grown in a more equitable economy. The equation of social equality with lowered rates of pesticide abuse is far from exact, but it is precise enough to ensure a healthier population and environment as a predictable result of more justice for rural people.

Are Traditional Technologies an Obstacle or a Resource?

The program of agricultural modernization carried out in Mexico over the last decades proposed that the "members . . . of the mission are the teachers and the nationals the learners" (Stakman, Bradfield, and Manglesdorf 1967: 316). If the matter at hand were the straightforward transfer of the technology of manufacturing silicon chips, as a hypothetical example, this statement might seem reasonable. But agriculture is different because its techniques have been worked out over thousands of years in association with the abilities and requirements of families, the culture of villages, and the character of the local biological and physical environment. The transformation of agriculture changes the relationships among people and the relationship between society and nature in far-reaching ways, many of which may be irreversible. To propose a simple, one-way relationship of teachers to learners in such a setting is ethnocentric and dangerous in the extreme.

It is particularly remarkable given the following: Mexico was the site of the domestication of a major share of the world's crops; Mexicans had developed highly productive agriculture within dozens of different cultural and ecological settings ranging from tropical rainforests to temperate deserts, from humid coastal plains to cool, high mountain plateaus; and a good deal of Mexico's agricultural production had for centuries been tied into international markets and waves of technical innovations diffused through those markets. On the other hand, the teachers from the United States were teaching the techniques of an American agriculture that was just emerging from nearly two decades of depression, strongly related to the problems created by overproduction, and the Dust Bowl, one of humanity's greatest natural calamities, directly related to the tragic application of inappropriate agricultural technologies (see Worster 1979 for an analysis of the role of technology in the agricultural depression of the 1920s and the creation of the Dust Bowl of the 1930s). The obvious arrogance involved needs little more discussion here in terms of its negative results. The lack of any serious research effort fifty years ago to see what indigenous Mexican techniques might have offered to the nation and to world agriculture is a tragedy, particularly to the degree that the successful promotion of agricultural modernization eliminated forever a huge range of genetic diversity in crops that simply can never be recovered. Along with the genetic diversity went many highly evolved cultural patterns and agricultural techniques that survive only in literature or in hopelessly fragmented form, patterns and techniques that might now have seemed quite useful.

Fortunately, the destruction of traditional techniques was far from complete, and a new generation of agricultural researchers is demonstrating the value of what remains. We have discussed the historical development of many of these technologies and their role in providing ecological stability and security to peasant farm families. It now appears probable that a new kind of agriculture can be constructed out of the combination of these older, conservative techniques and new ecological insights about agriculture.

Miguel Altieri is an entomologist and pest management specialist who teaches at the University of California at Berkeley. With Ford Foundation grants, he is establishing a program for research on alternative agriculture in Latin America. His book, *Agroecology: The Scientific Basis of Alternative Agriculture* (1987), aims to lay out the principles for the integration of ecological sciences with traditional agricultural systems and new ideas for more environmentally benign ways of growing food. Altieri confesses that there are major problems about how to proceed in such an effort because there are fundamental conflicts between the way Western science views the world and the systems of knowledge and practice of traditional agriculturalists. In *Agroecology*, Altieri's colleague, economist Richard Norgaard, writes,

> In the absence of a consensus about epistemological beliefs, agroecologists have resorted to pragmatism. Western knowledge is not rejected, for the mechanical world view has given us many insights, and conventional agricultural explanations help the agroecologist understand traditional systems as well. At the same time, agroecologists are receptive to the explanations of traditional peoples. Traditional knowledge may not survive Western tests. Traditional knowledge may not generate testable hypotheses; when it does, the hypotheses may be refuted; and the knowledge—typically contained in myths and social expectations—may not even be internally consistent. But traditional knowledge has survived the test of time—the selective pressures of droughts, downpours, blights, and pest invasions—and usually for more centuries than Western knowledge has survived. (1987: 25)

Altieri proposes a "coevolutionary" view of human culture and nature in dynamic interrelationship through time. He envisions an anthropological approach to agronomic research, seeking to identify agricultural practices that work well within the ecological and social setting where they are found. Scientists can study such practices, not necessarily with a view to changing them or improving upon them for the people who are using them satisfactorily, but in order to develop insights about adapting to externally imposed change or about the possible adaptation of the techniques studied for use in other settings. It is interesting to note that Altieri has reached similar conclusions to those of Carl Sauer decades earlier, advocating that research for the improvement of agriculture should begin with the anthropological study of the peoples involved. Altieri now has a much broader base of research into traditional systems to draw upon than was available to Sauer. This research includes work he has done in Mexico along with such pioneers in the study of traditional Mexican systems as Le Tourneau, Gliessman, Gómez Pompa, Trujillo, Hernández Xolocotzl, and Wilkins. He can also point to the well-known failings of decades of work within the framework of the program of agricultural modernization that Sauer was attempting to analyze and criticize at its inception.

Altieri and Stephen Gliessman of the University of California at Santa Cruz have already amassed a body of experimental data showing the value of traditional agricultural techniques used in Mexico and adapted to California farms and gardens. They have obtained excellent production results with interplanting of various vegetables, attributing much of the high-yield results to the control of insects and weeds achieved by "polyculture" plantings. Insect pests attracted to cabbages under ordinary circumstances do not appear in troublesome numbers when the cabbages are interplanted with beans, for example. Some plants emit chemicals which discourage the growth of weeds around them—such "alleopathic" plants can be interplanted with other crops that do not emit such chemicals in order to achieve effective weed control for both crops. Such techniques have been known to some gardeners and farmers for centuries, in Europe as well as in Mexico, but Mexico, with its extraordinarily high cultural and ecological diversity and long cultural history, seems to be a particularly fruitful area for researchers interested in incorporating traditional techniques into modern agriculture. Gliessman writes, "Improvements upon these traditional systems, or the development of new or alternative systems for the future, will involve the integration of ecological and cultural knowledge. Only in this manner can agriculture establish a truly sustainable base" (Jackson, Berry, and Colman 1984).

Wes Jackson's approach to ecological agriculture research is closely related to the approach of Altieri and Gliessman. In discussing the "unifying concept for sustainable agriculture," Jackson writes,

> The use of "the"—instead of "a"—"unifying concept" . . . was not accidental. I chose it because I believe that a truly sustainable agriculture will be directly keyed to nature, which already has a well-understood unifying concept of its own. I am speaking of the unifying concept of biology as discovered by Darwin and which was later coupled with the discovery that the DNA-RNA hereditary code is universal. Essentially all the natural ecosystems of the earth are many times more complex than our most elaborate agricultural ecosystems, and if there can be a unifying principle for the diverse natural biota, why shouldn't there be one for agriculture, especially if nature is our standard? (Jackson 1987: 119)

Jackson's assumption is that agriculture will be most stable, with stable or growing soil fertility, low requirements for fossil fuel or other imported energy sources, and a high degree of diversity among wild species as well as crops, when humans grow their food in patterns mimicking nature as closely as possible.

Thus nature, not culture, is Jackson's primary standard for the design of new agricultural systems. His own work involves the attempt to breed high-yielding perennial prairie plants to be grown in associations somewhat like those found on the prairie, eliminating the need for plowing, cultivation, and pest control as now practiced, with a constant plant cover that would protect and enhance soil fertility, as in the natural prairies. Jackson recognizes the importance of studying traditional systems for clues about how agriculture can be best integrated with nature's design. He follows the work of Gliessman, Altieri, and others in Latin America closely and has developed an ongoing dialogue with the Amish people and other traditionalists in the United States. "Some could argue, of course, that agriculture is so much a product of human manipulation that a unifying concept will have to be *invented*. I don't think so. I think it is only to be *discovered*. There are enough examples of good farming that we recognize as being in harmony with nature that we can look to them" (1987: 119).

Jackson believes, however, that "biological information is subject to Darwinian selection and is more reliable than cultural information." As Jackson makes clear in other writings, saying as much is to assert that human purposes can only be properly served in an enduring way when human will is subjected to what he takes to be a discoverable wisdom of nature that can be inferred from the study of ecology and from relatively harmonious traditional systems embedded in cultures that strongly constrain human pride and desire. For that reason, Jackson has maintained a fascination with Mennonite and Amish agriculture, agriculture he believes is relatively sustainable and based on powerful religious principles that ordinarily keep the rules of sustainability from being broken. Interestingly, many of these constraining principles, in very different forms, are those we have discussed earlier as part of the "technology of security" in the Indian "closed corporate communities" of Mexico and Central America.

Another way to recognize the value of traditional systems is often called the "farming systems" approach. Teams of researchers, nearly always including social scientists as well as technicians and natural scientists, attempt to define the problems related to agriculture faced by peoples or regions. Researchers start from the presumption that whatever technologies are in use have some rationale within the overall social and economic context. That the use of these technologies can be understood and explained does not mean that they are the most desirable ones available. The value of technologies has to be weighed in terms of a defined set of identified problems and goals for improvement. Although this is often not genuinely the case, farming systems research relies on the idea that there is significant participation by the people of the region under study in defining what the problems are and what would be considered desirable goals for the future. Then, agricultural researchers can study both traditional and nontraditional technologies, either indigenous to the region, imported, or newly invented that might be useful to meet the determined goals. The farming systems approach has become very popular in the International Centers for Crop Improvement in Los Baños, in the Philippines, in West Africa, and at the East-West Center at the University of Hawaii. It is an approach very explicitly worked out as a way of meeting the criticisms that have been launched against the Green Revolution and other agricultural modernization campaigns. It has been most popular with the institutions and agencies historically involved in such campaigns, such as the International Rice Research Institute at Los Baños, making some hopeful that a more enlightened approach has taken over the citadels of expertise and making other observers skeptical about the possible use of the farming systems approach as a more elaborate political masking of otherwise discredited programs. The short history of the farming systems method makes it difficult to judge (Butler Flora 1988; Altieri 1987: 50–58).

There are a number of unresolved questions in the farming systems approach. Does it take into sufficient account the overall context of the national and international economy? Does it allow for a critical approach to national development policies or the policies of international investors, development banks, and governments, or does it merely operate to facilitate such policies? What are the mechanisms for local participation? Who designs and operates such mechanisms? Are the researchers who carry out the work able to bridge the gaps between the concepts and languages of Western science and the concepts and languages of traditional peoples? The questions are all, in a broad sense, political.

The political nature of the questions that immediately come to mind about the farming systems approach really should help to make clear that all research on traditional systems and their incorporation into modern economies raises political issues. For example, the general principles laid out by Altieri and Jackson are different, even though they agree on the value of studying traditional systems. Jackson holds that the unifying concept of biology that he says comes from Darwinian evolutionary theory is clearly the starting point for all meaningful investigation. Altieri proposes a coevolutionary theory in which the information coming from non-Western cultures is assumed to have possible adaptive value even where its implied claims can be disproven by scientific methods. Altieri's colleague, economist Richard Norgaard, clarifies this in his assertion that "agroecology has a different epistemological basis than most of Western science." The differences in epistemology— the philosophical systems we use to decide what we know and how we know it—"could lead to an unhealthy conflict and cause a potentially fruitful new way of thinking to be 'nipped in the bud.' At the same time, the differences could lead to some exciting lines of questioning and highly productive research" (Norgaard in Altieri 1987: 27).

Lying buried not far beneath the surface of such intellectual reflections are some very complicated political issues. To explore some of them briefly, recall the *lama y bordo* system of the ancient Mixtecs and the *chinampa* agriculture of the Aztecs. The *lama y bordo* system enriched valleyland terraces of the Mixtec nobility as the nobles' serfs captured the eroded soils washed down from the steep, deforested slopes worked by the commoners. This system apparently supported a highly sophisticated culture known for its exquisite arts and ability to dominate neighboring people with a minimum of warfare, and supported it for five hundred years before the Spanish Conquest. The heritage of this system can be seen in the

Mixtec farmers who still work to stabilize eroding soils by beginning with terraces at the bottom of the slope rather than by the more common method of stabilizing the upper portions of the slope first, much to the chagrin of local soils technicians trained in agricultural schools. It is clear that the ancient Mixtec system was based on an extreme degree of class stratification and social inequality, and it is clear that the productive potential of the system was unusually vulnerable to collapse as a result of changes introduced into this system by the Spanish Conquest. Does the fact that this system existed for half a millennium or more before the Conquest and that it still survives in the minds and methods of some Mixtec farmers indicate that it had positive adaptive value worth studying and perhaps adapting to modern circumstances? As long as we know as little as we do about it, the very existence of the method over a long time makes it worth studying as a cultural artifact, but unless we discover some startling and unknown rationale for the system we can probably sympathize with the agronomists who would like to do away with it once and for all. The urgent crisis of the region does not permit a simple wait-and-see attitude that would be most convenient to the academic researcher but perhaps very expensive to the people of the region.

The *chinampa* system is a different case. It represents an unusually productive form of agriculture, perhaps the most productive per unit of land area of any known. Researchers have discovered in it a truly remarkable ecological ingenuity. *Chinamperos* make use of dozens of elements in their environment to achieve high yields with very few, if any, required imports from outside the area, although the present occasional use of some pesticides and chemical fertilizers indicates that *chinamperos* can and do introduce changes they consider desirable. *Chinampas* have survived five centuries of conquest by Spaniards who understood the system poorly, valued it little, and did a great deal to attack it directly and undermine it indirectly. It would seem that a system with such high productivity, ecological stability, and historical survival value deserves not only study but support and promotion.

Even with such admirable systems as *chinampas*, however, some cautions are appropriate. *Chinampas* do not seem to pass the test of mimickry of wild nature proposed by Wes Jackson. They require the creation of new fields out of lakes or marshland, the removal of most wild vegetation, constant management of water, and intense dedication to annual crops for human consumption. While the interplanting systems do, as Altieri and Gliessman have pointed out, mimic the natural vegetation of the replaced natural ecosystem to a consid-

erable extent, and while materials native to the environment pre-
dominate as tools and fertilizers, *chinampas* definitely demand extensive reshaping of the physical and biological environment. How
much, we might want to ask Jackson, is too much when it comes to
human manipulation? That has yet to be defined. Research into
chinampas is useful for raising and exploring that question within a
specific cultural and ecological context.

The most highly productive and extensive *chinampas* we have
known in historical times began in the lakes of the Valley of Mexico
during the flowering of the Toltec city-states. So far as we know, the
Aztecs did most to develop *chinampas* to the productive heights for
which they are famous. For all its political success and artistic mag-
nificence, Aztec society was cruelly imperialistic over neighboring
and distant peoples and was based on a rigid and brutally enforced
system of class privilege. Are the political characteristics of the
Aztec system relevant to an understanding of the subsistence base of
the society, primarily *chinampa* agriculture?

Surely so. Immediately one notices that the systems most similar
to *chinampas*—the intensive paddy agriculture of East and South-
east Asia—also arose in societies with exceptionally rigid class
structures. Is it possible that the ingenuity of such highly productive
systems is the product of tightly organized, authoritarian rule capable of enforcing greater and greater production out of peasants
who found no other way to resist the pain of such demands? Perhaps
the great monuments common to such societies—the pyramids and
palaces—are a testimony to the ability of a well-organized, con-
fident elite to extract tremendous surpluses of labor and goods out of
a peasant population, and the ecological ingenuity of such peasants a
testimony to the myriad ways different cultures have of dealing with
oppressive situations. It is also possible, of course, that the excep-
tional productivity of the systems gave rise to the possibility of leisure and labor specialization that led to highly stratified, rigid social
systems (see Geertz 1963).

Archaeologists have discovered in recent years that classical
Mayan civilization, which underwent a catastrophic and very rapid
collapse in the tenth century A.D., was fed from an agricultural
system that included raised beds on swampy ground, resembling
chinampas. Some contemporary peasants of Mayan or closely related ethnic identity practice a variant on *chinampa* agriculture
along the Mexican Gulf Coast. Enthusiastic researchers have concluded that the historical importance of *chinampa* agriculture may
have been much greater than previously thought.

Here, too, some disturbing questions come to mind. Classical Mayan civilization did, after all, undergo one of history's most spectacular collapses, for reasons archaeologists have long speculated had something to do with the interaction of rigid social structures with the ecological limits of the Mayan production system. Was the resilience of the agricultural system pushed very far by the rulers, obsessed with their pyramids and their great intellectual achievements and always demanding more of the peasants and the land the peasants worked? Could this have led to the development of an amazingly sophisticated form of agriculture, survivals of which later made their way to highland central Mexico? Perhaps such highly sophisticated systems developed under great social pressure over centuries are finally subject to unusually rapid collapse when some ecological event—drought, hurricanes, or pest outbreaks—or social disturbance pushes them a little too far. Such a view would be entirely consistent with and supportive of the most influential theory of the Mayan collapse as laid out by J. Eric S. Thompson in 1954 and revised in 1968, before the role of raised fields in Mayan agriculture was properly appreciated.

The promotion of *chinampas* by agricultural researchers in recent years has met with some disappointing results at times. A project on the Gulf Coast fell apart because of inadequate planning and, especially, because of the insensitivity of the researchers with regard to a number of issues that were crucial to peasants but not so important to the researchers. The labor required for the system made it difficult for the farmers to supplement their incomes with wage labor in town as they had been accustomed to. Inadequate research into markets led to exploitation by brokers taking advantage of the peasants' inability to transport the crops to market. Design of the canals and raised beds made access difficult at first and later cut off the flow of water, reducing the accumulation of silt and weeds needed to fertilize the gardens. In another project, the lake selected for constructing *chinampas* was too deep for proper construction of garden beds. The dominance of the researchers in the project led to an exclusion of participation and active advice from the peasant participants. While the critical report on these projects written by Mac Chapin makes a compelling case for the actual failure of the projects versus the successes claimed by project designers, Chapin blurs the distinctions between what went wrong with the technical design and what was wrong from the beginning with the processes of participation and decision making. The sour experiences by no means disprove the potential usefulness of *chinampas*, but they do caution against the possibility that great errors will be made when, in Chapin's words, "we become blinded by the beauty of a conceptual model and lose our bearings" (Chapin 1988).

I cannot emphasize too strongly that I believe that *chinampa* agriculture holds great promise for Mexico in the near and distant future, especially as a more ecologically stable, healthier way to grow vegetables intensively. But as a sophisticated agricultural system based on the constant manipulation of the environment and the use of an extraordinary amount of information, *chinampas* demand thoughtful inquiry into both the social and technical requirements of the system. It is also possible that these requirements are too strict to allow for the development of the technique far beyond its current use. The point of my historical reflections here is to suggest reasonable caution and to warn against romanticizing traditional techniques such that their application is insulated against critical thought. Ultimately, if researchers promote traditional techniques in the absence of effective critical scrutiny, as Green Revolution crops were promoted, the results will discredit rather than honor technologies of great potential value.

Reflection on the modern value of ancient technologies serves to emphasize that it is also a mistake to attempt to divorce technological research from debate and the political process. Critics of traditional techniques should not be defined as "utterly unworthy," the term used by Green Revolution researchers to define their critics (see below).

Different ways of thinking about traditional technologies also sometimes reveal very fundamental differences in the way scientists and social scientists conceive of their relationship to traditional peoples. People practicing the many varieties of traditional agriculture are not necessarily "happy peasants" who simply want to be left alone, nor are they necessarily simply "oppressed peoples" who only await the opportunity for liberation. I have tried in this book to present a picture of one traditional people, the Mixtecs, who happen to be those who suffer most from the abusive use of pesticides. This picture is not a simple one. There are Mixtecs who believe in the most ancient kind of magic embodied in "living crosses," there are other Mixtecs who continue to honor the Catholic saints and the Virgin, there are born-again Protestant Mixtecs, there are Mixtecs who believe strongly in the value of mutual help organizations, and there are Marxist-Leninist Mixtecs who hope that their own struggle as workers is part of a global project for the liberation of the world proletariat. These people are rural landless migrant workers, urban wage laborers, and peasants. There are hundreds of thousands of illiterate Mixtecs, and there are Mixtecs in highly technical professional jobs—one I know is an expert on the disposal of urban toxic wastes, and another is a graduate of Harvard Law School who founded and directs a Spanish-language radio station in the San Joaquin Valley notable for its close coverage of the problems of pesticide poisonings experienced by California farm workers.

As peasants, Mixtecs have so far been studied mostly as a disaster case, although Gene Wilken has found that many Mixtecs are highly sophisticated farmers who conserve soils, water, and energy very successfully under very difficult circumstances. So far as we know, the world's most important grain crop—maize—was first domesticated in the Tehuacán Valley of the Mixteca, quite likely by the ancestors of these modern people who today harvest our tomatoes in Culiacán, Baja, and the San Joaquin Valley. Nonetheless, the Mixtecs, like most traditional peoples, continue to be treated as helpless peasants, cheap labor, and objects of study and concern. They are only occasionally seen as a dynamic people with their own strategies of adaptation, their own diverse beliefs and desires.

When agricultural researchers explore the value of traditional technologies, it is vital that they keep in mind that the people who are keeping these technologies alive are often involved in a process of sweeping historical change, and that traditional peoples themselves are doing a great deal to shape the outcome of that change. The process of adaptation, social and technological, does not present a static array of "cultural characteristics" but a moving, diverse, and willful group of people who can act, speak, and innovate. As fellow human beings or as scholarly researchers or agricultural technologists, it is vital to remember that traditional peoples and Westernized people are involved in historical change together.

In a deeper sense, there are not "traditional technologies" and "nontraditional technologies" but various traditions of technological development and change. The one we call "modern" has a long history in Europe and the United States. It is taken as definitive of present, "modern" reality because it is part of the dominant world trend of commercial, capitalist development over the last several hundred years and it has displaced others. But during this period, other traditions, including the Mexican peasant agricultural technologies, have been changing and developing and adapting to present-day circumstances insofar as possible, just as the traditions of Europe and the United States have been changing. We may make a serious error if we try to appropriate "traditional" technologies into our own technological realm without recognizing the integrity of the peoples and cultures who embody distinct traditions. Part of such a recognition is an understanding that other traditions, like our own, are dynamic.

Traditional technologies of the kind of greatest interest to researchers for the most part are not museum pieces awaiting resurrection by agricultural scientists. As Wilken has so convincingly shown, they are actively and skillfully practiced by tens of thousands of peasant farmers in Mexico and Central America, in spite of the tendency of agricultural modernization efforts to undermine them. And, as Wilken also emphasizes, these technologies are not static. Peasants using them constantly change their practices in order to exploit new opportunities or accommodate to new pressures.

The most important agricultural researchers involved in developing traditional technologies are not found on the staffs of universities, research institutes, or field stations—they are instead to be found working in the fields their ancestors cleared generations or centuries ago. Exploration into the value of traditional technologies belongs on the agenda of research institutions. But more important, respect for the value of the labor and knowledge of traditional agriculturalists belongs on the agenda of governments. Only by promoting agricultural policies consistent with the survival of traditional farmers can valuable technologies used by them really flourish. Fortunately, such traditional farmers have shown that many of them can achieve precisely the kind of high productivity with low resource and financial expenditure that the economies of many modern nations desperately need.

It is important that scientists and technicians investigate the value of traditional technologies, but it is more important, from both a scientific and an economic perspective, that the traditional peoples be protected from thoughtless ruin by promotion of technologies that cannot endure. It is also vital that the chronic crisis of traditional regions trying to survive in an economy hostile to them not be simply exploited for their ability to produce cheap labor for areas like the vegetable fields of the Culiacán Valley, founded on an unsustainable and damaging technology that deepens class and regional distinctions.

Should Agriculture Serve as the Instrument for Industrialization?

There are few nations in the world that will, or can, seriously contemplate a future without extensive industrialization within their own economies. Predominantly agricultural economies without any significant industrialization are simply too weak and vulnerable in the face of industrialized nations. Even if a nation wants to rely primarily on agriculture for its income, it will be able to greatly increase the value of its agricultural products to the extent that it can do the processing and packaging now universally expected for many products bought by urban consumers. The local manufacture of agricultural tools and implements can insulate local farmers from exploitive marketing practices of international oligopolies or monopolies. And most nations want to go beyond that, to a manufacturing sector providing a share of the large array of goods people want and expect, and exploiting any special advantages of resources or labor the country may have to export and earn the hard currencies necessary to participate in the international economy.

All of that taken for granted, the role of agriculture in promoting industrialization can and does vary widely. The policy followed by Mexico sought to squeeze farmers very hard to gain savings that could be invested in industry, as we have seen. Within agriculture, Mexico's policies deliberately favored ways of farming that required the chemicals, machines, and fuels provided by industry as a way of encouraging industrial growth. Dozens of nations have followed similar policies, from the Soviet Union to Brazil, from Taiwan to Ethiopia. While the results are mixed, some fairly clear choices have been defined by these experiences.

The worst experiences with using agriculture as a basis for industrialization have occurred in Africa. Newly independent Africans hoped that industrialization would make them economically as well as politically independent. Instead, the failed attempts at industrialization have mostly created economic bankruptcy, with ailing agriculture, little industrialization, and heavy environmental damage. The nations stretched along the southern border of the Sahara desert are a distressing example.

Livestock-herding peoples of the Sahel, the semiarid borderland between the southern Sahara and the humid lands farther south, had worked out grazing patterns adapted to the climate over many centuries. They spent the dry season at the southern end of their range or along major rivers with peasant farmers, bartering meat, hides, and cheese for grazing rights, grain, and hay. As the rains moved northward, they followed the greening grass. Reaching the far northern part of their range, they returned south through pastureland that had recovered from the passing herds earlier in the season. With the end of the wet season, they and their cattle were back again with the peasant farmers.

The colonial period disrupted these patterns in a number of ways.

A vastly increased demand for slaves by the Europeans set tribe against tribe. Arbitrary boundaries laid across the pasturelands by competing European powers upset the evolved logic of the migration routes. Colonial powers enforced taxes that required herders to run larger herds, leading to overgrazing. European business interests did all they could to encourage cash crops to displace the barter economies of peasants and herders.

As independent states displaced the colonial empires in the 1950s and 1960s, the new governments moved to encourage industrialization. They set food prices favorable to urban settlers but devastating to farmers and herders. They found financial help from development banks and European, American, and Soviet aid programs to build dams and irrigation schemes. The irrigation water would encourage commercial farming, and the dams would provide energy for minerals processing and modern farming technologies and to industrial enterprises associated with modernized mines and farms.

The herders found themselves excluded from the best land as it was turned over to modern farming, and the barter economy with peasants was further disrupted. Taxes increased to help finance the new development schemes, but governments did their best to hold down traditional food prices. Herders put more and more cattle on the land to try to compensate for their losses, leading to further overgrazing. Governments, with the advice of Western experts, replaced the watering sites that had been taken from the herders with wells, but by now the situation was so desperate that herds of livestock crowded around the wells and ruined the surrounding grasslands. Cattle died after drinking their fill, because the journey outward from the wells to fresh grass was too long. The extreme pressures created by this situation exacerbated old tensions between tribes and regions, and rebellions, civil strife, and, in some cases, decades-long wars made the situation much worse. Increasing desperation led to increased grazing on lands that remained available, and the Sahara desert advanced southward (Swift 1973; Ware 1975; Sterling 1974).

In the Sahel case, the decision to make agriculture serve as an instrument for industrialization was not the original cause of the troubles. But it was a misconceived solution that made the problem worse rather than, as governments hoped, resolving the difficulties established by the colonial heritage. The advantages of the new industrial enterprises and modernized agricultural regions were overwhelmed by national crises and the drastic loss of previously productive land, perhaps irreversibly. Many climatologists believe that the damage was so generalized that it changed weather patterns

for the worse over much of central Africa, and there is wide agreement that much of the grazing land is beyond recovery without huge
new expenditures. By attempting to industrialize on an agricultural
base, African countries dealt a fatal blow to much of their agriculture and gained little in modernized agriculture or industry.

In the Soviet Union, the squeeze placed on agriculture succeeded
in helping to build an industrial economy. Prices and credit policies
kept life bearable for urban workers who had little to spend on food.
An emphasis on machine and chemical dependence in agriculture
gave incentive for industrial growth. But planners treated farming
as they would an industrial enterprise. The regime saw traditional
knowledge and culture as largely reactionary, and industrial models
for farm management prevailed. Organizational and managerial failures in Soviet farms became routine. Planners, frustrated by lag-
ging farm production, pushed such massive frontier development
schemes as the "Arid Lands Project" of the 1960s, and they rushed to
emulate the farm technology of the United States during the same
period. Ironically, they succeeded in emulating a piece of American
farm history, creating a huge Soviet Dust Bowl with tremendous
economic loss and environmental damage. Just as important, there
was little incentive to continue to produce food when crop prices
were held so low, and the farm economy languished. The former
minister of agriculture, Mikhail Gorbachev, is attempting to rem-
edy this situation. The government is allowing food prices to rise
rapidly, and it is allowing farmers much more freedom to respond to
demand where it is found and to make their own farm management
decisions. Gorbachev has apparently concluded, as have many internal and external critics, that the industrial economy cannot move
forward successfully without a healthier farm economy capable of
reliably providing the food and fiber the cities and factories must
have to function smoothly. In the Soviet Union, the decision to
make agriculture serve the purposes of industrialization was not
nearly so disastrous as in many African nations, but it has proved
very troublesome (Eckholm 1976; Medvedev 1987; Gorbachev 1987:
96–97).

Taiwan presents yet a different kind of case. The Japanese won
Taiwan as a prize of war in 1896 and soon incorporated it into their
plans for the industrialization of Japan. They established a well-
financed agricultural research program in Taiwan. The Japanese tax
system captured a large share of the farmers' earnings. By the 1920s,
they had doubled the production on Taiwanese farms and doubled
the rate of taxation, while providing substantial exports to feed Japa-

nese industrial workers. When the Nationalist Chinese regime fled mainland China to Taiwan, they were able to build on the productive base established by the Japanese.

The Nationalist general, Chen Cheng, upon becoming governor of Taiwan under the new regime enacted a serious land reform program within three months of assuming office. In less than four years, the regime decreed a more ambitious "land-to-the-tiller" reform. The defeat of the Nationalists by the Communists on the mainland was the most important influence on both the Nationalist government and its sponsor, the U.S. government. It was obvious that land reform was essential to the political and economic stability of the regime. Within a few years of the reform, researchers found that Taiwanese agriculture was as productive as that in any but three other nations, one of which was Japan, just coming out its own reform pushed through by the American occupation administration under Douglas MacArthur.

The Taiwanese were determined, however, to make agriculture serve the purposes of industrialization, as elsewhere. But the security and relative degree of equality among the majority of peasant farmers under the reform laws eased the pain of credit, price, and tax squeezes for industrialization. In addition, Taiwan used the draft and recruitment into one of the world's largest standing armies to relieve unemployment during the first two decades of the reform and industrialization process. Crucial to their ability to do so, and to so many other elements of their success, was the foreign assistance program of the United States—during the 1950s, Taiwan sometimes received a fifth of the total U.S. foreign assistance budget. The political commitment of the United States to the Nationalist regime made for stability and secure investments, further aided by the trading policies of the United States, deliberately keeping U.S. markets open to Taiwan's goods. The hard-handed dictatorship of the Nationalists allowed for the close coordination of agricultural and industrial policies without effective complaints about the unequal terms of the peasants' relationship to the urban economy. The security of landholdings and rural incomes and the availability of many basic health and social services encouraged poor people to have small families, giving Taiwan one of the earliest successful family planning programs, keeping the high economic growth rates from being eroded by overly rapid population growth (Tai 1974; King 1977).

Taiwan is an example of economically successful use of agriculture as a base for industrialization. But it is an example whose success is dependent on a large number of factors that could be duplicated in only a few nations and that many nations would reject if given the choice. The strong financial and political support of the U.S. government in the wake of a massive political defeat in China is a circumstance that has not been duplicated outside of a few East Asian nations strongly influenced by the same experience. In any case, it shows that if the intent is to successfully drain agricultural earnings into industrialization without creating serious bottlenecks to continued growth, thoroughgoing land reform is best seen as a requirement, not an obstacle.

It has to be admitted from the global experience of the last century that peasants will almost always have to accommodate themselves to the demands made by others in the drive for industrialization. Their interest in free, autonomous communities that Eric Wolf and others cite as central to peasant ideals will be necessarily compromised. But the terms of the compromise are very diverse and subject to negotiation.

The first point is the obvious one of reasonable distribution of land, access to credit and markets, and security of land tenure. Without resolving this issue, the classic problems of colonial economies will prevail—high concentration of earnings and the flight of capital to other economies, poor conservation of natural resources, dependency on other economies for basic food imports, and poorly developed internal markets for both agricultural and industrial goods. If the basic issue of land reform is reasonably resolved, agriculture can contribute a great deal to industrial growth without destroying the nation's resource base, its rural culture, or its own economic potential. Also at issue is whether the knowledge and experience of peasant producers is to be honored. While every agricultural researcher knows a great deal about agriculture, no one knows as much about the specific problems and possibilities of particular pieces of ground as peasant farmers. To ignore or eliminate this knowledge is like giving away the nation's minerals to encourage imports of metal tools or chopping down all the nation's forests to make it easier to see where roads should go. A basic resource is squandered as though it has no value. And because this resource is contained within the culture and character of a significant portion of the nation's people, the waste of knowledge is an indignity that erodes confidence and good will. Any tendency to increased social distance due to economic policies will be strengthened as peasants must bear contempt on top of injustice.

Another issue at stake is the control of crop choice and the degree of crop diversity. In David Ronfeldt's excellent study of a town in Morelos called Atencingo, we see a case that has been repeated with some variations throughout Mexico. Land reform in Atencingo gave land from a sugar estate to peasant-controlled *ejidos*. Most of the peasants chose to work this land in small plots, growing a wide variety of grains and vegetable crops for subsistence needs and nearby urban markets. The diversity gave them protection against crop failure due to weather or pest outbreaks and against poor prices for single crops. It also gave the peasant families a good diet. Soil enhancement and protection from crop pests were easier to achieve with diversity. Peasant farmers knew and understood local markets and were capable of making good decisions about what to grow and when and where to market to obtain good prices, in sharp contrast to their total lack of control in the national and international sugar market.

Unfortunately, the former owner of the estate, a North American named Jenkins, accommodated to the expropriation of the land but wanted the peasants to continue to grow sugar in order to protect his investment in the local sugar mill. He from time to time prevailed upon the Mexican government to see things his way, arguing that the nation needed to boost exports and that the peasant division of the land was inefficient. Eventually the government itself, after nationalizing the sugar industry, made the same arguments. Peasants in Atencingo have fought for decades for the right to a balanced, diverse agriculture consistent with the needs of their families and communities. The government has vacillated, stirred up internal divisions in the community, and bolstered the power of local political bosses willing to represent the government's point of view. As a result, there is neither healthy, diverse peasant agriculture nor a vigorous sugar plantation. Had the government been willing to grant the wishes of the majority of the peasants, it would have successfully enlisted their energies and enthusiasm in strengthening the domestic Mexican market on both the supply and demand sides. They could have had rural prosperity and food for urban people with relatively high, year-round employment. Instead, they achieved a depressed, divided community with low productivity and seasonal employment. The government chose the logic of monocrop and export dependency over the logic of diversified crops and strong rural communities (Ronfeldt 1973).

If the drive for industrialization creates severe regional inequalities, more damage is done. This is especially clear in the case that is the centerpiece of this book—the brutal exploitation of the poverty and environmental ruin of one region to promote reckless, environmentally destructive growth in another—the relationship that ties Culiacán to the Mixteca. The problems of the poorer region deepen, and the very cheapness and abundance of labor flowing out of the

poor region encourage a cavalier attitude toward the health and loyalty of workers. Because workers are abundant and desperate, they
are considered expendable. And because the agrarian reform process
has been so corrupted in the richer region, land eventually becomes
expendable, too. A government network of direct and indirect subsidies to the wealthy farmers of the developing regions and production of crops for a foreign market that does not depend on the pur-
chasing power of local people complete the picture. The wealthy
farmers can make profits high enough to guarantee their ability to
diversify their holdings beyond the land that produces the profits,
and guarantees that they can do so rapidly. In such conditions, such
local problems in the region as pollution of water and soil, soil salin-
ity, and the buildup of pest resistance can be largely ignored—the
trick is to make profits quickly and run. Another region will then
fall into decline, as has happened in one agricultural frontier after
another in Mexico and elsewhere. This strategy certainly can aid in
directing new investment into industry, but at a high and unnecessary cost. To the extent that it makes it more difficult to create a
healthy rural economy and hard to feed urban workers, the strategy,
as we have seen, helps to defeat the drive for industrialization.

Agricultural development can certainly aid in the drive for industrialization, but the lesson repeated by dozens of nations and scores
of scholarly studies is that it can only do so when requirements internal to the needs of agriculture are met. Family security and strong
rural communities keep the exodus to the city from swamping improvements in economic growth with excessive social costs. Most
countries will be better off by avoiding heavy dependency on foreign
markets and by providing a diverse, balanced diet for the population
out of local resources.

Policies that depend on environmentally destructive technologies
are self-defeating, certainly in the long run and often in the short
run. Degraded land and water and uncontrollable pests undermine
future growth possibilities. When growth strategies poison nature
and sizable numbers of the people of a country, the confusion between ends and means becomes hopelessly distorted. Isn't the purpose of growth to bring health rather than disease?

When growth leads the way to stagnation, dependency, squander-
ing of resources, and increasing inequality, there is no reason for the
nation to continue to pursue growth. Under such circumstances,
only dominating elites benefiting from their intermediary role in negotiating the terms of further dependence and from the growing social inequalities have an interest in maintaining the existing policies and political relationships. They will find themselves, as does

the ruling party of Mexico, in an increasingly defensive and dangerous position, relying more and more on their foreign supporters.
Their foreign backers will increasingly have to answer questions of a
type they have had to deal with in Vietnam and Central America,
with results that have pleased no one.

At present, Mexico has clear choices, as represented by the candidates in the 1988 presidential election. It can continue with the
present policies under the ruling party. Or, Mexico could choose to
emphasize even more strongly the export-oriented, highly concentrated economic growth favored by the right, with a much closer integration with the United States, politically and economically. Or,
Mexicans could decide to turn to reducing glaring inequalities and
turn the productive capacity of the nation more directly to the task
of feeding, clothing, and housing the people of Mexico.

To choose this last path would be difficult and expensive, particularly because it would earn the immediate hostility of Mexico's private and international creditors, who clearly want some combination of the present policies with those proposed by the right. Among
these policies are those dedicated to squeezing the majority of the
nation's poor, and especially farmers, to help finance further growth
in industry, mineral exports, and agricultural exports. Historically,
Latin American nations that have attempted to challenge their creditors and remove the creditors' local allies from power—Guatemala,
Chile, the Dominican Republic, Nicaragua—have faced coups or invasions backed by the U.S. government. Only Cuba has survived
such pressures for a long time, and at the cost of strong dependence
on another superpower that exacts its own price. Nicaragua's experiment is now being severely tested. Many Mexicans are looking to
the experience of Nicaragua, whose population is that of a single
Mexico City suburb, for inspiration and guidance. But Mexico will
have to make its own decision, a decision that could affect the United
States far more powerfully than anything else that has happened or
is likely to happen in Latin America. The citizens of the United
States will bear the responsibility for the reaction of their government to the inevitably disturbing results of any path Mexico
chooses.

Is Agriculture Vitally Linked to Wild Nature?

When Avila Camacho announced new policy directions to emphasize productivity over distribution and industrial growth over
the needs of agriculture itself, no particular thought was given to the

changes such policies might introduce into the natural environment of Mexico. When agricultural researchers experimented on new crop varieties and when extension workers and chemical salesmen promoted pesticides and fertilizers, they did not worry much about the relationship of crops to the wider environment.

Part of the charm of chemical dependence is the fantasy that it grants the chemical user a high degree of control over the plants and animals in the field. Secondary pest outbreaks, increasing pesticide resistance among pests, the disappearance of diseases, parasites, and predators that formerly controlled pests, and the spiraling need for more and more chemicals eventually interrupt the fantasy. The public health and environmental effects and the high financial costs make it clear that the fantasy is very expensive. Unfortunately, the result is not always sober reassessment. Like addicts, people are often inclined to simply spin out new fantasies, in this case based on the illusion of ultimate human power over nature. Many agricultural researchers are now working to elaborate such new fantasies, while others seek to establish agriculture more firmly within the limits set by nature.

Outside of a few science fiction writers and overenthusiastic promoters of the odd scheme to feed humanity from hydroponic greenhouses, no one openly defends the proposition that agriculture has no vital link with wild nature. But agricultural researchers and economic planners, judging from their behavior, sometimes seem to forget agriculture's dependency on the physical and biological environment.

A crop seed contains a vast amount of information encoded into a genetic message that can be passed on to the next generation of plants. This information derives from the wild ancestors of the plant, with millions of years of adaptive experience in the environment. The information also comes from the experience of the plant's ancestors as a human crop—for most of the major crops, six to ten thousand years of experience. The seed's ancestors have survived droughts, floods, frosts, molds, mites, insects, birds, mammals, and innumerable combinations of circumstances of every sort. Humans have selected seeds from the plant at some point, cross-bred them with other plants, deliberately and accidentally, and nourished and protected the seed as far as possible. In this long history, far more seed varieties have been lost to natural circumstance or human choice than have survived to still serve as human crops. The information in seeds is very complex and strongly reflects the story of human interaction with nature over millennia.

Each seed contains a kind of survival formula that is adapted both to nature and to human needs and preferences. Human cultures, in turn, are partially adapted to the survival needs of the crops we cultivate. For thousands of years, people have been breeding plants consciously and unconsciously. Until the relatively recent past, plant breeding was in most cultures an activity carried on primarily by millions of individual farmers on the lookout for especially good seeds and for opportunities to breed desirable qualities into the seeds they encountered. The process by which some were selected and others rejected was rather precisely as complex as the interaction of the whole culture with all of nature in the region.

More recently, and especially in the twentieth century, plant breeding has become a profession. Like most professions, it has established its special rights and privileges on the basis of some demonstrated ability to do things faster and better than they could be done by others. Plant breeders all over the world have reshaped human crops with an unprecedented rapidity, adapting them to the requirements of farmers, governments, or corporations. Plant breeders have been especially helpful in the project of adapting crops of interest to European peoples in their growing dominance over environments foreign and sometimes hostile to European seedstocks. The influence of rapidly growing, specialized global commercial markets has been especially strong, moving plant breeders to emphasize the high yields that make participation in such markets profitable. But whenever plant breeders strongly select for one quality, they will deemphasize or eliminate other adaptive qualities of the seeds. The drive for more productive seeds has created, for the most part, seeds more vulnerable to failure.

Plant breeders responded to the rapidly changing needs of an increasingly commercialized and universalized world agricultural economy. But plant breeding can be painfully slow—each new crop variety must demonstrate over generations its ability to survive and prosper, first under experimental conditions and then under the actual conditions of farmers' fields and human society. The advent of a whole new range of chemical products to fertilize crops and protect them from pests provided a way to respond to the quickening pressures of the world economy and world politics. Problems with old varieties could be quickly solved with chemicals, as could anticipated problems with new varieties. Soon, it became obvious that new varieties could be designed specifically to be part of what Green Revolution researchers call a "production package." The seeds would be specifically adapted to the use of the chemicals.

Not all agricultural researchers are plant breeders. Some specialize in various aspects of the manipulation of crops in the field.

Irrigation experts, inventors of agricultural machinery, experts in crop nutrition and disease, agricultural economists, and other specialists soon followed the lead of the plant breeders in using chemicals to solve old problems and, especially, to redesign crops and cropping systems to adapt to the use of chemicals.

Agricultural policy makers followed in the same path. A key change made possible by pesticides in particular was the ability to grow pest-vulnerable crops on vast acreages under highly productive humid tropical and subtropical conditions. Previously, large acreages of such crops as cotton and tomatoes could be grown only in the arid or semiarid tropics, but with chemicals it appeared to be easy to extend monocrop export agriculture to whole new frontier regions. Forests fell to the ax, and agribusiness moved in. The same chemicals used to protect crops seemingly would provide protection against malaria and other epidemic disease that flourished when the forest was cut and insects came down out of the canopy to prey on humans, breeding in the stagnant pools of water left in the wake of deforestation (Desowitz 1987: 46–58).

Governments, investors, and researchers transformed agriculture into an activity newly dependent on chemicals. In terms of natural evolution of organisms and ecosystems and in terms even of the traditionally long periods required for the evolution of human crops, this adaptation to chemicals came very fast—a matter of years or a few decades. We now know a large number of problems associated with this abrupt change. We know that the severe changes introduced into the populations of crops, their pests, and the organisms that ordinarily would keep pests in check have gotten out of human control. We lose a slightly larger share of human crops to pests than we did before chemical dependence (30 to 40 percent). Although the total crop from chemical-dependent world crops is larger than it was without chemicals, it is clear that chemical dependence is only one way to higher productivity, and a costly and dangerous one. We have seen the spiral of increasing chemical dependence that occurs when no sharp break is made with largely chemical-dependent methods, digging us in even deeper.

Research now indicates that with little or no reliance on synthetic agricultural chemicals, farmers in and out of the tropics can attain high levels of productivity without excessive pest losses and with improved soils and resource protection. It appears from the existing research that many farmers can raise their profits in the process. The combination of traditional techniques and new research can lead the way out of chemical dependence.

A recently completed report from the National Academy of Sciences concludes that the chemical dependency of agriculture and large increases in demand for irrigation water have arisen in response to research, trade, and commodity policies that have placed production of the greatest volume of food ahead of environmental, social, or rural economic concerns. Commodity, tax, and economic policies have favored capital over management as a means to increase production. The objective has been to increase total production of food and feed by reducing and simplifying biological interactions. In contrast, the report found, alternative systems cultivate and take advantage of the biological and natural resources available on the farm.

The NAS Board of Agriculture researchers found that many farmers were beginning to swim against the stream of national farm policies—a growing number of farmers in all regions of the United States are producing most crops using methods that involve little or no use of synthetic chemicals. These farms are more diversified and take advantage of local, on-farm resources. The report notes that the "scientific basis for some of these practices and systems is not understood, but they work."

Perhaps most surprising, the report found that successful alternative farms are almost always profitable with little or no government support. In many cases, the use of alternate methods results in higher yields. In most cases, per unit production costs decline a great deal, about $80 per acre for Midwestern corn and soybean producers, but for fruit and vegetable producers, costs do not show such decline. The report concludes that with adequate government support of farming systems research and reform of commodity program policies, farmers and society will reap the benefits of alternative systems. Benefits will come in the form of far less environmental degradation, more consistent farmer income and production levels, and lower government program costs (National Academy of Sciences 1989).

We must give up some things in the process. Most notably, we have to dedicate agriculture to crop diversity and severely reduce our reliance on monocrops. It will mean that the production of such winter luxury crops in a few regions like Culiacán and Florida must be sharply cut back and the remaining production scattered across more regions and larger areas to allow for diversity. The luxury of winter tomatoes may come at a higher price, and consumers can choose to pay it or return to dietary patterns more adjusted to the season, patterns universal less than a generation ago. We will have to settle for somewhat lowered world production of crops grown as monocrops that are pest vulnerable—cotton especially. But these crops have long bedeviled farmers with low prices because they are produced around the world in superabundance, with dozens of governments dedicated to bailing out farmers with expensive price support programs for these crops. Some crops, particularly rice, that have been adapted to chemical dependence present problems more difficult than others because of their importance in the diets of so many poor people and because traditional rice varieties and technologies have been lost. It is reasonable to surmise, however, that synthetic pesticide use can be reduced sharply worldwide in the next decade or two. As in Indonesia in the last few years, governments of rice-producing nations are moving to reduce pesticide dependence because of the troubling effects of pest resistance and secondary pest outbreaks, as well as more specific environmental and human health dangers. With generalized reduction in pesticide use, specific worries over pesticide dangers can be limited to very localized and particular issues, like ordinary occupational health or community sanitation problems, rather than as a broad problem of human health and generalized environmental decay.

Not everyone is happy with this prospect, however. At the same time that executives in large pesticide-manufacturing companies announce that they themselves see the end of the pesticide age ahead, they herald new technologies that promise to repeat the same old errors. In particular, they hope that genetic engineering can shortcut the processes of adapting human crops to nature and to human culture even more rapidly than pesticides did.

Razzle-dazzle press agentry aside, most genetic engineering projects proposed for agriculture are simply ways to do what classical plant and animal breeders have done in the past. The big attraction is speed and economy. New varieties and species can be produced with great rapidity and put on the market. The ability of genetic crosses to reproduce successfully over generations is not always an issue—the greater the dependence on laboratory manufacture of the organism, the more profitable the sales.

Theoretically, biotechnologists could design plants that were resistant to pests and thus reduce pesticide use. So far, however, most bioengineering in agriculture, chemical company propaganda to the contrary, has been aimed at producing crops resistant to increased herbicide applications or creating predators on crop pests that are themselves resistant to insecticides. A much-heralded project to produce "Ice-Minus," a bacteria meant to retard frost formation on strawberries, failed. Experiments revealed that while the genetically altered bacteria did retard frost formation, the naturally occurring bacteria from which the genetic material was taken for the new strain did a better job. (For two critical examinations with different perspectives, see *Development: Journal of the Society for International Development* 4 [1987] and *Development Dialogue* 1–2 [1988], edited by Cary Fowler, Eva Lachkovics, Pat Mooney, and Hope Shand.)

There is a reason that bioengineering techniques are proving slow in producing the miracles promised from them. Bioengineering removes the researcher and promoter farther from the limitations imposed by wild nature and the special conditions of the agricultural field, farther than they were carried by synthetic pesticides. The distance between flights of human fantasy and the real complexities of ecosystems becomes ever greater. Classical breeders learned relatively quickly if a new variety was at all viable in the environment, because breeding techniques required the reproduction of successive generations of living plants or animals in the field. Ultimately, genetic engineers must make their creations pass through the same test, but it is harder for them to anticipate whether designs will succeed, because the question of viability in nature does not derive from the long history of the organisms used.

As geneticist and plant ecologist Wes Jackson writes,

> My opposition to many of the grand claims of biotechnology is not because they are new, but because they are painfully old. They are part of the recent acceleration of the fall that came with the industrialization and chemicalization of agriculture. The ancient split between humans and nature widened during this epoch. . . .
>
> There is one new possibility in the biotechnology grab bag—our ability to transplant genes from one organism to another remotely related creature. Little else in that bag is really new. Cloning is old and easy, as anyone who had had anything to do with a strawberry patch knows. Tissue-culture techniques allow us to cut corners considerably, but we should remember that cloning is simply monoculture in spades. (Jackson 1987: 20)

Many critics have pointed out that the almost certain result of biotechnology is the further narrowing of our genetic base in living, adapting crops. So we narrow the adaptability of crops at the same time we introduce new varieties that break sharply with old, proven ones.

This is not the place to make the whole, complicated case against embracing biotechnology as the solution to the current dilemmas of agriculture. Other authors have excellent work concentrating on this theme. It is also important to recognize that genetic engineering will almost certainly make a variety of valuable and enduring contributions to stable, productive agriculture. But it must be kept in perspective as a tool among many tools, not as a sweeping solution to the agricultural dilemma.

What we can see more clearly from the story of agricultural modernization in Mexico is that biotechnology promoters routinely repeat the same kind of intellectual and political errors that were made in designing Mexican agriculture for pesticide dependence. They assume that the technology is neutral in its social and ecological effect, they ignore the wisdom of traditional approaches to farming, they make agriculture more dependent on the industrial sector, they minimize or deny the vital link of crop plants with the environment, and they mask the drive for profits in the mistaken idea that low productivity is the basic cause of hunger and poverty. Finally, they ignore other available solutions that tend to correct these errors rather than repeat them and that do not rely on technological miracles.

The insistent use of the word "miracle" in agricultural modernization campaigns is extremely revealing. Science originally won its intellectual prestige through the rejection of the idea of miracles. But publicists and grant seekers with commercial and political motives find the rules of scientific reasoning inconvenient. Everyone must remind themselves that belief in miracles in the natural world is comforting but dangerous. Agriculture, as a human activity deeply rooted in the natural world, must survive within the constraints of natural laws and the requirements established by human culture.

Biotechnology promoters intend to take agriculture farther from the tests applied by the conditions of the field and of broader nature. They want to remove agricultural research and the reproduction of crops even farther from the wisdom of practicing farmers and the slow processes of adaptation through natural and cultural evolution. We need instead to move in the opposite direction, toward the readaptation of agriculture to the complexity of nature and the requirements of healthy human beings and healthy human communities.

Are Technologies Neutral?

It is obvious that agricultural modernization had profound effects on Mexican agriculture, Mexican rural life, and the rates and patterns of growth of Mexico's cities. The balance of wealth and power in Mexico was profoundly affected by the country's program of agricultural modernization. As we have seen, it was conceived in a highly political context and specifically addressed itself to issues of political alliances, consolidation of power, and the relationship of Mexico to the United States. Many of the agricultural scientists and technicians who took part in the design of modernization may not have been fully aware of all the political implications. But their own writing makes it clear that they had some understanding of the role of the program with respect to the distribution of wealth and power, and they knew that their vision competed with the views of others.

The Green Revolution researchers, in recounting their own history, wrote in 1967, "Such data as are available in Mexico indicate that the increased wealth produced by the improvement of agriculture in the past 20 years has gone largely to the upper income groups. Even though the minimum wage was legally increased . . . the level of real wages (corrected for inflation) for the country as a whole is still below that of 1939, and wages on farms are usually lower than in cities" (Stakman, Bradfield, and Manglesdorf 1967: 214–215). They go on to quote Ramón Beteta, once secretary of the treasury, who wrote, "We industrialists are treating farmers as an underdeveloped nation. We sell high to them and buy cheap from them and the inevitable result is, in the long run, that they cannot acquire the products turned out by industry." As discussed in previous chapters, the tendencies noted here by the Green Revolution modernizers themselves have been thoroughly documented by one scholar after another and by government studies undertaken at various times over the decades. Rich farmers prevailed over poor farmers, and urban interests prevailed over rural—it is no accident that Mexican cities have, on average, doubled in size every ten years since World War II.

But most remarkable is that after citing official concern in Mexico over these results of the agricultural modernization program, the Green Revolution researchers end with the pious wish that the benefits of the research program should be more equitably distributed. "Research and extension have contributed much to Mexico, but they still have a big job to do, contributing still more to many more Mexicans" (Stakman, Bradfield, and Manglesdorf 1967: 215). Did these men appreciate their own sleight of hand? Do they see that the inequitable distribution and the starving of agriculture for the sake of industrial growth was a result, number one, of the high financial cost of the technologies they had created; number two, of the deliberate dependency of the new technologies on industrial inputs in order to stimulate industrial growth; and, number three, of the political program promoting growth rather than redistribution in which their research had played an absolutely key and premeditated role?

When President Avila Camacho had announced in 1941, with Vice-President Wallace standing at his side, that henceforth redistribution of land would be de-emphasized in favor of growth in productivity for the sake of rapid industrialization, and when Wallace then immediately called upon the agricultural scientists to formulate a research program consistent with the direction of the new Mexican government and its American allies, they seemed to know what they were doing. And he and the Rockefeller Foundation got the research they envisioned and financed. This is politics, politics ably designed to shape technology and society together toward particular purposes.

Public relations campaigns, especially of the aggressive and grandiose kind used to promote chemical-based agriculture since 1945, are meant to sell something. When such advertising techniques are aimed not just at the actual purchasers of the seeds, chemicals, and equipment, we can be sure that they are selling not just a commercial product but that they are meant to create a general political climate and specific public policies favorable to the sale of the merchandise. And indeed, much of the literature arguing for agricultural modernization is quite candid about this—it is considered absolutely necessary that the governments of countries switching over to chemical-based agriculture be fully committed to the change. As the architects of the Green Revolution wrote in *Campaigns against Hunger*, "Cooperation is essential within each mission and between the missions and the countries in which they operate. The primary obligation for insuring cooperation rests with the members of the mission, for supposedly they are the teachers and the nationals are the learners." But what is to be done if cooperation is not forthcoming?

"When persuasion fails, it is time to use compulsion." Hunger and ignorance and disease are so tragic that anyone who obstructs or delays their alleviation commits a gross misdemeanor against those who suffer. . . . A man who plays a lone hand may be useful where cooperation is not essential, but he is intolerable in a cooperative program and should be eliminated from the mission unless he has only a limited technologic task to perform.

Institutions too must cooperate. But many countries are so highly and capriciously institutionalized that it is difficult to learn what all of the institutions and their functions are, and still more difficult to get them working cooperatively. . . .

But effective cooperation cannot really be compelled, nor can it be exorcised at pleasure. The ideal solution would be to convince everyone that the cause of fighting hunger and igno-

rance is so worthy that it is utterly unworthy to hinder the
cause. (Stakman, Bradfield, and Manglesdorf 1967: 316)

It is precisely this spirit that guided the programs of agricultural
modernization carried out in country after country—the programs
of persuasion were not simply aimed at showing the advantages of a
given technology but at identifying anyone who opposed the programs as "utterly unworthy." This was true internally, where dis-
senters who complained too loudly of the social or environmental
consequences were eliminated from the research and education programs, and externally, where alliances with elites, political parties,
and governments facilitated the elimination of political opposition.
(Jennings 1988 gives a detailed account of this process, especially
with regard to the treatment given to researchers and professors who
questioned the goals and methods of the Green Revolution program.)
    What was wrong here was not the creation of an ideological and
political program that sought to prevail over opposition. Such is the
nature of political life. The problem arises when people try to hide
behind the mask of scientific or technological neutrality to disguise
the "persuasion" and "compulsion" involved in promoting given
technologies, and when others are taken in by such claims.
    The principle that the technologies must be examined for their
social and environmental consequences is now widely recognized,
in law and practice as well as in theory. Laws such as the U.S. National Environmental Policy Act require environmental impact statements for many kinds of projects. Court cases have led to the extension of the NEPA principle to the funding or promotion of specific
technological programs even where specific projects have not been
identified. The U.S. Congress has required the Agency for International Development to submit environmental impact reports for
projects, and the World Bank also has a process for evaluating environmental effects of projects. Congress also established the Office of
Technology Assessment, with the realization that whole technologies, as well as particular projects, must be better understood in
order to foresee their effects on the public and the environment.
Governments around the world, including the Mexican government,
recognize that adoption of new technologies reshapes society to
some degree.
    For example, Congress has itself undertaken and commissioned
studies to assess the possible results of genetic engineering technology. In all these studies, investigators ask what effect the technology
will have on the environment, who will benefit, and who will suffer.
While in this area, as in others, there is widespread disagreement,

debates can be carried on much more intelligently now that we have recognized at the highest levels of public policy that technologies are not neutral (National Academy of Sciences 1972; USGAO 1984).

The researchers, scholars, and farmers who have been experimenting with ecological agriculture usually intend that the techniques they are developing will have effects on society and the environment. Their research is inspired by what they see as the negative societal and environmental results of chemically dependent agriculture, and so it is obvious to them that any technology they develop must be responsive to more than narrow technical or economic concerns. Wes Jackson, a prominent exponent of a new, sustainable form of agriculture, believes that the effort to create a better agriculture is far-reaching and unfamiliar to human culture that has for millennia lived with agricultural technologies that have been destructive of natural diversity and land. In *New Roots for Agriculture* (1980), he presents a vision of ecological agriculture founded on a generalized reorganization of rural society. He believes that the present destructive technologies are the product of capitalism, but that socialist societies have shown little or no ability to develop better ecological alternatives. Jackson proposes that we may even require the creation of a new vocabulary or language to bring a sustainable agriculture into being. "That language will only come as we discover the proper relationship of people and land in a modern setting, as we assess not only the tools and techniques but the social, political, economic, and religious arrangements suitable for a highly populated sun-powered planet" (Jackson 1984: xiv–xv).

There is a strong sense among most of those doing research into alternative agricultural techniques that what they are doing is closely linked to even wider projects for the improvement of human life. Richard Thompson, an organic farmer in Iowa whose success is being imitated by many other conventional, commercial farmers, sees his conversion away from chemical-dependent agriculture as part of his conservative Christian faith. He served as a campaigner for the fundamentalist Christian presidential candidate Pat Robertson in the 1988 Iowa primaries. Anthropologists working to analyze the diffusion of organic techniques among farmers in Ohio have noted that those ready to switch over to a low dependence on chemicals seem to do so primarily from a moral perspective usually deriving from a Christian interpretation of the "stewardship" responsibility of humans for earth and its creatures (Thompson 1988; Esainko and Wheeler 1987).

On the other hand, many ecological farmers see their work as a revolt against the most destructive aspects of capitalist technological development and against the modern emphasis on large, centralized organizations, both capitalist corporations and socialist bureaucracies. Such farmers often actively organize against U.S. intervention in the Third World and for nuclear disarmament and more liberal domestic social policies. These people would normally characterize themselves as liberals, leftists, or progressives and would be nervous with the conservative Christian faith of people like Richard Thompson. But in spite of the different visions that guide people interested in lower chemical dependence, in the United States it is clear that an interest in alternative agriculture is strongly correlated to a wider commitment to strong moral ideals, usually including a desire for greater social equality and generalized restraint on environmentally destructive technologies, products, and policies (Jackson 1980, 1987; Callicott and Lappé 1988).

To some, this may seem ridiculously grandiose, and many find occasions to ridicule the high hopes of the alternative agriculture movement. But everyone should remember that the most recent program for the modernization of world agriculture, begun in Mexico in 1941, made claims to be part of a wide movement of societal change, proposing not only to drastically change methods of agricultural production but to provide the basis for industrialization and urbanization in countries that had seemed resistant to modernization along these lines.

The way we treat the land on which we live, the way we grow our food, and the principles by which we organize our cultures and societies to grow and distribute food are not trivial matters. Agriculture is the basic sustenance of us all and our most significant form of interaction with wild nature. It is still the principal activity of most human beings on earth, and even in the United States, where only two in a hundred are farmers, one in four works in jobs that are part of the agricultural economy. Agricultural technologies must be judged with the realization of their profound effects on nature and society.

The Questions Are Political

What we have seen by looking at our assumptions about agriculture is that there are many choices to be made toward a healthier agriculture and better human communities. These choices are essentially political, although every kind of human creative endeavor will be necessary in the search for the right decisions and the ways to carry them forward. What is clear is that the barriers are not the obstacles thrown up by cruel nature but the adversities created by unresolved human conflict.

There are some simple truths that underlie all we have said here. Chemical companies want profits, banks want to lend money and collect interest, governments want to grow and to please the powerful, large countries like to dominate smaller ones, and scientists and other professionals like to have jobs and feel important. Traditional ways and people have little power and are usually easily dominated by more aggressive, industrial peoples. Everybody needs food, and most of us apparently like to fantasize about technologies that will always save us from all our problems. We admire nature, but we would like it to conform to our convenience. We will have to begin to examine much more closely the way these simple things cause us some very complicated and perhaps ultimately tragic problems.

In 1952, Carl Sauer wrote an essay for a conference and book called *Man's Role in Changing the Face of the Earth*. His words could hardly have been more prophetic. He began by describing the "continuous and cumulative" spread of European culture, "borne by immediate self-interest . . . but sustained also by a sense of civilizing mission." He spoke of the drive for control of markets and resources and of some of the benefits this had brought to distant places. "We also wish to be benefactors by increasing food supply and by diverting people from rural to industrial life, because such is our way, to which we would like to bring others."

"The road we are laying out for the world is paved with good intentions, but do we know where it leads?" Sauer pointed to the rapid depletion of resources and the promotion of environmentally damaging production techniques. He questioned the careless promotion of genetically uniform monocrops and of the changes that came with them.

> Industrialization is recommended to take care of surplus populations. We present and recommend to the world a blueprint of what works well with us at the moment, heedless that we may be destroying wise and durable native systems of living with the land. The modern industrial mood (I hesitate to add intellectual mood) is insensitive to other ways and values. . . . What we need more perhaps is an ethic and aesthetic under which man, practicing the qualities of prudence and moderation, may indeed pass on to posterity a good Earth. (Sauer 1969: 68)

10
Points North

*I*n July 1988, I took a trip to Tijuana to find the family of Ramón González. I had heard that Ramón's mother had moved with her children to Colonia Obrera in the border town, and I wanted to know the news of the family since I had spoken with her in their Oaxacan village of San Jerónimo Progreso in 1984. It also seemed possible they would have further information or ideas on the death of Ramón in the fields of Culiacán. Luckily, my anthropologist friend from Riverside, Michael Kearney, who knows the Mixtec community very well, was free to help me find Guadalupe and her children. But, as always, Michael had friends to look up and errands to run on the way. Following along with him would teach me a lot about the concrete details of the lives of people who provide the labor for agribusiness in California and Mexico and who live in the peculiar world of a transborder culture.

Rugged, rocky hills rise up west of the booming real estate developments around Sun City and Rancho California, about halfway between Riverside and San Diego on the way to Tijuana. The striking geometric patterns of avocado groves have been laid across the hills with little regard for their natural shape. Michael and I drove along twisty two-lane roads into these hills to try to find a Mixtec farm worker who had lived in Michael's house from time to time. Roberto Barrera, the young Mixtec man, had been expecting some documents from the U.S. Immigration and Naturalization Service and had given Michael's address to the INS because it was the only stable one he had north of Oaxaca. He had told Michael that he would probably be working in the avocado groves near Rancho California when the documents arrived. Michael had received the papers from INS, and so we went looking for Roberto.

The first farm worker we encountered was taking a scythe to some weeds at the border of an orchard. He didn't know Roberto, but suggested a couple of camps he might be living in. The second man we encountered was a very dark, handsome man with a bushy black mustache, carrying a pair of pliers to reset wires holding up young trees and repair fence lines. He knew Roberto from the camps. Unfortunately, he reported that Roberto had left for Oaxaca three weeks earlier. On the off chance we would find Roberto, and simply to take a look at the camps, we drove to the *campamento* along Via de los Robles.

There were no signs leading to the *campamentos*. Each one was well hidden among groves of oak trees down in the steep arroyos and canyons formed by the hills. We made several futile sorties into bottomland woods where we saw beer cartons laid out on the ground for sleeping, cans, bottles, and other signs of recent habitation. At a road closed with a locked cable extended across it, we noticed the tracks of a lot of heavy boots and followed them up the road by foot. After another false turn or two up arroyos, we found the *campamento* concealed perfectly by an oak grove.

The *campamento* consisted of about a dozen squarish shelters made by stretching sheets of clear or black plastic across a frame made of sticks and pieces of cast-off lumber. It was likely an all-male camp, empty at this hour during the work week. At the bottom of the camp, where we entered, was a wooden shed put together as a shower, using plastic pipes from the drip irrigation system for the avocados. The men had cleared a piece of ground under the trees for a basketball court and had mounted an improvised hoop and backboard on a sturdy pole. They had made a variety of stove designs out of fifty-gallon drums and large tins. Some of the designs included ovens. In two hammocks made out of plasticized fiber fertilizer bags, Mexican pulp novels lay awaiting the return of the readers. Beds ranged from cardboard on the ground to mattresses supported by a frame laid out on four forked-stick corner posts, as in villages in Oaxaca. Most of the designers of the *casitas de plástico* had taken advantage of the drip irrigation system on the surrounding hills to install water and a sink made from a bucket or large tin can. There were a few, but very few, personal belongings neatly arranged in most of the houses, usually including a minimal set of dishes and eating utensils. One man had left his dishes unwashed that morning before heading for the groves, in contrast to his neighbors, who had their forks, spoons, and cups laid out to dry. There were no latrines, but simply paths leading across the hill away from the camp, sometimes littered with a few pieces of toilet paper. In one shelter, a drain from the plastic-bucket sink led to a small raised bed with a few tomato and geranium plants.

By the time we walked back to the car in the late morning the temperature was rising quickly to pass one hundred degrees. Somewhere, the men from the camp worked hidden to us in the avocado groves. Signs along some of the roads leading into the groves warned visitors not to enter the orchards. "Help us protect our trees—help prevent root rot. Please do not enter the groves." For those who had to enter, there were in some places buried plastic pails with copper sulfate to scrape onto shoe soles.

Arriving in Tijuana, we went to visit Victor Clark Alfaro, executive director of the Bi-National Center for Human Rights and a friend of Michael's. Victor's offices are up well-worn, wooden stairs above a corner store near downtown Tijuana. The front meeting room is decorated with posters reflecting some of Victor's concerns— a poster detailing the UN Declaration on the Rights of Children, decorated with funny cartoons featuring Mafalda, the hyperintellectual, anxiety-driven six-year-old Argentine cartoon figure. Another poster asks, "Who killed 'Gato' Félix?" with a picture of the prominent Mexican journalist recently assassinated on the streets of Mexico City, one of twenty-five to thirty journalists murdered in the last six years in Mexico. (Culiacán's state of Sinaloa accounts for nearly half of them.) Other posters concerned human rights congresses in Mexico and the formation of the Academy of Human Rights by internationally known Mexican scholar Rudolfo Stavenhagen. Others commemorated celebrations in Nicaragua or announced academic conferences dealing with farm workers and rural issues.

Victor's secretary directed us into his plain, tiny office to await his return. Victor was happy to meet Michael, and after an embrace and a few pleasantries, got down to business discussing recent meetings, the problems of various migrant families known to both men, and recent political events affecting the border. We told Victor of our visit to the camp in the avocado groves, and he went immediately to his files to bring out a front-page story from Tijuana's *El Heraldo*, published two days earlier.

"They Live in Caves," the red-letter banner declared across the whole page, with a subtitle that said, "The inhuman conditions of undocumented workers in the United States." The long article, accompanied by more than a page of photographs, described the foxholes and caves that migrant tomato pickers use as houses in northern San Diego County. An interview with a tomato grower who owns a farm and large packing shed, Allen Yosiguchi, included the grower's contention that whatever conditions they saw were temporary because the tomato growers intended to move their operations across the border into the San Quintín area of Baja. Aside from labor costs, the construction boom in San Diego County, according to Yosiguchi, was sending the price of land beyond the reach of farmers. In spite of Yosiguchi's driving the reporter and a group of Mexican university students off his land, the reporter and photographer had somehow managed to show graphically the incredibly primitive and unhealthy conditions of farm worker life in the area. (The same conditions have also been documented by North American reporters, including Lynne Walker, writing for the *San Diego Union,* and a writer for the widely read Sunday supplement *Parade,* who declared that in covering ten wars he had never seen refugee camps as miserable as these.)

Michael and Victor discussed the case of a small group of farm workers who had gone to work in Oregon. A young Tzotzil man from Chiapas had joined some Mixtecs on the journey, and they had all become good friends. In Oregon, after a day's labor in the fields, the apparently healthy Tzotzil had lain down to rest. Within a couple of hours, the man's companions realized he was no longer breathing, and then that he was dead. An autopsy by a local physician failed to discover any explanation for the death—there were no signs of a heart attack, stroke, or previously unknown illnesses or organic defects. The doctor also said there were no toxins in his blood. While his Mixtec friends had considered that pesticides were a possible cause of his death, because pesticides had been recently used in the fields where the man worked, they decided that the autopsy results ruled out pesticides.

I explained that it was possible that pesticide poisoning might not have been detected by an ordinary autopsy. Diagnosis of pesticide poisoning through organophosphates would normally be accomplished by analyzing acetylcholinesterase levels, not by looking for the toxins themselves. The death previous to analysis, breakdown of the toxins in the body, the costs and unreliability of tests for toxins by labs not especially experienced in pesticide analysis, and the complications concerning individual variations in the chemicals that regulate nervous system functioning made it quite possible that even the most rigorous autopsy aimed at investigating the hypothesis of pesticide poisoning would fail to establish any clear conclusions. I concluded by saying that pesticide poisoning could certainly not be ruled out, although there was also no way of establishing pesticides as the cause so long after the death.

Victor had stored in the closet of his office the traditional Tzotzil clothing the deceased young man had left with him for safekeeping. The brightly embroidered blouse, the loose white pants, and the shoulder bag with elaborate leather decorations were neatly folded into a clean, cheap canvas gym bag. Michael volunteered to see that they were returned to the man's companions, but Victor said they would certainly be returning to his office to get them, so it was best he kept them.

Victor is a stocky, short man with dark, neatly trimmed hair and a closely clipped black beard and mustache. He was dressed in a dark polo shirt and jeans and wore horn-rimmed glasses. Everyone we met in the next twenty-four hours greeted him enthusiastically. According to Michael, the state government considers him a very irritating thorn in its side. He works with cases of arbitrary arrest and police brutality, legal problems of poor neighborhoods, and all the intricate bureaucratic, legal, and personal difficulties of people who cross the border to earn their living. Victor was born and raised in Tijuana of a family that in spite of the Anglo name Clark has been thoroughly Mexican for generations. Like Michael Kearney, he is an anthropologist—for reasons I never asked about someone had written across the top of a small, movable blackboard in his office the sentence "Anthropology deals with observable behavior and its results." Victor is deeply involved with the politics of the streets and neighborhoods of Tijuana, not as a political figure but as someone who looks for ways to help people obtain respect for their basic rights. His center is financed by personal contributions and foundation grants. He is deeply involved with the Mixtec community and knows it well.

In the afternoon, the three of us set out on the streets of Tijuana to see if we could track down Ramón's mother. Most of the Mixtec women in Tijuana work selling small, home-made decorative items to tourists on the streets. We parked in a lot downtown and sought out the first group of Mixtec women we could find. On the first corner, we found three Mixtec women seated with two children, selling necklaces and bracelets made from colorful yarn. We described the woman we were looking for as "the widow of Irineo González," and they immediately knew whom we meant. They described the house where she was living with her family, but there was some disagreement among the three about the whereabouts of her house. They also assured us that she would not be returning to her house until early evening, because she would be somewhere on the streets selling things.

We visited with several such groups of women, all of whom knew Victor and most of whom knew Michael. The Mixtec women are easily recognizable as Mixtecs by the dress and apron they wear over polyester pants. Most of them are well under five feet tall, and many have children with them. Victor discussed with most of them the troubles they were having because the municipal government was not letting them sell on the choicest blocks along Avenida Revolución until after six in the afternoon. Only those who are members of the PRI-affiliated street vendors union are allowed to sell in the more desirable spots. The Mixtec women didn't belong to the official union but instead had four separate, competing organizations of their own, and Victor suggested in our conversations with the women that they would be able to resolve their problems more easily if they were to unify the four groups into one.

We ran into a woman who had lived in Michael's house in Riverside during her pregnancy and through the birth of one of her children. Other women we met had lived with Michael's family for a time. Most of the women knew of Irineo's widow, but the disagreements about where she lived multiplied with each conversation. When Victor and Michael had decided we had done all we could until evening to find Guadalupe, Victor decided to take us to the border area where much of his work is focused. From the border we could reach a much deeper understanding of some of the problems faced by the migrant workers whom I knew from Culiacán, San Quintin, and Oaxaca.

A four-lane street defining the northern edge of Tijuana runs east and west from the downtown area, becoming a freeway as it leaves the older part of the city and heads out toward the beaches. On the north side of this street is a high cyclone fence that is the border with the United States. There are breaks in the fence where people have cut through or where groups have simply pushed it down, and there are places where there is no fence. One of the places where there is no fence is the municipal headquarters of the Mexican ruling party, PRI, and we walked through its parking lot onto the concrete-lined southern levee of the Tijuana River. While the legal border is the fence, the border enforced by *la Migra*, the U.S. Border Patrol, is the northern levee, also lined with concrete, of the Tijuana River. Mexicans attempting to cross the border illegally stand or sit on the southern levee or down in the mostly dry riverbed itself, watching for breaks in the vigilance of the patrols of *la Migra* strung out along the top of the northern levee in passenger vans or all-terrain motorized tricycles or on foot. At times of heavy traffic across the border, the patrols also use helicopters, one of which crashed a few years ago after being hit by a hail of stones thrown by people along the southern levee top.

Human rights concerns in Tijuana are nowhere more concentrated than along this strip, an area Victor knows very well. At night there would be hundreds of people standing along the levee top, but as we walked along with Victor in the afternoon, there were only a few dozen. Many of them knew Victor and greeted him cheerfully. From time to time, we would stop to talk with people, hearing their stories or their concerns. Most who were waiting to cross on that day were from various rural regions of Mexico and were headed for the harvest season in the California fields. All those we encountered that afternoon had been across before. One toothless, elderly man with bow-legs and a marked limp was waiting to cross to go to Stockton, California, where he intended to work picking fruits and nuts as he had done in many other years.

There are some consistent problems beyond the obvious ones in this area. Many of the people seated here, from groups of young men to whole families with children and babes in arms, have hundreds of dollars in cash with them to pay a "coyote" to help guide them to the other side or simply for living expenses while they try to find work and wait for a paycheck in the United States. These *pollos*, or chickens as everyone here calls them, in contrast to the coyotes who live off them, have two immediate worries: can they get across the border, and can they do so before teenage gangs or armed thieves come along the levee top to beat and rob them? Aside from that, they also have to worry that as large groups of them sometimes storm the border, they may be turned back by the border patrol and be badly hurt in the melee, or be thrown en masse into the street at the southern side of the levee. There, according to Victor and some of the people we met along the levee, they may be beaten, robbed, and perhaps arrested by members of the Tijuana police force, or beaten and robbed by neighborhood gangs.

The area along the southern levee is a kind of no-man's-land, not part of Mexico but not policed by the United States. People smoke marijuana and buy and sell it in this area because they need not fear arrest on the north side of the fence. One short, fat man who knew Victor talked to us about the drugs. One of his eyes was swollen shut and he was cut and scratched on his face and hands from a recent fight.

"No, there's just *mota* [marijuana] around here now. Not much hard stuff. You're not seeing a lot of heroine or morphine right now."

"Crack?" Victor asked.

"No, not that. Maybe a little heroin. Not much now."

We spoke to a young, attractive woman dressed in the fashionable youth style of 1988 of black blouse, black jeans, and black high-top sneakers. She told us she had a legal work permit for the United States and had been working in a restaurant in San Diego but had visited Tijuana, become drunk, and had her work permit and money stolen. She had been working in the red light district across the street to get money to pay a coyote and was hoping to cross over the levee top into the United States that night.

One man in his twenties was dressed in the "crazy *vato*" fashion of California cities, with a pompadour hairdo, baggy pants, black shoes, and a shirt buttoned at the collar but not below and splayed into a V over a white T-shirt. The shirt was a prison shirt. We seldom understood what he shouted to us from time to time—he was apparently badly drunk or stoned. The majority of the *pollos*, especially the family groups, did their best to shy away from him and the other people along the levee who looked less than trustworthy. People like the *vato* and those apparently dealing drugs added an atmosphere of menace to the palpable anxiety of most of the *pollos* who after all just wanted to get to the other side.

Victor pointed to some apartment buildings on the U.S. side, perhaps half a mile from where we stood. "Some coyotes have apartments rented in those buildings. They will give a special deal of one hundred dollars to guide a person from here to one of those apartments, where they can stay a night before moving north." The usual price for the coyotes is in the neighborhood of $250.

Most of the crossings are made at night, eluding the border patrols and the infrared scopes they use to locate people in the darkness. From what people told us, they rely heavily on the diversion of the patrols by a first person or group, opening a gap that can be penetrated by those who are alert and quick. Sometimes, heavily drugged people who have no particular desire to cross the border are used as decoys. As we stood there in the late afternoon, a few young men made runs for it as they spotted places where the patrols seemed inattentive. They jogged across planks laid over the raw sewage flow of the Tijuana River and ran as fast as they could up the steep slope of the northern levee. In every case we witnessed, the patrols hurried to meet them, and the young men turned around. They would probably have to wait until nightfall for success.

Victor spent a long time talking to a thin man with a handsome, heavily lined face in his late forties or fifties who was pushing a two-wheeled cart selling oranges and cantaloupes to the people along the levee top. He had once worked in the United States as a stevedore but had experienced a lot of trouble and decided it was not worth crossing again. He and Victor discussed the complex politics of street-vending unions and the competition between the PRI-affiliated union and the independent unions that sometimes associated themselves with one or another of the small leftist parties. Like the Oaxacan women, this man was in a group that PRI had excluded from the more lucrative spots in Tijuana, but he thought a deal had been cut that would soon allow his groups to return to those spots. In the meantime, he was working the border. Virtually everything the man said was punctuated with smiles, sly comments, or jokes.

Victor asked him, "Did you vote?"

He smiled and replied, "No, I don't vote."

"You should always vote, it's important," Victor said.

"It doesn't make any difference."

"But look, in this election on July 6, Cárdenas [the opposition candidate to PRI's Salinas] nearly won. He probably did get the majority of the votes."

The man grinned and asked simply, "Is he president?" Like many Mexicans, he apparently believed that PRI would always enforce its own presidential candidate on the nation whatever the vote.

The toothless old man headed for Stockton said that he had voted. Victor congratulated him. The man added, "I'm for Cárdenas, but I voted for Salinas. It's the union," he said, patting an official, plasticized card in his shirt pocket that indicated his membership in a PRI-controlled organization. Victor nodded his understanding.

There is considerable debate about whether the newly restrictive Simpson-Rodino Immigration Bill will, as it intends, reduce the flow of immigration over the border. Initial signs after two years are inconclusive. Recalling that nearly half of Mexico's population is under the age of fifteen, that Mexico must create a million jobs a year in a chronically stagnant economy to keep unemployment from rising even further, and that Mexico's population of 85 million is expected to double in less than thirty years, it seems obvious that there is no reasonable way to stop an epic migration of Mexicans into the United States. The scene at the border only confirms this suspicion, given the palpable mixture of hope, desperation, and determination among the people waiting at the shores of the Río Tijuana.

As we walked back toward the end of the levee, we encountered two young women who called out to Victor. They turned out to be people who work for him. One was a lawyer. They were carrying clipboards with questionnaires inquiring of people whether they had been mistreated or arrested in this area—the survey was meant to uncover abuses by the Tijuana police or border patrol, and Victor explained that he was accumulating data. "If we can document these things with enough solid information, we will use it to bring the abuses to public attention, trying to force some changes."

We had to return to Victor's office for a meeting scheduled there. The meeting was between the *mayordomía* of villages in the Mixteca now located in Tijuana and a Catholic priest. The *mayordomía* are men chosen to be responsible for community functions, including the celebration of important religious holidays. They were planning for the celebration on August 4 of the festival of Nuestra Señora la Virgen de Nieves, an important occasion in the Mixteca Baja that takes place in the town of Nieves.

One of the striking things about the migration of Mixtec people out of their homeland north to Mexico City, Culiacán, Tijuana, and California is that the Mixtecs seem to have been stimulated by the migration to reassert their ethnic identity. Michael Kearney and Carole Nagengast (Nagengast 1988) emphasize the rise of Mixtec ethnicity as a feature of the Mixtec migration, pointing out that the Mixtecs are not at all anxious to have their culture lost or subsumed in the larger cultures that now surround them as they migrate. They are insistent on maintaining the strength of their community ties to help them in their travels and to present a solid front to an outside world with which they must deal but which is often hostile to their interests. Mixtecs in the San Joaquin Valley of California, for example, have been reluctant to join the United Farm Workers union as individuals but have carried on discussions with the union through their own, separate Mixtec organization. The Mexican government has apparently begun to recognize the persistence of Mixtec identity—PRI arranged through a Tijuana town councilor to have a Mixtec Child Care Center built in Colonia Obrera in Tijuana, finished just in time for the presidential candidate Salinas de Gortari to inaugurate it personally on a campaign trip in early summer, 1988.

The Mixtec men arrived ahead of time at Victor's office and the priest shortly thereafter. The Mixtec men wore work clothes but looked strikingly neat and well groomed relative to other people on the hot, dusty streets of Tijuana. The priest was a tall, light-skinned man who towered over the Mixtecs. He was dressed in the cheapest kind of huarache sandals, polyester pants, and a shirt worn open to his sweaty belly. His face, way of talking, glasses, and manner all indicated a style popular with studiedly unpretentious Mexico City intellectuals. We later learned that he was a Jesuit from Mexico City who had no parish, made his living as a plumber, and had a small bookshop specializing in "liberation theology" that he managed to keep open a few hours a week. He also gave mass and heard confessions in the penitentiary.

The priest readily agreed to say mass, but to the obvious surprise of the Mixtec men insisted, in a friendly, casual way, on having the men tell him in detail about the festival. What did it mean? They said it was a thanks for the good fortune of the year and especially for the food they had been given by the grace of God. Which villages

celebrated it and where were they located? What was the role of men
and women in the celebration? Was there food, drink, music? He returned repeatedly to the question of music as part of the mass. The
men said at first that there was no music in this celebration. Insistent questioning revealed that there was usually a band at home, but
they despaired of putting one together in Tijuana. First, they said
there was no one who knew how to play—the best musicians were
over the border in California. "But," the priest said, "people are not
going to get happy without some music, surely." Then, it seemed
there were people in Tijuana who could play a guitar and a clarinet,
but no one had instruments. The priest promised to look for instruments that could be loaned for the day. Through each detail of the
day's celebration, and with frequent restatement and affirmation and
reaffirmation, the men worked out every detail of the festival with
the priest over a period of more than an hour. They seemed very satisfied with him when the discussion was over.

At the end of the meeting, we asked if the men of the *mayordomía* knew Ramón's mother, Irineo's widow, the woman called
Guadalupe. "Of course, she lives in Colonia Obrera, section three."
After a little discussion, it was universally agreed that we could find
her after nine o'clock or so in her family home at the top of the hill
in a house right by a utility pole with election posters tacked to it.

Before dark, we found Guadalupe's house, just as they said, near
the pole with election posters for Cuauhtemoc Cárdenas. Her daughter, who remembered my face from four years earlier when we had
spoken in San Jerónimo, said that her mother would be back about
nine-thirty. We should come back then.

Taking advantage of the time, we went to visit friends of Michael's.
At dusk, the neighborhood set on top of a long ridge on the southern
edge of Tijuana seemed pleasant. There were children and dogs wan-
dering the streets. Most houses, while small and crude by middle-
class standards, had solid walls and roofs built of rough-cut lumber.
Electric lines ran to most of the houses, and many had some arrangement for water within the house. There was no sewage system, but
there were no open streams of sewage as I had seen in many slums
throughout Mexico and other countries in Latin America. Latrines
seemed well-enough planned to keep the smells of human waste to a
minimum. There were a few dim street lamps along the dirt streets.
A voice from a loudspeaker urged people to come to the movies—a
truck containing a projection booth formed one side of a movie theater, with the other sides made of canvas like a circus tent.

The first friend of Michael's we visited was a young man sitting in his two-room house making yarn parrots for sale to tourists. He and his young son both worked without missing a beat as we chatted with them. Before we came they had been watching television, and as we talked they occasionally took a glance at the shows detailing the emotional lives of fabulously rich people and the commercials advertising soft drinks and automobiles that were apparently bought only by other fabulously rich people. A girl of three or four slept on a blanket thrown over a plastic couch.

In an adjoining house, another friend of Michael's worked at the same task, making parrots. His report card from a private business school showed he was making very good grades. There was a color television and a bed with a mattress in the room he worked in—it seemed to me that the people on his television were even more fabulously rich, probably because of the color. His father returned after nine o'clock after working twelve or fourteen hours as a gardener. Within two or three minutes of coming into the house, he also set to work making brightly colored little yarn parrots. His father was planning to make a visit back to San Jerónimo soon to work on his house there. Michael had rented the house from him the previous summer and said it was in good condition, but the man insisted it needed major repairs. The time came to visit Guadalupe and her family, so we took our leave.

Guadalupe had returned at about nine-thirty, and she and her children were all at home. Their house was mostly surrounded by a cyclone fence, and the children showed us the way past the fence at one side of the small, dry yard. Guadalupe's oldest son had, with some friends of his, just finished building the house of rough-cut lumber, spending about U.S. $800 for all the materials. It was simple but fairly solid and tight, and larger than many others in the *colonia*. The family's only chair was a plastic dinette chair with the back missing, and there were a few small electric wire spools with little plastic-covered pillows used as stools. There was a little embarrassment and joking about where we would sit, but of course we had to sit on the best seats available while Guadalupe sat on the floor in a corner with her three-year-old daughter cuddled in her lap.

Guadalupe remembered a lot of details of my visit with the family shortly after her husband's death, when she was pregnant with the youngest child and with no adult children living with her to help her through the crisis. She soon became frustrated with trying to speak Spanish, however, as Mixtec is her first language, and began to refer the questions to her oldest child in the house, Catalina.

Catalina is twenty years old, a young woman with lovely eyes, a beautiful warm smile, and a careful, thoughtful way of talking. She works at a restaurant in downtown Tijuana helping to prepare food. She seems remarkably self-confident. At one point in the conversation, she asked for some clarification about who Victor was, and when she heard his name clearly, said, "Oh, yes! Victor Clark! I have heard you on the radio many times. You have said some very fine things about us Mixtecs. I want to tell you how much we thank you for that. You know, many times people think that we are stupid burros, perhaps because our Spanish is not always so good. But that's not true, we are . . . well, it's just not true, and it is very good that you say the things you do. We are very grateful for that."

Guadalupe and her children had been in Tijuana for two years. They had come because they thought they could do better economically, and the decision was made easier by the fact that they had been in Tijuana before and that three of the older children were already living in Tijuana when the rest came in 1986. The oldest of Guadalupe's family was Cipriana, a twenty-four-year-old recently married and living with her husband. Ramón, had he lived, would have been twenty-six. Juana, twenty-two, lived with Guadalupe and Catalina, as did Asunción, María, Juventina, Rosa, Felipe, Modesto, Vernadina, and Teresa, ranging in age from seventeen down to three. Four of the children were in primary school. Aurelio, nineteen, had recently left to work in California, the family didn't know exactly where, but suggested the Rancho California avocado camps as one likely place. "But who knows?"

We asked whether they had any new information about Ramón's death. Catalina said they did not.

I said, "The last time we talked, some people had said he died of pesticide poisoning, and others that he died of a blow during karate practice. Do you have any ideas about that now?"

Catalina replied forcefully, "No, none of that is true. They killed him. They just made up those stories. We don't have any idea how he died."

Michael asked, "Did he have any enemies? Was there some reason someone would want to kill him?"

"No, everyone was his friend," Catalina said, perhaps a little defensively.

Michael said, "Yes, that was my memory of him. He always conducted himself well and never made any occasion for fights."

I asked, "My understanding was that most of his companions thought he died of pesticides, but that some thought it was a blow while practicing karate . . ."

Catalina interrupted, "No, that's all just stories, they just made it up, there's nothing in any of that. They just told us that. He didn't have anything to do with karate or any of that."

Michael asked, "Then why do you feel so sure that these stories about the pesticides are invented?"

Catalina suddenly looked a little nonplussed. "What is the word you are using? Pesticide? What is that?"

I explained, "The poisons they use in the fields to kill weeds or insects."

"Yes," she said, "I understand. But, no, he certainly didn't die of those. That's what they said, but it is not true."

Michael asked, "Are you saying then that you know that it is not true, or are you saying that in your opinion it is not true? Is this something you and Guadalupe have decided as your opinion?"

"Yes, that's our opinion. No one else told us it wasn't true. *They* tried to tell us it was the poisons. But we decided no, those were stories they made up, because it could not be true. They said he took a bath in the canal after working in the fields and then died of the poisons. A person cannot die of those poisons."

I asked, "Did someone tell you that?"

"No," Catalina answered, "but we know that. It's ridiculous to think one could die of those poisons that are just for using on the fields. After all, I was there in the camps in Culiacán one time when I was a small child, and I didn't die. My mother didn't die. My father didn't die of them. How could Ramón have died of them when we didn't?"

I explained that one could die of the poisons they used in the fields. I also said that some people might die of them when others might not. "While I cannot say what caused your brother's death, I can say that it is certainly possible that a person can die of those poisons. Many people do. It is possible that your brother did."

Catalina looked at me very intensely for a moment and said, "I didn't know that. I would never have believed that. I didn't realize that."

Catalina spoke quietly to her mother in Mixtec. Guadalupe's expression darkened, and she closed her eyes for a moment.

After more conversation about what kind of person Ramón was and about other family matters, we took our leave. The parting was warm, and they invited us back at any time and thanked us for our interest in them.

We walked to my car, parked in the yard of a friend of Victor's and Michael's. There we met a man who was a teacher and principal of a primary school for six hundred children in another *colonia*. Leaning up against his battered old Datsun in his shorts, flip-flops, and T-shirt, the man asked us to stay and talk awhile. We were soon discussing various matters in the Mixtec community. I asked him if he was Mixtec.

"Yes," he said cheerfully. "Bone and flesh. Down to the last hair on my head, I am Mixtec."

We told him of our conversation with Guadalupe and Catalina.

"That's very interesting," he said. "Just day before yesterday I was talking with some people from the agency for adult education. They had prepared a draft for a booklet on health for use in the Mixtec community. They wanted our comments. They had listed some of the major health problems of the Mixtecs and asked if they had left out anything. One of the other teachers said they should have a section on pesticide-caused health problems, because there seemed to be a lot of them and people didn't have much information about it—there is a lot of confusion and misunderstanding. I lent my support to what the man said. But they got all excited and said no, no, that wouldn't do. They couldn't put that in, it might damage the whole project.

" 'We can't deal with that,' they said. 'It's just too politically hot.' "

On returning to Sacramento, I learned that Cesar Chavez, leader of the United Farm Workers union, was nearing the end of his fast. He said that he was fasting to call attention to the dangers of pesticide use in California's fields. As a way of putting pressure on agribusiness to move toward more conservative use of pesticides, the union was asking the public to boycott table grapes, a tactic that had helped to win contracts for the union in the 1970s. Chavez's doctors were urging him to end the fast soon, as his body was showing signs of possibly permanent damage from lack of nutrition. Chavez already has serious medical problems—his back is badly damaged from the use of the short-handled hoe. Growers had insisted for decades that if workers used long-handled hoes they would not see all the weeds in a field or they might injure crop plants. Only after an intensive UFW campaign in the 1970s did California finally outlaw the use of the short-handled hoe that had crippled Chavez and many other farm workers.

On the thirty-third day, Chavez ended his fast with a mass attended by various celebrities and public figures, including the Rev. Jesse Jackson, who had just finished his campaign for the Democratic nomination for president of the United States. Some observers claimed that the UFW was using the boycott to gain new contracts and not to call attention to the danger of pesticides. Others ridiculed the participation of Jackson and others in the mass as cynical manipulation of the public's sentiments. But for the first time since the appearance of Rachel Carson's *Silent Spring* in 1962, journalists were writing about the dangers of pesticides to farm workers and the environment on the front page of most of the nation's major newspapers, and experts were debating the issues on national radio and television programs.

On September 14, 1988, Dolores Huerta, co-founder and vice-president of the UFW, was carrying a UFW flag in a demonstration in San Francisco to protest presidential candidate George Bush's denunciation of the grape boycott. A video tape later showed that as the police began to move people, Huerta, who is fifty-eight years old, five feet tall, and weighs 110 pounds, asked a policeman where she should go to avoid trouble. As she moved away, a policeman struck her in the side with his nightstick, breaking two of her ribs and rupturing her spleen. The ensuing outrage left many, including San Francisco's mayor, who was a long-time political ally of the UFW, openly wondering how this kind of mistreatment could occur, and many were further outraged when the policeman was exonerated of blame by an official inquiry (*Sacramento Bee*, September 17, 1988, p. 1).

The farm workers were forcing the American public, not for the first time and with painfully repeated risks to themselves, to examine some hard truths.

References

Adkisson, Perry L., George A. Niles, J. Knox Walker, Luther S. Bird, and
Helen B. Scott
    1982 "Controlling cotton's pests: A new system." *Science* 216 (April 2).
Ahluwalia, M. S., N. Carter, and H. Chenery
    1978 "Growth and poverty in developing countries." World Bank Staff
        Working Paper no. 309. Washington, D.C.: World Bank.
Ahmed, Dr. Arif Jamal Mohamed (National Agricultural Research Council,
Sudan)
    1988 Interview, May 6.
Akesson, N. B., W. E. Yates, and P. Christensen
    1978 "Aerial dispersion of pesticide chemicals of known emissions, par-
        ticle size, and weather conditions." Manuscript. Cited in Pimentel
        and Perkins 1980.
Alanis, Gilberto López, Laura E. Alvarez Tostado, Arturo Retamoza Gu-
rrola, Yolanda Ponce Conti, César Cisneros Puebla, Norberto Gaxiola Ca-
rrasco, Manuel López Alvarez, and Mario A. Lamas Lizarraga
    1982 *Monografía del Estado de Sinaloa.* Culiacán: UAS.
Albert, Lilia
    1977 "Contaminación por plaguicidas organoclorados en un sistema de
        drenaje agrícola del estado de Sinaloa." *Protección de la Calidad
        del Agua* 3 (1).
    1980 "Organochlorine pesticide residues in human adipose tissues in
        Mexico: Results of a preliminary study in three Mexican cities."
        Archives of Environmental Health, September–October.
    1981 "Organochlorine pesticide residues in human milk samples from
        Comarca Lagunera, Mexico, 1976." *Pesticides Monitoring Journal*
        15 (3).
    1983 Interview, October 12.
    1986 Interview, June 15.
Allen, Patricia, and Debra Van Dusen, eds.
    1988 *Global Perspectives on Agroecology and Sustainable Agricultural
        Systems.* Proceedings of the Sixth International Scientific Con-
        ference of the International Federation of Organic Agricultural
        Movements. Santa Cruz: University of California.

Alonso, Jorge
1984 El estado mexicano. 2nd ed. Mexico City: Nueva Imagen.
Altieri, Miguel A.
1987 Agroecology: The Scientific Basis of Alternative Agriculture. Boulder, Colo.: Westview.
Altieri, Miguel A., Deborah Letourneau, and James A. Davis
1983 "Developing sustainable agroecosystems." Bioscience 33 (1): 45–49.
Anderson, Edgar
1978 Plants, Man, and Life. Boston: Little, Brown.
Astorga Lira, Enrique, and Clarisa Hardy Raskovan
1978 Organización, lucha, y dependencia económica: La Unión de Ejidos Emiliano Zapata. Mexico City: Nueva Imagen.
1984 Interviews, February 18, March 12, 15.
Avery, Dennis T.
1985 "Central America: Agriculture, technology, and unrest." Department of State Bulletin, January.
Aviles González, Mayra
1984 Interview, February 20–21.
Barberán, José, Cuauhtémoc Cárdenas, Adriana López Monjardín, and Jorge Zavala
1988 Radiografía del fraude: Análisis de los datos oficiales del 6 de julio. Mexico City: Ed. Nuestro Tiempo.
Barkin, David, and Billie Dewalt
1984 "El sorgo y la crisis alimentaria mexicana." Problemas del Desarrollo mexicano. Mexico City: Instituto de Investigaciones Económicas, UNAM.
Barkin, David, and Timothy King
1978 Desarrollo económico regional: Enfoque por cuencas hidrológicas. Mexico City: Siglo XXI.
Barkin, David, and Blanca Suárez
1978 El complejo de granos en México. Mexico City: Centro de Ecodesarrollo.
1983 El fin del principio: Las semillas y la seguridad alimentaria. Mexico City: Océano.
1985 El fin de la autosuficiencia alimentaria. Mexico City: Océano.
Barnett, Paul G.
1988 "Survey of research on the impacts of pesticides on agricultural workers and the rural environment." Paper delivered at the Research Review Conference of the Working Group on Farm Labor and Rural Poverty, California Institute for Rural Studies, May 7.
Barney, Gerald O.
1980 The Global 2000 Report to the President. Elmsford, N.Y.: Pergamon.
Bassols Batalla, Angel
1982 Recursos naturales de México: Teoría, conocimiento, y uso. 14th ed. Mexico City: Editorial Nuestro Tiempo.

Beadle, George
   1972 "The mystery of maize." *Field Museum Natural History Bulletin*
      43: 2–11.
Benítez Zenteno, Raúl, ed.
   1980 *Sociedad y política en Oaxaca 1980.* Oaxaca: Instituto de Inves-
      tigaciones Sociológicas, Universidad Autónoma de Oaxaca.
Benítez Zenteno, Raúl, and Julieta Quilodrán, comps.
   1983 *La fecundidad rural en México.* Mexico City: El Colegio de México.
Berry, Wendell
   1978 *The Unsettling of America: Culture and Agriculture.* New York:
      Avon.
Blair, Aaron
   1979 "Leukemia among Nebraska farmers: A death certificate study."
      American Journal of Epidemiology 110: 264–273.
Boardman, Robert
   1986 *Pesticides in World Agriculture.* New York: St. Martin's.
Boletín de Epidemiología
   1981 "Intoxicación por plaguicidas." 1 (18).
Borah, Woodrow Wilson
   1943 *Silk Raising in Colonial Mexico.* Berkeley: University of Califor-
      nia Press.
Borlaug, Norman
   1986 "Accelerating agricultural research and production in the Third
      World." *Agriculture and Human Values* 3 (3): 5–14.
Bottrell, Dale G., and Perry L. Adkisson
   1977 "Cotton insect pest management." *Annual Review of Entomology*
      22: 451–481.
Bourjilay, Vance
   1973 "One of the Green Revolution boys." *Atlantic Monthly* 231 (2):
      66–76.
Brown, M., R. García, C. Magowan, A. Miller, M. Moran, D. Pelzer, J. Swartz,
   and Robert van den Bosch
   1978 *Investigation of the Effects of Food Standards on Pesticide Use.*
      Environmental Protection Agency Contract 68-01-2602, NTIS no.
      PB-278 976. Washington, D.C.: EPA Office of Pesticide Programs.
Bull, David
   1982 *A Growing Problem: Pesticides and the Third World Poor.* Ox-
      ford: OXFAM.
Burmeister, Leon
   1983 "Selected cancer mortality and farm practices in Iowa." *American
      Journal of Epidemiology* 118: 72–77.
Butterworth, Douglas
   1975 *Tilantongo: Comunidad mixteca en transición.* Mexico City: In-
      stituto Nacional Indigenista.
Calderón Vega, Salvador
   1984 Interview, February 6.

California Assembly Office of Research
    1985 *The Leaching Fields: A Nonpoint Threat to Groundwater*, March.
California Department of Food and Agriculture
    1978 *Report on Environmental Assessment of Pesticide Regulatory Programs.*
Callicot, J. Baird, and Frances Moore Lappé
    1988 "Marx meets Muir: Toward a synthesis of the progressive political and ecological visions." In Allen and Van Dusen 1988.
Calva, José Luis
    1988 *Crisis agrícola y alimentaria en México: 1982–1988.* Mexico City: Fontamara.
Carabias, Julia, and Victor Manuel Toledo
    1983 *Ecología y recursos naturales.* Mexico City: Ediciones del Comité Central del PSUM.
Cárdenas Solorzano, Cuauhtémoc
    1988 Interview. See Reding 1988.
Carneiro, Joao
    1983 Interview, December 3.
    1984 Interviews, February 12, 22.
Carrasco, Pedro
    1976 "La sociedad mexicana antes de la conquista." In El Colegio de México 1976.
Carson, Rachel
    1962 *Silent Spring.* New York: Houghton Mifflin.
Castaños, Carlos Manuel
    1981 *Testimonios de un agrónomo.* 2nd ed. Mexico City: Universidad Autónoma de Chapingo.
Castro, Dr. Yolanda
    1987 Interview, February 17.
Cecena Cervantes, José Luis, Fausto Burgueno Lomeli, and Silvia Millan Echeagaray
    1974 *Sinaloa: Crecimiento agrícola y desperdicio.* Mexico City: UNAM, Instituto de Investigaciones Económicas.
Center for Rural Affairs, Marty Strange, ed.
    1984 *It's Not All Sunshine and Fresh Air: Chronic Health Effects of Modern Farming Practices.* Walthill, Nebr.
CEPAL (Comisión Económica para América Latina)
    1982 *Economía campesina y agricultura empresarial.* Mexico City: Siglo XXI.
Chapin, Georgeanne, and Robert Wasserstrom
    1981 "Agricultural production and malaria resurgence in Central America and India." *Nature* 293 (September 17): 181–185.
Chapin, Mac
    1988 "The seduction of models: Chinampa agriculture in Mexico." *Grassroots Development* 12 (1).

Chevalier, François
  1972 *Land and Society in Colonial Mexico.* Berkeley: University of California Press.
CIAPAN (Campo de Investigaciones Agrícolas del Pacífico Norte)
  1984 "Breve historia del control químico del gusano alfiler en el Valle de Culiacán." (Under the supervision of Ing. Mayra Aviles González, who provided additional information on pesticide resistance in cotton and other vegetable pests in an interview on February 21, 1984.)
Clements, Charlie
  1981 "Pesticide use and farmworker health in the Salinas Valley: A case study." Master's thesis, University of Washington.
Cockcroft, James D.
  1983 *Mexico: Class Formation, Capital Accumulation, and the State.* New York: Monthly Review Press.
El Colegio de México
  1976 *Historia general de México.* Mexico City: El Colegio de México.
Collier, George, and Daniel Mountjoy
  1988 "Pesticide Use among Tzetzales in Highland Chiapas." Draft paper and personal communication.
Compton, Paul, and Billie Dewalt, eds.
  1985 *El sorgo en sistemas de producción en América Latina.* Mexico City: INTSORMIL.
Cook, Roberta, and Ricardo Amon
  1988 "Competition in the fresh vegetable industry." In Moulton et al. 1988.
Cook, Sherburne F.
  1949 "Soil erosion and population in Central Mexico." *Ibero Americana* 34.
Cook, Sherburne F., and Lesley Boyd Simpson
  1948 "The population of Central Mexico in the sixteenth century." *Ibero Americana* 31.
Cook, Sherburne F., and Woodrow Borah
  1968 "The population of the Mixteca Alta, 1520–1960." *Ibero Americana* 50.
Corillo Sánchez, Dr. José Luis
  1983 Interview, December 4.
Crosby, Alfred
  1973 *The Columbian Exchange: Biological and Culture Consequences of 1492.* Westport, Conn.: Greenwood Press.
Cross, Harry E., and James A. Sandos
  1981 *Across the Border: Rural Development in Mexico and Recent Migration to the United States.* Berkeley: Institute of Governmental Studies, University of California.
Damante Bilello, Dr. Luis Jorge
  1983 "Investigación de un brote de intoxicación por plaguicidas." Paper

presented at the Thirty-seventh Reunión Nacional de Salud Pública, Guanajuato, November 23.

Danbom, David B.
1979 *The Resisted Revolution: Urban America and the Industrialization of Agriculture, 1900–1930.* Iowa City: University of Iowa Press.

David, Wilfred L.
1986 *Conflicting Paradigms in the Economics of Developing Nations.* New York: Praeger.

Desowitz, Robert S.
1981 *New Guinea Tapeworms and Jewish Grandmothers: Tales of Parasites and People.* New York: Norton.

Díaz, Lilia
1976 "El liberalismo militante." In El Colegio de México 1976.

Díaz Ramo, Patricia
1987 Personal communication.

Dittrich, V., S. O. Hassan, and G. H. Ernst
1985 "A case study of the emergence of a new primary pest." *Crop Protection* 4 (2): 161–176.

Dover, Michael
1985 *A Better Mousetrap.* Washington, D.C.: World Resources Institute.

Dover, Michael, and Brian Croft
1984 *Getting Tough: Public Policy and the Management of Pesticide Resistance.* Washington, D.C.: World Resources Institute.

Doyle, Jack
1985 *Altered Harvest: Agriculture, Genetics, and the Fate of the World Food Supply.* New York: Viking.

Drum, David
1987 "Boomtime in Baja." *California Farmer*, September 19: 60–61.

Duoll, John, Curtis D. Klaasen, and Mary O. Amdur
1980 *Toxicology: The Basic Science of Poisons.* New York: Macmillan.

Duran, Celeste, and Jim Mitchell
1984 News broadcasts on pesticide use in Mexico and residues on Mexican produce sold in the United States. Channel 2, Los Angeles, February 26–28.

Eckholm, Erik P.
1976 *Losing Ground: Environmental Stress and World Food Prospects.* New York: Norton.

Esainko, Peter, and Valerie Wheeler
1987 "Cultural factors in the diffusion of organic agricultural techniques." Presentation to Association of Research Scholars, California State University, Sacramento.

Esteva, Gustavo
1982 *La batalla en el méxico rural.* 3rd ed. Mexico City: Siglo XXI.

Farris, Nancy M.
   1984 *Maya Society under Colonial Rule.* Princeton, N.J.: Princeton
      University Press.
Fazal, Anwar
   1986 Remarks at meeting of the Pesticide Action Network, Ottawa,
      Canada, June 3.
Feder, Ernst
   1976 *El imperialismo fresa: Una investigación sobre los mecanismos
      de la dependencia en la agricultura mexicana.* Mexico City: Edi-
      ciones Campesino, Sec. de Reforma Agraria.
Feenstra, Gail W.
   1988 *Who Chooses Your Food: A Study of the Effects of Cosmetic Stan-
      dards on the Quality of Produce.* CalPirg Report, August.
Fernández Flores, Jesús
   1984 Interview, February 24.
Fertimex
   1981 *Plan de desarrollo de Fertimex en la producción, formulación y
      comercialización de insecticidas.* Mexico City: Gerencia General
      de Planeación y Desarrollo, Gerencia de Planeación, Subgerencia
      de Programación.
Flannery, Kent
   1983 "Precolumbian farming in the Valleys of Oaxaca, Nochixtlán,
      Tehuacán, and Cuicatlán: A comparative study." In *The Cloud
      People: Divergent Evolution of the Zapotec and Mixtec Civiliza-
      tions,* ed. Kent Flannery and Joyce Marcus. New York: Academic
      Press.
Flint, Mary, and Robert van den Bosch
   1981 *Introduction to Integrated Pest Management.* New York: Plenum.
Flora, Cornelia Butler
   1988 "Farming systems approaches in international technical coopera-
      tion in agriculture and rural life." *Agriculture and Human Values*
      5 (1–2).
Florescano, Enrique
   1971 *Origin y desarrollo de los problemas agrarias de México, 1500–
      1821.* Mexico City: Era.
Frisbie, R. E.
   1981 "Pest management systems for cotton insects." In *CRC Hand-
      book of Pest Management in Agriculture,* vol. 3, ed. David Pi-
      mentel. Boca Raton, Fla.: CRC Press.
Galinat, W. C.
   1971 "The origin of maize." *American Review of Genetics* 5: 447–448.
Geertz, Clifford
   1963 *Agricultural Involution: The Processes of Ecological Change in
      Indonesia.* Berkeley and Los Angeles: University of California
      Press.
George, Susan
   1976 *How the Other Half Dies.* London: Penguin.

Gibson, Charles
 1964 *The Aztecs under Spanish Rule.* Stanford, Calif.: Stanford University Press.
Gips, Terry
 1987 *Breaking the Pesticide Habit.* Minneapolis: International Institute for Sustainable Agriculture.
Gliessman, Stephen R., and Moisés Amador Alarcón
 1980 "Ecological aspects of production in traditional agroecosystems in the humid lowland tropics of Mexico." In *Tropical Ecology and Development,* ed. José Furtado. Kuala Lumpur, Malaysia: International Society for Tropical Ecology.
 1984 "An agroecological approach to sustainable agriculture." In Jackson, Berry, and Colman 1984.
Gliessman, Stephen R., E. R. García, and A. M. Amador
 1981 "The ecological basis for the application of traditional agricultural technology in the management of tropical agro-ecosystems." *Agroecosystems* 7: 173–185.
Goldenman, Gretta, and Sarojini Rengam
 1986 *Problem Pesticides, Pesticide Problems.* San Francisco: Pesticide Action Network North America, International Organization of Consumer Unions.
Gómez Pompa, Arturo
 1978 "An old answer to the future." *Mazingira* 5: 50–55.
González, Guadalupe Pérez
 1984 Interviews, January 12, 13.
 1988 Interview, July 15.
González, Catalina
 1988 Interview, July 15.
González, Héctor Miguel
 1987 Interview, February 6.
González, Luis
 1976 "El liberalismo truinfante." In El Colegio de México 1976.
González Pacheco, Cuauhtémoc
 1983 *Capital extranjero en la selva de Chiapas, 1863–1982.* Mexico City: UNAM.
Gorbachev, Mikhail
 1987 *Perestroika: New Thinking for Our Country and the World.* New York: Harper.
Griffin, Keith
 1974 *The Political Economy of Agrarian Change.* Cambridge, Mass.: Harvard University Press.
 1987 *World Hunger and the World Economy: And Other Essays in Developmental Economics.* London: Macmillan.
Grindle, Merilee Serrill
 1977 *Bureaucrats, Politicians, and Peasants in Mexico: A Case Study in Public Policy.* Berkeley: University of California Press.

Guenzi, W. D.
    1974 *Pesticides in Soil and Water.* Madison: Soil Science Society of America.
Gutelman, Michel
    1983 *Capitalismo y reforma agraria en México.* 9th ed. Mexico City: Era.
Hale, Charles
    1968 *Mexican Liberalism in the Age of Mora, 1821–1853.* New Haven: Yale University Press.
Hamilton, Nora
    1982 *The Limits of State Autonomy.* Princeton, N.J.: Princeton University Press.
Hammond, Norman
    1982 *Ancient Mayan Civilization.* New Brunswick: Rutgers University Press.
Hansen, Michael
    1986 *Escape from the Pesticide Treadmill: Alternatives to Pesticides in Developing Countries.* Mt. Vernon, N.Y.: Consumers Union.
Harrison, Paul
    1987 *The Greening of Africa.* New York: Penguin.
Hassig, Ross
    1985 *Trade, Tribute, and Transportation: The Sixteenth Century Political Economy of the Valley of Mexico.* Norman: University of Oklahoma Press.
Hayes, Wayland J., Jr.
    1982 *Pesticides Studied in Man.* Baltimore: Williams and Wilkins.
Henderson, J.
    1968 *Legal Aspects of Crop Spraying.* Circular 99. Champaign-Urbana: University of Illinois Experiment Station.
Hernández Xolocotzl, Efraím
    1977 *Agroecosistemas de México: Contribuciones a la enseñanza, investigación y divulgación agrícola.* Chapingo, Mex.: Colegio de Posgraduados.
Herring, Hubert
    1962 *A History of Latin America.* New York: Knopf.
Hewitt de Alcántara, Cynthia
    1976 *Modernizing Mexican Agriculture: Socio-Economic Implications of Technological Change, 1940–1970.* Geneva: United Nations Research in Social Development.
Hinshaw, Robert
    1975 *Panajachel: A Guatemalan Town in 30 Year Perspective.* Pittsburgh: University of Pittsburgh Press.
Hoar, Sheila, et al.
    1986 "Agricultural herbicide use and risk of lymphoma and soft-tissue sarcoma." *Journal of the American Medical Association* 256: 1141–1147.

Huntington, Samuel P.
    1968   *Political Order in Changing Societies.* New Haven: Yale University Press.
Jackson, Wes
    1980   *New Roots for Agriculture.* San Francisco: Friends of the Earth.
    1987   *Altars of Unhewn Stone: Science and the Earth.* Berkeley: North Point.
Jackson, Wes, Wendell Berry, and Bruce Colman, eds.
    1984   *Meeting the Expectations of the Land: Essays in Sustainable Agriculture and Stewardship.* Berkeley: North Point.
Jennings, Bruce H.
    1988   *Foundations of International Agricultural Research: Science and Politics in Mexican Agriculture.* Boulder, Colo.: Westview.
Jiménez Moreno, Wigberto
    1966   "Mesoamerica before the Toltecs." In Paddock 1966.
Kahn, E. J., Jr.
    1984   *The Staffs of Life.* Boston: Little, Brown.
Kahn, Ephraim
    1976   "Pesticide related illness in California farm workers." *Journal of Occupational Medicine* 18 (October): 693–696.
Karim, M. Bazhul
    1986   *The Green Revolution: An International Bibliography.* New York: Greenwood.
Kearney, Michael
    1983   Interview, September 9.
    1984a "Integration of the Mixteca and the western U.S.-Mexico Region via migratory wage labor." Conference on Regional Aspects of U.S.-Mexican Integration. Center for U.S.-Mexican Studies, University of California, San Diego, May.
    1984b Interview, March 10.
    1986   "From the invisible hand to the visible feet: Anthropological studies of migration and development." *Annual Review of Anthropology* 15: 331–361.
    1987   Interview, September 30.
    1988   Interviews, July 16–18.
    1989   "Mixtec political consciousness: From passive to active resistance." In *Rural Revolt in Mexico and U.S. Intervention,* forthcoming from Center for U.S.-Mexican Studies, University of California, San Diego.
Kearney, Michael, and Carole Nagengast
    1988   "Anthropological perspectives on transnational communities in rural California." Prepared for California Institute for Rural Studies, Davis, July 15.
King, Russell
    1977   *Land Reform: A World Survey.* Boulder, Colo.: Westview.

Kissinger, Henry, et al.
 1984 *Report of the Bipartisan Commission on Central America.* Washington, D.C.: U.S. Government Printing Office.
Kistner, William
 1986 "Scrutiny of the bounty: The chemical fog over Mexico's farmworkers." *Mother Jones,* December 28.
Krauze, Enrique
 1987a *Plutarco E. Calles.* Mexico City: Fondo de la Cultura Económica.
 1987b *Emiliano Zapata.* Mexico City: Fondo de la Cultura Económica.
Kunkel, John H.
 1965 "Economic autonomy and social change in Mexican villages." In *Contemporary Cultures and Societies of Latin America,* ed. Dwight B. Heath and Richard N. Adams. New York: Random House.
Kurzman, Dan
 1987 *A Killing Wind: Inside Union Carbide and the Bhopal Catastrophe.* New York: McGraw-Hill.
Lambert, John
 1981 "The ecological consequences of ancient Maya agricultural practices in Belize, Central America." Presented at the Symposium on Prehistoric Intensive Agriculture in the Tropics, Australian National University, Canberra, August 1981.
Leopold, Aldo
 1953 "The Land Ethic." In *A Sand County Almanac.* New York: Ballantine; repr. 1974.
Levine, R. S.
 1985 *Assessment of Mortality and Morbidity Due to Unintentional Pesticide Poisonings.* Geneva: World Health Organization.
Lightstone, Ralph
 1987 Remarks at symposium on NAS Delaney Paradox study, Goethe House, sponsored by the Pesticide Education and Action Project and the Center for California Studies, Cal State University, Sacramento, August.
Lipton, Michael
 1977 *Why Poor People Stay Poor.* London: Temple Smith.
Lira, Andrés, and Luis Muro
 1976 "El siglo de integración." In El Colegio de México 1976.
Lockhart, James
 1969 "Encomienda and hacienda: The evolution of the great estate in the Spanish Indies." *Hispanic American Historical Review* 49: 411–429.
Loera, Rogelio
 1988 *La información sobre plaguicidas en México.* Cuadernos de Divulgación INIREB no. 33. Xalapa, Veracruz: Instituto Nacional de Investigaciones sobre Recursos Bióticos.

Lomelí, Arturo

1983a "El dilema de los plaguicidas: Esperanza de liberación y progreso o pesadilla contaminadora?" *La Voz del Consumidor* 1 (3).

1983b Interview, October 10.

1984 Interview, March 28.

1986 Interview, July 1.

López Alanis, Gilberto, et al.

1982 *Monografía del Estado de Sinaloa.* Culiacán: Universidad Autónoma de Sinaloa, Instituto de Investigaciones de Ciencias y Humanidades.

López Ballieau, Cynthia

1988 Interview, November 15.

Maggio, Frank

1987 Interview, February 8.

Manglesdorf, Paul C.

1974 *Corn: Its Origin, Evolution, and Improvement.* Cambridge, Mass.: Belknap Press of Harvard University Press.

Marco, Gino G., Robert M. Hollingworth, and William Durham

1987 *Silent Spring Revisited.* Washington, D.C.: American Chemical Society.

Marcus, Joyce, and Kent Flannery, eds.

1983 *The Cloud People: Divergent Evolution of the Zapotec and Mixtec Civilizations.* New York: Academic Press.

Mares, David

1987 *Penetrating the International Market: Theoretical Considerations and a Mexican Case Study.* New York: Columbia University Press.

Martínez, Ifigenia

1988 Interview. See Reding 1988.

Martínez Muñoz, Marco Antonio

1982 "Infraestructura y organización para el buen uso de plaguicidas en México." Master's thesis, Universidad Autónoma de Chapingo.

Martínez Saldaña, Tomás

1980 *El costo social de un éxito político.* Chapingo, Mexico: Colegio de Postgraduados.

1986 "Historia de la agricultura mexicana." Presented at INIREB, June 18, Primer Taller Nacional sobre Riesgos de Plaguicidas, Xalapa.

Medellín Leal, Fernando, ed.

1978 *La desertificación en México.* San Luis Potosí: Universidad Autónoma de San Luis Potosí.

Medvedev, Zhores A.

1987 *Soviet Agriculture.* New York: Norton.

Meier, Barry

1987 "Poison produce." *Wall Street Journal,* March 26, 1ff.

Merchant, Carolyn
  1980 *The Death of Nature: Women, Ecology, and the Scientific Revolution.* New York: Harper and Row.
Metcalf, Robert L.
  1982 "Historical perspective of organophosphorous ester-induced delayed neurotoxicity." *Neurotoxicology* 3 (4): 269–284.
Meyer, Michael C., and William L. Sherman
  1983 *The Course of Mexican History.* 2nd ed. New York: Oxford.
Meyerhoff, A., and Lawrie Mott
  1984 "Pesticide residues: Swallowing the government line." *Sierra,* July–August: 21–27.
Monsivais, Carlos
  1976 "Notas sobre la cultura mexicana en el siglo XX." In El Colegio de México 1976.
Moore, Monica
  1986 Interview, January 28.
Morales, Dr. Luis
  1987 Interview, February 6.
Morelos González, José
  1984 Interview, February 24.
Morgado, Dr. Javier
  1983 Interview, October 12.
Moses, Dr. Marion
  1988 *A Field Survey of Pesticide-Related Working Conditions in the U.S. and Canada: Monitoring the International Code of Conduct on the Distribution and Use of Pesticides in North America.* San Francisco: Pesticide Education and Action Project.
Mott, Lawrie
  1986 *Pesticide Registration: An Evaluation of EPA's Progress.* San Francisco: Natural Resources Defense Council.
Mott, Lawrie, and Karen Snyder
  1988 *Pesticide Alert: A Guide to Pesticides in Fruits and Vegetables.* San Francisco: Sierra Club Books.
Mouat, Andrew
  1986 Interview, June 19.
Moulton, Kirby, David Runsten, Roberta Cook, James Chalfant, and Ricardo Amon
  1988 *Competitiveness at Home and Abroad, Report of a 1986–87 Study Group on Marketing California Specialty Crops: Worldwide Competition and Constraints.* Berkeley: University of California Agricultural Issues Center.
Muñoz Ledo, Porfirio
  1988 Interview. See Reding 1988.
Nabhan, Gary Paul
  1982 *The Desert Smells Like Rain: A Naturalist in Papago Indian Country.* San Francisco: North Point.

1989   *Enduring Seeds: Native American Agriculture and Wild Plant Conservation.* San Francisco: North Point.

Nagengast, Carole
   1988   "Mixtec ethnicity: Social identity, political consciousness and political activism." Forthcoming in untitled volume ed. Arturo Gomez Pompa. Boulder, Colo.: Westview, [1991].

Nagengast, Carole, Michael Kearney, and Rodolfo Stavenhagen
   1989   "Human rights and indigenous workers: The Mixtecs in Mexico and the United States." Forthcoming in untitled volume ed. Arturo Gómez Pompa. Boulder, Colo.: Westview, [1991].

Nakayama, Antonio
   1983   *Sinaloa: Un busquejo de su historia.* Culiacán: Universidad Autónoma de Sinaloa.

National Academy of Sciences
   1972   *Genetic Vulnerability of Major Crops.* Washington, D.C.: NAS Press.
   1986   *Pesticide Resistance: Strategies and Tactics for Management.* Washington, D.C.: NAS Press.
   1987   *Regulating Pesticides in Food: The Delaney Paradox.* Washington, D.C.: NAS Press.
   1989   *Alternative Agriculture.* Washington, D.C.: NAS Press.

National Geographic Magazine
   1945   "Your new world of tomorrow," October.

Nations, James
   1984   "The Lacandones, Gertrude Blom, and the Selva Lacandona." In *Gertrude Blom: Bearing Witness,* ed. Alex Harris and Margaret Sartor. Chapel Hill: University of North Carolina Press.

Nations, James, and Jeffrey Leonard
   1986   "Grounds of conflict in Central America." In *Bordering on Trouble: Resources and Politics in Latin America,* ed. Andrew Maguire and Janet Walsh Brown. New York: Adler and Adler.

Nicolson, Harold
   1935   *Dwight Morrow.* New York: Harcourt, Brace and Company.

Norgaard, Richard
   1987   "The epistemological basis of agroecology." In Altieri 1987.

Olson, Leon
   1986   "The immune system and pesticides." *Journal of Pesticide Reform,* Summer.

Ornelas, Rene Jiménez, and Alberto Minujin Zmud
   1984   *Los factores del cambio demográfico en México.* Mexico City: Siglo XXI.

Ottawa Declaration
   1986   Fate of the Earth Conference, June 5, Ottawa, Canada. Pesticide Action Network, San Francisco and Penang, Malaysia.

Owens, Clarence, and Emiel Owens
 1982 *Atlantic Coast Migrant Stream Pesticide Study.* Environmental
  Protection Agency.
Paddock, Joe, Nancy Paddock, and Carol Bly
 1986 *Soil and Survival: Land Stewardship and the Future of American
  Agriculture.* San Francisco: Sierra Club.
Paddock, John
 1966 *Ancient Oaxaca: Discoveries in Mexican Archeology History.*
  Stanford: Stanford University Press.
Palerm, Angel
 1973 *Obras hidráulicas prehispánicas en el sistema lacustre del Valle
  de México.* Mexico City: Instituto Nacional de Antropología e
  Historia.
Palerm, Angel, and Eric Wolf
 1972 *Agricultura y civilización en Mesoamérica.* Mexico City:
  SepSetentas.
Palerm, Juan Vicente
 1987 "Transformations in rural California" (anonymously written re-
  search report on Palerm's work). *UC Mexus News,* nos. 21–22
  (Fall 1987–Winter 1988).
Pare, Luisa
 1983 *El proletariado agrícola en México: Campesinos sin tierra o pro-
  letarios agrícolas.* 5th ed. Mexico City: Siglo XXI.
Payer, Cheryl
 1974 *The Debt Trap.* New York: Monthly Review.
Pérez, Andrés
 1984 Interview, February 24.
Perkins, John
 1983 *Insects, Experts, and the Pesticide Crisis.* New York: Academic
  Press.
Perlman, Michael
 1977 *Farming for Profit in a Hungry World.* Montclair, N.J.: Allanhold,
  Osmun and Company.
Pesticide Education and Action Project (now Pesticide Action Network,
 North American Regional Center)
 1987 Meeting on the Delaney Clause and Pesticide Regulatory Policy,
  Sacramento, Goethe House, August 30.
Pimentel, David, and John Perkins
 1978 "Environmental aspects of world pest control." In *World Food,
  Pest Losses, and the Environment,* ed. D. Pimentel et al. Boulder,
  Colo.: Westview.
 1980 *Pest Control: Cultural and Environmental Aspects.* AAAS Sym-
  posium Series 43. Boulder, Colo.: Westview for AAAS.
Posadas, Florencio
 1982 "El proletario agrícola en el estado de Sinaloa." Master's thesis,
  FLACSO, Mexico City.

Postel, Sandra
    1988 "Controlling toxic chemicals." In *State of the World, 1988*, ed. Lester Brown et al. New York: Norton.
Ragsdale, Nancy N., and Ronald J. Kuhr
    1987 *Pesticides: Minimizing the Risks.* Washington, D.C.: American Chemical Society.
Rama, Ruth, and Raúl Vigorito
    1979 *El complejo de frutas y legumbres en México.* Part of a series, Transnacionales en América Latina. Mexico City: Nueva Imagen.
Ravicz, Robert S.
    1965 *Organización social de los Mixtecos.* Mexico City: Instituto Nacional Indigenista; repr. 1980.
Reding, Andrew
    1988 "The democratic current: A new era in Mexican politics." *World Policy Journal*, Spring: 323–363. Includes interviews with Cuauhtémoc Cárdenas Solorzano, Ifigenia Martínez, and Porfirio Muñoz Ledo.
Reed, Nelson
    1964 *The Caste War of Yucatan.* Stanford, Calif.: Stanford University Press.
Repetto, Robert
    1985 *Paying the Price: Pesticide Subsidies in Developing Countries.* Washington, D.C.: World Resources Institute.
Restrepo, Iván, and Susana Franco
    1988 *Naturaleza muerta: Los plaguicidas en México.* Mexico City: Oceano.
Reynolds, Clark
    1987 Keynote address at the Annual Meeting of the Pacific Coast Council on Latin American Studies, Tempe, Arizona, October 12.
Ritchie, D., ed.
    1985 *Fungicide and Nematicide Tests.* Vol. 40. St. Paul, Minn.: American Phytopathological Society.
Rojas Rabiela, Teresa, ed.
    1974 *Nuevas noticias sobre las obras hidráulicas prehispánicas y coloniales en el valle de México.* Mexico City: Sep/Inah.
    1983 *La agricultura chinampera.* Mexico City: Universidad Autónoma Chapingo.
Ronfeldt, David
    1973 *Atencingo: The Politics of Agrarian Struggle in a Mexican Ejido.* Stanford, Calif.: Stanford University Press.
Rostow, W. W.
    1959 *The Stages of Economic Growth.* Cambridge: Cambridge University Press.
    1971 *Politics and the Stages of Growth.* Cambridge: Cambridge University Press.

Rudd, Robert L.
1970 *Pesticides and the Living Landscape.* 2nd ed. Madison: University of Wisconsin Press.

Runsten, David, and Kirby Moulton
1988 "Competition in Processing Tomatoes." In Moulton et al. 1988.

Sailer, Reece J.
1981 "Extent of biological and cultural control of insect pests of crops." In *CRC Handbook of Pest Management in Agriculture,* ed. David Pimentel. Boca Raton, Fla.: CRC Press.

Salinas de Gortari, Carlos
1982 *Political Participation, Public Investment, and Support for the System: A Comparative Study of Rural Communities in Mexico.* RR-35. San Diego: Center for U.S.-Mexican Studies.

Sánchez, Arturo
1984 Interview, February 9.

Sanderson, Steve E.
1981 *Agrarian Populism and the Mexican State: The Struggle for Land in Sonora.* Berkeley: University of California Press.
1986 *The Transformation of Mexican Agriculture: International Structure and the Politics of Agrarian Change.* Princeton: Princeton University Press.

Sanidad Vegetal
1982 *Plan nacional fitosanitario.* Mexico City.

Sauer, Carl O.
1955 "The agency of man on the Earth." In *Man's Role in Changing the Face of the Earth,* ed. Wm. L. Thomas, Jr. Chicago: University of Chicago Press.

Sax, N. Irving
1979 *Dangerous Properties of Industrial Materials.* New York: Van Nostrand.

Schapsmeier, Edward L. and Frederick H.
1970 *Prophet in Politics: Henry A. Wallace and the War Years, 1940–45.* Ames: Iowa State University Press.

Secretaría de Agricultura y Recursos Hidráulicos
1977 *Estado agrológico de reconocimiento de las zonas bajas del Estado de Sinaloa.* Mexico City: SARH.

Secretaría de Programación y Presupuesto
1981 *El sector alimentario en México.* Mexico City.

Secretaría de Programación y Presupuesto, Centro para el Desarrollo Rural Integral
1984 *Elementos para formulación de un programa de desarrollo rural para la mixteca oaxaqueña.* Mexico City.

Sheridan, David
1981 *Desertification of the United States.* Washington, D.C.: Council on Environmental Quality, U.S. Government Printing Office.

Shilling, Elisabeth
1938 "Los 'jardines flotantes' de Xochimilco." In *La Agricultura Chinampera*, ed. Teresa Rojas Rabiela. Chapingo: Universidad Autónoma Chapingo.

Sinkin, Richard
1972 "Modernization and reform in Mexico: 1858–1876." Ph.D. dissertation, University of Michigan.

Spores, Ronald
1967 *The Mixtec Kings and Their People.* Norman: University of Oklahoma Press.
1984 *The Mixtecs in Ancient and Modern Times.* Norman: University of Oklahoma Press.

Stakman, E. C., Richard Bradfield, and Paul Manglesdorf
1967 *Campaigns against Hunger.* Cambridge, Mass.: Belknap Press.

Stavenhagen, Rodolfo
1975 *Social Classes in Agrarian Societies.* New York: Doubleday.

Sterling, Claire
1974 "The making of the sub-Saharan waste-land." *Atlantic Monthly*, May: 98–105.

Stimman, M. W.
1980 *Pesticide Application and Safety Training.* Publication no. 4070. Davis: Division of Agricultural Sciences, University of California.

Strange, Marty, ed.
1984 *It's Not All Sunshine and Fresh Air: Chronic Health Effects of Modern Farming Practices.* Walthill, Nebr.: Center for Rural Affairs.
1988 *Family Farming: A New Economic Vision.* Lincoln, Nebr., and San Francisco: University of Nebraska Institute for Food and Development Policy.

Stuart, James, and Michael Kearney
1981 *Causes and Effects of Agricultural Labor Migration from the Mixteca of Oaxaca to California.* La Jolla: Program in U.S.-Mexican Studies, University of California, San Diego.

Swezey, Sean, and Rainer Daxl
1983 *Breaking the Circle of Poison: The Integrated Pest Management Revolution in Nicaragua.* San Francisco: Institute for Food and Development Policy.

Swift, Jeremy
1973 "Disaster and a Sahelian nomad economy." In *Drought in Africa*, ed. David Dalby and R. J. Harrison. London: Centre for African Studies.

Tai, Hung-chao
1974 *Land Reform and Politics: A Comparative Analysis.* Berkeley: University of California Press.

Tatum, L. A.
1971 "The southern corn leaf blight epidemic." *Science* 171: 1113.

Tax, Sol
   1965 "World view and social relations in Guatemala." In *Contemporary Cultures and Societies of Latin America*, ed. Dwight B. Heath and Richard N. Adams. New York: Random House.
Taylor, William B.
   1972 *Landlord and Peasant in Colonial Oaxaca.* Stanford, Calif.: Stanford University Press.
Thompson, J. Eric S.
   1968 *The Rise and Fall of Mayan Civilization.* 2nd ed. Norman: University of Oklahoma Press.
Thompson, Richard
   1988 Interview, February 16.
Thrupp, Lori Ann
   1988 "Pesticides and policies: Approaches to pest-control dilemmas in Nicaragua and Costa Rica." *Latin American Perspectives* 15 (4).
Todorov, Tzvetan
   1984 *The Conquest of America.* Trans. from the French by Richard Howard. New York: Harper and Row.
Toledo, Alejandro
   1983 *Como destruir el paraíso: El desastre ecológico del sureste.* Mexico City: Oceano.
Toledo, Alejandro, Julia Carabias, Cristina Mapes, and Carlos Toledo
   1985 *Ecología y autosuficiencia alimentaria.* Mexico City: Siglo XXI.
Trujillo Arriaga, Javier
   1983 Interview, October 13.
   1985 Interview, December 10.
   1987 "The agroecology of maize production in Tlaxcala, Mexico: Cropping systems effects on arthropod communities." Ph.D. dissertation, University of California, Berkeley.
University of California Division of Agricultural Sciences
   1980 *Pesticide Application and Safety Training.* Sale Publication 4070.
UNFAO (United Nations Food and Agriculture Organization)
   1985 "International code of conduct on the distribution and use of pesticides." C 85/25-Rev. 1, November, Twenty-third Session, Rome.
UNIDO (United Nations Industrial Development Organization)
   1980 "Report on the use of pesticides in Latin America." UNIDO/IDO. 353, May 19.
UNWHO (United Nations World Health Organization)
   1973 *Safe Use of Pesticides.* 20th Report of the WHO Expert Committee on Insecticides, Tech. Rept. Series No. 513. Geneva.
USDA (United States Department of Agriculture)
   1980 *Report and Recommendations on Organic Farming.* Washington, D.C.: U.S. Government Printing Office.
USEPA (United States Environmental Protection Agency)
   1983 "U.S.-Mexico Cooperative Environmental Agreement." August 14.

1987a *Unfinished Business: A Comparative Assessment of Environmental Problems.* February.

1987b *Agricultural Chemicals in Ground Water: Strategic Plan.* June.

USFDA (United States Food and Drug Administration)

n.d.   "MOU regarding foreign commerce between Mexico and the United States." Guide 7156b.01.

USGAO (United States General Accounting Office)

1984   *Commercial Biotechnology: An International Analysis.*

1986   *Pesticides: Better Sampling and Enforcement Needed on Imported Food.* GAO/RCED-86-219.

Ussher, Ricardo García

1984   Interview, February 8.

van den Bosch, Robert

1978   *The Pesticide Conspiracy.* New York: Doubleday.

Vanderwood, Paul J.

1981   *Disorder and Progress: Bandits, Police and Mexican Development.* Lincoln: University of Nebraska Press.

van Heemstra, E. A. H., and W. F. Tordoir

1982   *Education and Safe Handling in Pesticide Application.* Amsterdam: Elsevier.

Vargas, Humberto

1984   Interview and field visits, February 9.

Vaughan, Christopher

1988   "Disarming farming's chemical warriors: Research brightens the dark underside of the Green Revolution." *Science News* 134: 120–121.

Velázquez, Luis Antonio

1984   Interview, February 16.

Verduzco Gutiérrez, José, and Guillermo Gaytan

1972   *Erosión, pobreza mexicana.* Mexico City: Subsec. Forestal y de la Fauna.

Vetorrazi, Gaston

1979   *International Regulatory Aspects for Pesticide Chemicals.* Boca Raton, Fla.: CRC Press.

Vitale, Luis

1983   *Hacia una historia ambiental en América Latina.* Mexico City: Nueva Imagen.

Wade, Nicholas

1974   "Sahelian drought: No victory for Western aid." *Science* 185: 234–237.

Walker, Lynne

1988   "The invisible work force: San Diego's migrant farm laborers from Oaxaca." *San Diego Union,* December 18; reprinted as special publication, March 10, 1989.

Walker, William O.

1981 *Drug Control in the Americas.* Albuquerque: University of New Mexico Press.

Wallace, Henry A., and William L. Brown

1956 *Corn and Its Early Fathers.* Lansing: Michigan State University Press.

Ware, Helen

1975 "The Sahelian drought: Some thoughts on the future." Special Sahelian Office, U.N. Food and Agriculture Organization, March 26.

Warman, Arturo

1978 . . . *y venimos a contradecir: Los campesinos de Morelos y el estado nacional.* Mexico City: Ediciones de La Casa Chata.

Wasilewski, Amia

1987 "The quiet epidemic: Pesticide poisonings in Asia." *IDRC Reports* 16 (1).

Wasserstrom, Robert

1983 *Class and Society in Central Chiapas.* Berkeley: University of California Press.

Wasserstrom, Robert, and Richard Wiles

1985 *Field Duty: U.S. Farmworkers and Pesticide Safety.* Washington, D.C.: World Resources Institute.

Watt, Kenneth

1973 *Principles of Environmental Science.* New York: McGraw-Hill.

Weatherwax, Paul

1954 *Indian Corn in Old America.* New York: Macmillan.

Weir, David

1987 *The Bhopal Syndrome: Pesticides, Environment, and Health.* San Francisco: Sierra Club.

1989 "Will the Circle Be Unbroken?" *Mother Jones,* June.

Weir, David, and Mark Shapiro

1981 *The Circle of Poison.* San Francisco: Institute for Food and Development Policy.

Wells, Miriam J., and Martha S. West

1988 "Regulation of the farm labor market: An assessment of the farm worker protections under California's Agricultural Labor Relations Act." Paper presented at the Conference of the Working Group on Rural Poverty, Fresno, May 7, California Institute for Rural Studies, Davis.

Wheeler, Valerie, and Peter Esainko

1987 "Value Construction and Choice in Ohio Agriculture." Lecture, California State University, Sacramento, October.

Wiles, Richard

1983 "The United States and the proliferation of pesticides in developing countries." M.A. thesis, California State University, Sacramento.

Wilken, Gene C.
    1987   *Good Farmers: Traditional Resource Management in Mexico and Central America.* Berkeley: University of California Press.
Wilkes, Garrison
    1972   "Maize and its wild relatives." *Science* 177: 1071–1077.
Williams, Robert G.
    1986   *Export Agriculture and the Crisis in Central America.* Chapel Hill: University of North Carolina Press.
Wolf, Eric R.
    1959   *Sons of the Shaking Earth.* Chicago: University of Chicago Press.
Wolterding, Martin
    1981   "The poisoning of Central America." *Sierra,* September–October: 63–66.
Womack, John Jr.
    1968   *Zapata and the Mexican Revolution.* New York: Vintage.
Worster, Donald
    1977   *Nature's Economy.* New York: Anchor.
    1979   *Dust Bowl: The Southern Plains in the 1930's.* New York: Oxford University Press.
Wright, Angus
    1985   "Innocents abroad—American agricultural research in Mexico." In *Meeting the Expectations of the Land,* ed. Wes Jackson et al. San Francisco: North Point Press.
    1986   "The politics of pesticide poisoning among Mexican farmworkers." *Latin American Perspectives* 13 (4): 26–59.
Yates, Paul Lamartine
    1981   *Mexico's Agricultural Dilemma.* Tucson: University of Arizona Press.

Index

Abandonment of land, 227–228
ABC Farms, 41, 201
Acetylcholinesterase (ACHE) inhibition by pesticides, 31–33, 206, 218
Aerial application of pesticides, 19–24, 59–60, 78–79
Africa, 265–267
Agent Orange, 141
Agrarian reform: and development banks, 204; importance in development, 238–239, 268–269; in Mexico, 151–153, 166–172, 175, 186–187, 241; in Taiwan, 268–269. *See also Ejidos*; Landholding
Agricultural modernization: assessment, 244–285; in Central America, 234–236; and culture, 263–264; global, 140–187, 284; and industrialization, 227–229, 264–272, 284; in Mexico, 67, 138–139, 140–187, 221; in Mixteca, 133; origins, 6; and population growth, 175, 234–235; and Sonoran dynasty, 166–168; theory, 222–243, 284
Agricultural productivity, 8, 174–177, 181–183, 246, 247–252; altered by pesticides, 275
Agricultural technology: assessment, 244–285; and culture, 263–264; negative consequences, xvi; neutrality, 246, 279–284; origins in Mexico,

140–187; reliance on nature, 272–279; for security, 153–166, 256; for surplus, 166–187; transfer, 60, 66–67, 253. *See also* Green Revolution; Pesticides; Sustainable agriculture; Traditional agriculture
Agroecology: the Scientific Basis of Alternative Agriculture (Altieri), 254–255
Agro-ecosystems, 64, 246, 254–255, 272–279. *See also* Sustainable agriculture
Agroexports. *See* Central America; Culiacán; Development theory; Export economy
AIDS, 244
Albert, Lilia, 67–69
Aldicarb, 16, 201
Aldrin, 17
Alleopathic plants, 255
Alternative Agriculture (National Academy of Sciences), 275–276
Altieri, Miguel, 254–255, 256, 258, 259
American Products Overseas Services, 73
Amish, 256, 257
Anderson, Edgar, 160–161, 179
Apatzingan, Michoacán, 42, 49
Arid Lands Project (Soviet Union), 267
Asparagus, 202
Astorga Lira, Enrique, 82, 138
Atencingo, Morelos, 269
Atropine, 41–42

Avery, Dennis, 234–235
Avila Camacho, Manuel, 171–174,
    185–186, 241, 272, 281
Aviles González, Mayra, 35–37
Avocados, 286–288, 298
Azinphos-ethyl (Bionex), 22
Aztec, 126, 158–159, 163, 260

Bacillus thuringiensis, 62, 213
Backpack application of pesticides,
    24–29
Baja California, Mexico, 40–42,
    202, 203, 288
Banrural, 73–76
Bayer Chemical, 22, 82
Beans, monocropped, 45. *See also*
    Triculture
Beteta, Ramón, 280
Bhopal, India, xii, 244
Bilello, Damante, 82–83
Bi-National Center for Human
    Rights, 288
Biotechnology, 277–279
Birds, 78
Black (bubonic) plague, 127–128
Boll weevil, 210
Bordeaux mixture, 141
Border crossing, 291–294
Borlaug, Norman: on importance of
    pesticides, 8; plant breeding re-
    search, 181–183
Bourbon reforms, 148
Bradfield, Richard, 174–177
Brazil, 204, 224, 231, 236, 265
Brenner, Anita, 168–169
Broccoli, 202
Bull, David, xii, 3
Bush, George, 301
Butterworth, Douglas, 108

Calderón, Salvador, 78–81
Calibration of pesticide application
    equipment, 20
California: agricultural research,
    178; cancer, 205; contamination
    of groundwater by pesticides,
    xii–xiii, 5; cosmetic standards

and pesticide use, 35; DBCP in-
    cident, 192; farm worker poison-
    ing, 5, 32, 205, 300; farm
    workers, 4–5, 286–289,
    300–301
California Department of Food and
    Agriculture, 78, 192, 197, 201
California Department of Health
    Services, 78
California Rural Legal Assistance,
    196
Calles, Plutarco Elías, 166–168,
    173
Campaigns against Hunger (Stak-
    man, Bradfield, and Man-
    glesdorf), 175, 280–281
Campbell's Soup, 203
Campo Gobierno (Villa Juárez), 53,
    54–58
Campo Patricio, 25, 27, 116
Campos, Sergio, 133
Cancer, 16, 22, 27, 39, 53; and
    DBCP, 192; and herbicides, 205
Canelos brothers, 41
Capital flight, 237–238
Carbamate insecticides, 4
Carbaryl (Sevin), xii
Cárdenas, Cuauhtémoc, 187,
    240–243, 252
Cárdenas, Lázaro, 94; presidency,
    168–171, 186–187
CARE, 218
Cargo system, 115, 118–119, 164
Caribbean, 203
Carson, Rachel, xi, xiii–xiv, 301
Castaños, Carlos Manuel, 215
Castle and Cook, 41, 76, 203
Cauliflower, 202
Centers for Crop Improvement, 257
Central America, 17, 40, 42, 70,
    124, 142; and agricultural mod-
    ernization, 234–236, 272; politi-
    cal instability, 243, 272
Centro de Investigaciones para el
    Desarrollo Rural Integral
    (CIDRI), 132, 136
Chahuixtle, 214–215

Chapin, Mac, 261–262
Chavez, Cesar, 5, 300–301
Chen Cheng, 268
Chevalier, François, 148
Chiapas, Mexico, 47–49
Chile, 203, 272
China, 236, 238
Chinampa: described, 158–160; *in-digenismo,* 168; peasant defense of, 145; potential for future, 258–262; and social systems, 163, 258–262
Chloredimiform, 201
Cholula, 126
Ciba-Geigy: DDT, 141; executives' viewpoint, xiv–xv; pollution of Rhine, xiii, 244
Científicos, 150
Circle of Poison, The (Weir and Shapiro), xi, 18
Clark Alfaro, Victor, 288–301
Clear Lake, California, 39
Closed corporate communities, 118–119, 164, 257
Clotts, H. V., 157–158
Clouthier, Manuel, 240, 252
Cochineal, 129
Coevolution of culture and agricultural technology, 255, 258
Colonialism: in Africa, 265–266; and agro-exports, 239; in development theory, 232–233
Colombia, 219
Columbian Exchange, The (Crosby), 146–147
Competitive advantage in pesticide use, 204
Comte, Auguste, 150
Comuneros, 101
Consumers, 189–190, 196–203
Consumer's Union Organization, International, 219–220
Corn. *See* Maize
Corporate landownership, 131, 147
Cosmetic standards, influence on pesticide use, 35
Costa Rica, Central America, 194

Costa Rica, Sinaloa, 51–52, 53, 58
Cotton: IPM, 209, 210; pest control, 37, 42–43, 44, 61, 62, 65; reducing production, 276–277
Credit, agricultural, 73–76, 183–184
Crosby, Alfred, 146–147
Cuba, 225, 272
Culiacán, Sinaloa: assasinations, 288; death of Ramón González in, 1–2; export region, 138, 270–271; favored by policy, 185; federal health services, 54–55; foreign investment, 73, 203; grower's perspective, 188–190, 222; history, 10–12; irrigation projects, 12; labor unions, 14–15, 194–196; land disputes, 12; Mixtec migration to, 132; model of agricultural modernization, 5–7, 245, 264; pesticide poisoning in, 51–64, 102–103, 115–118, 120; pesticide residue lab, 196; pesticide use in, 10–40, 49; relation to Mixteca, 270–271; university activity, 81, 83
Cupravit, 22–23, 27

Dairy products, residues, 43, 68
Daniels, Josephus, 170, 174
Darwin, Charles, 256
Daxl, Ranier, 218
DBCP, xii–xiii, 192, 194, 196
DDT, xi; invention and significance, 140–142; toxicity compared to parathion, 16–17
DDVP, 14
Debt: in Mexican economy, 137–139, 204, 229–231, 241, 249; and rural poor, 249
Debt peonage, 144
Debt Trap, The (Payer), 230–231
deforestation: and disease vectors, 42; in Chiapas, 47–49
Del Monte Foods, 203
Democratization, 242
Dengue fever, 42

Dermatitis, pesticide caused, 31
Devaluation of currency, 229–231
Development theory, 222–243
Díaz, Porfirio, 130–131, 150–151
Díaz Ordaz, Gustavo, 242
Dieldrin, 17
Disease and colonial population decline, 127–128
DNA-RNA, 256
Dole products, 203
Dominican Republic, 272
Dow Chemical Company: DBCP, xiii; 2, 4, 5-T, 141
Durango, 25
Dust Bowl, 175, 253

East-West Center, University of Hawaii, 257
Echeverría, Luis, 242
Ecological agriculture. *See* Sustainable agriculture
Economic development theory, 222–285
Economic threshold in pest management, 136
Effective demand for food, 227–228, 248–249
Egypt, 40
Eight Deer, 122, 123, 124
Ejidos: attacks on, 153; creation, 151–153, 167–168, 171–172; corruption of, 12, 44, 79, 171, 270; defined, 12, 93; in Mixteca, 93
El Bajío, 43–44, 49, 75
El Dorado, Sinaloa, 53
El Salvador, 17
Enclosure Acts, 232
encomienda, 144
Endosulfan, 16, 27
Endrin, 17
England as model of development, 231
Environmental consequences of pesticide use: in Culiacán, 38–40; general, xi–xiii. *See also*

Agricultural technology; Pesticides
Environmental degradation, global, 244
Epistemology, 258
Eradication of pests, 209–210
Ethiopia, 265
Europe as model of development, 232
Exiles in Mexican politics, 81–82
Export Agriculture and the Crisis in Central America (Williams), 235
Export economy, agricultural: in Central America, 234–236; development banks, 74, 204, 222; failure, 235–239, 245, 272; Mexican policy, 74, 222–224, 227–230, 252, 270–271; origins, 142. *See also* Culiacán
Export economy, industrial: in Mexico, 228, 238

Farming systems research, 257–258
Farm workers: as backpack pesticide applicators, 24–33; in California, 4–5, 286–289, 300–301; global estimates of numbers poisoned by pesticides, 3–4; growers' attitudes, 189; growing numbers in U.S., 4–5; health problems in Culiacán, 30–33, 103; housing and living conditions in Culiacán, 12–15, 115–118; Mexican government position on poisonings, 66; pesticide poisonings in California, 5, 32, 205; pesticide poisonings in Culiacán, 15–33, 51–64, 102–104, 115–118, 120, 298–299; pesticide poisonings in Florida, 205–206; pesticide poisonings in U.S., 205–206; unions, 191–196, 300–301
Fazal, Anwar, 220
Feder, Ernst, 43

Federal Insecticide, Fungicide, and Rodenticide Act, 197–198
Feed grains, 43–45, 228
Félix, "Gato," 288
Ferrell, J. A., 173
Fertilizers, 181–185
Fertimex (Fertilizantes Mexicanos), 70–71, 75
Fisheries, and pesticide use, 38, 79
Fitofilo, 60
Florescano, Enrique, 145–146, 148
Florida, 202–203
Food self-sufficiency, 227–228
Ford Foundation, 254
Foreign investment, 53, 189; and Díaz, 150–151; and liberalization, 229–231; petroleum, 169–170; and Sonoran dynasty, 166–167; and U.S.-Mexican relations, 249–250
Fosdick, Raymond, 173
French invasion, 130
Fresno, California, 25
Fungicides, 22–23, 27, 215

Galinat, Walter, 180
García, Benito, 14–15, 195
García Escobar, Alfonso, 65
Guasave, Sinaloa, 196
General Foods, 203
Genetic diversity: and Green Revolution, 178, 179–183, 214–215; and traditional agriculture, 155
Genetic engineering, 210, 277–279
Genetics, 174
Germany, 173
Gibson, Charles, 144–145, 148, 152
Gliessman, Stephen R., 148, 255, 256, 259
Gómez Pompa, Arturo, 148, 255
González, Irineo, 2, 94–95, 96, 108–109, 290
González, Ramón, 51; ancestors, 163; death, 1–2, 298–299; family in Tijuana, 286, 290–291, 296–299; home village, 87–121;

life, 2; significance of death, 9, 138–139, 187, 298–299
González y González, Guadalupe, 100, 104–105, 118–121, 139; in Tijuana, 186, 290–291, 296–299, 297–299
Gorbachev, Mikhail, 267
Grain, colonial regulation, 145–146
Grape boycott, 194, 220, 300–301
Great Depression, 175, 241
Greenhouse effect, 244
Green Revolution: consequences, 6–7, 65, 66, 224; controlling research, 281–282; critique, 244–285; in Culiacán, 5–8; and distribution of land and income, 279–281; farming systems, 257–258; in Mexico, xv, 6–8; neutrality, 279–284; origins, 6–8, 66, 142–153, 166–187; seeds, 156, 181–183, 274–275, 251; success, 224, 281
Green Uprising, 233–234
Groundwater contamination, xii–xiii, 5, 192, 196
Growing Problem: Pesticides and the Third World Poor, A (Bull), xii
Growers, Mexican. *See* Culiacán
Growers, U.S., opposition to imports, 202–205
Guadalupe, Catalina, 296–299
Guasave, Sinaloa, 196
Guatemala, 155, 272
Guaymas, Sonora, 207
Guthion, 16

Hacienda, origins, 148
Harvard University, 231, 234, 240, 263
Health system, Mexican federal, 54–55
Herbicides: and cancer, 205; and genetic engineering, 210. *See also* Paraquat; 2, 4, 5-T
Hermosillo, Sonora, 114

Hernández Xolocotzl, Efraim, 148, 255

Hidalgo, Miguel, 147

Hinshaw, Robert, 115

Huajaupan de León, Oaxaca, 90, 91–92

Huamantla, Tlaxcala, 44, 64–65

Huerta, Dolores, 301

Huichol, 42, 144

Hunger, 248–249. *See also* Food self-sufficiency; Green Revolution

Huntington, Samuel, 231, 233–238

Ice-Minus, 277–278

Imperial Valley, California, 203

Income distribution, Mexico, 226–229, 241, 280

Independence Wars, 130, 147

India, 186, 219, 231, 236

Indigenismo, 168

Indonesia, 40, 186, 204, 219, 231, 236; reducing pesticide use, 277

Industrial Bio-Test Labs, 198

Industrialization: global, 285; in Mexico, 224

Industrialization and agriculture, 236, 246, 264–272; in Africa, 265–267; in England, 232–233; in Mexico, 171–172, 178, 184, 222, 227–228, 280; in Soviet Union, 267; in Taiwan, 267–269

Inflation, 137

Instituto Nacional de Investigaciones sobre Recursos Bióticos (INIREB), 68

Instituto Nacional Indígena, 97

Instituto Nacional de Investigaciones Agrícolas (INIA): creation, 177; in Culiacán, 35, 62; in Mixteca, 133–136; perspective, 64–65, 66, 67

Integrated pest management (IPM), 36–37, 63, 207, 208–216; in citrus, 209; in cotton, 43, 209; in Culiacán, 49, 61–62, 78, 209; in

grains, 49; in soy, 209; in sugarcane, 49, 209; in wheat, 61, 209

Inter-American Development Bank, 203, 222

Internal markets, importance in development, 237–243, 250–252

International Monetary Fund: development theory, 230–231, 252; in Mexico, 72–76, 137–138, 203, 204, 222, 229–230, 252

International Rice Research Institute, 257

Interplanting, 36–47, 153–156, 160–161, 255

Iowa, 178, 205

Iran, 231

Irapuato, Guanajuato, 43

Irrigation, 12, 157–158, 161; in Africa, 266; required by new seed varieties, 182; and World Bank, 171

Jackson, Wes, 161, 255–257, 258, 259, 278, 283

Jalisco, 144

Japan, 224, 267–268

Jennings, Bruce, 177–179

Jesuit landholdings, 148

Jesuit priest, 295–296

Juárez, Benito, 131, 147–148

Kansas, 205

Kearney, Michael, 1–2, 41, 94–101, 107–108, 115, 286–301

Kenya, 219

King Eight Deer, 122, 123, 124

Kissinger report, 235

Lacondones, 47

Labor unions, 191–196; in California, 192–194; in Culiacán, 14–15, 194–196

Laguna Verde nuclear plant, 68

La Laguna, Mexico, 43, 49, 65, 68

Lama y bordo system, 123, 258–259

Landholding, 30, 144, 147; concentration, 248–249, 279–281; and population decline, 146. *See also* Culiacán; *ejido*

Lathrop, California, 192

Leukemia, 30

Liberalism, 131, 147–148; and Porfirio Díaz, 150–151

Liberalization of economy, 229–230

Licensing, 229

Lightstone, Ralph, 196

Limited good, 164

Livestock management, 174, 176; in Sahel, 265–266

Locusts, African, 214

Lomelí, Arturo and Lili, 69–70, 72

Los Baños, Philippines, 257

Lymphoma, 205

MacArthur, Douglas, 268

Madrid, de la, Miguel, 85

Maggio, Frank, 73, 76, 203

Magón, Flores, 152

Maize, 43, 142, 145; domestication, 263; and Green Revolution, 184–185; and plant breeding, 173; in triculture, 153–158

Malaria, 39, 42–43, 127, 275

Malathion, 16; in malaria control, 39

Malaysia, 40, 219, 220

Maneb, 27

Manganese ethylenebisdithiocarbamate (Manex), 22–23

Manglesdorf, Paul: on Carl Sauer, 179; Green Revolution research, 174–177; on triculture, 153–154

Man's Role in Changing the Face of the Earth (Sauer), 285

Mares, David, 195

Marxism, 81

Maya: *chinampas*, 160, 260–261; collapse, 143, 261; history,

124–125; resistance, 144; social stratification, 143, 163

Mennonites, 257

Metcalf, Robert, xii

Methamidophos (Tamaron), 16, 22, 201

Mexican Agricultural Program, 176–187. *See also* Green Revolution

Mexican border with U.S., 291–294

Mexican National Agricultural University, Chapingo, 64, 66–67, 133; and Rockefeller Foundation, 177

Mexican Revolution (1910–1920), 131–132

Mexican state, role in economy, 236

Mexican Supreme Court, 170

Mexican Union of Vegetable Growers, 73

Mexico: economic liberalization, 229–230; federal health system, 54–55; relation to U.S., xvii, 170–175, 243, 249–250

Mexico City, 63–64, 88–89, 132, 295; *chinampas*, 159–160; protest movement, 224–225

Millan, Jorge, 52–55

Mining, 130

Missionaries, 107, 110, 113, 116

Mixteca, La: description and history, 87–139; environmental decline, 123–145, 258–259, 270–271; housing, 97; labor surplus, 194; landholdings, 97–98, 118–120, 132–133; and national government, 101; Porfirio Díaz, 131; relation to Culiacán, 270–271; resource policies, 243. *See also* Soil erosion

Mixtecs: in border region, 286–289, 290–291, 294–300; diversity, 262–263; farm workers in Baja California, 40–41; farm workers in California, 131, 286–289;

farm workers in Culiacán, 1, 9, 87–139; health education, 299; labor surplus, 194; language, 124; market women, 290–291; in Mexican Revolution, 131–132; migration, 97, 106–108, 132–133; migration and ethnic identity, 295; pesticide education, 135–136, 299; Porfirio Díaz, 131; religion, 92–93, 99–100, 111–114, 295–296; as skillful farmers, 263; social stratification, 122–123, 163, 258–259; solidarity, 108, 110–111, 124, 295–296; stereotypes, 87; as victims of pesticide poisoning, 1, 5, 51–64

Modern agricultural dilemma, 244–285; defined, 245

Modernization, 185–186; theory, 232–243, 245. *See also* Agricultural modernization

Monoculture, 34, 36–37, 63, 67; as barrier to IPM, 213–216; and genetic engineering, 278; reduction for sustainability, 276–277; result of economic policies, 252, 270

Montelargo, Sinaloa, 21, 78

Morales, Luis, 29–31

Morales, Vicente, 115

Moroleón, Sinaloa, 21–22, 60, 117

Morrow, Dwight, 167

Mueller, Paul Hermann, 140–141

Murray, Douglas, 218–219

Mutations, 22

Nabhan, Gary, 157

Nagengast, Carole, 295

National Academy of Sciences, Board on Agriculture, 199, 275–276

National Indigenous Institute, 97, 113

Nationalization: and denationaliza-

tion, 229–230; of pesticide industry, 70–72; of petroleum, 169–170

National Union of Vegetable Growers, 12, 188

Natural resource management, 242–243

Natural resource pricing, 229–231

Nature, relation to agriculture, 246, 256–257, 272–279. *See also* Agro-ecosystems; Sustainable agriculture

Nebraska, 205

Nervous system damage, 23, 31–33

New Roots for Agriculture (Jackson), 283

Nicaragua, 209, 218–219, 243, 272

Nieves, Oaxaca, 93, 295

Nigeria, 204, 231, 236

Nochixtlán, Oaxaca, 133–136

Nogales, Sonora, 108

Noorgard, Richard, 254, 258

Nutri-Clean, 199

Oaxaca, 87–139. *See also* Mixteca, La

Obregón, Alvaro, 166–167

Olympic games protest, 224–225, 239

Organic agriculture. *See* Sustainable agriculture

Organochlorine insecticides: implications of use, 16. *See also* Pesticides, persistent

Organophosphate insecticides: as acetlycholinesterase inhibitors, 31–33; as central nervous system poisons, 31–33, 198, 206; consumer poisoning, 82–83; early research, 4; examples in use, 22, 27, 41, 44; farm worker poisoning, 4, 51–64; implications of use, 16–19; neurotoxicity, 198. *See also* Pesticides, nonpersistent

Ottawa Declaration, 220

Owens, Clarence and Emiel,
205–206
Ozone depletion, 244

Packing plants, 34–35
Papago, 157–158
Pakistan, 186
Pan American Health Organization,
69
Panneta, Leon, 203
Paraquat, 16; applied by farm workers, 25–27; peasant use, 48–49;
residues, 201; toxicity to farm
workers, 4, 26–27; use in drug
control, 4
Parasitic wasps, 209, 213
Parathion: origins, 141; poisoning,
55, 82; residues, 201; toxicity
compared to DDT, 16–17; in
use, 45. *See also* Organophosphate insecticides
Paris green, 141
Partido de Acción Nacional (PAN),
240
Partido Revolucionario Institucional (PRI), 136, 137, 138, 291,
293–294; and C. Cárdenas, 240,
294; and L. Cárdenas, 169; and
farm worker unions, 194–196
Patents, 229
Payer, Cheryl, 230–231
Peasants: abandonment of land,
229; in Central America,
234–235; and crop diversity,
153–166, 269–270; dispossesion
and hunger, 248; dispossesion in
England, 231–232; dispossesion
in Mexico, 143–150; households, 178; income, 227; strategies of resistance, 144–146,
151–152; traditional technologies, 143, 153–166; use of pesticides, 45–49
Pelicans, xi, 16
Pemex (Petróleos Mexicanos),
71–72

Penang, Malaysia, 219, 220
Peregrine falcon, xi, 16
Perennial polyculture, 256–257
Pérez, Andres, 51–53, 58
Perkins, John, 208
Peru, 40, 229
Pesticide Action Network (PAN),
219–220
Pesticide residues: consumer reaction, 189–191, 196–203; in
dairy products, 43, 68; in Mexican foodstuffs, 68; official Mexican position, 61, 66; safety, 190,
196–203; in shift to nonpersistents, 18–19
Pesticides: application rates, 34, 35;
cost, 33, 59, 207; dependence of
new seeds, 182–185; drift,
20–22, 23, 24; ecological significance, 273–279; epidemiological
analysis, 30–33, 53–54, 58; farm
worker poisonings, 22–49,
51–64, 66, 205–206, 298–299;
invention, 140–142; legal controls, 59–63; neurotoxicity, 198;
peasant use, 45–49; suppression
of immunity, 198
Pesticides, nonpersistent: characteristics and examples, 16–17;
consequences of substitution for
persistents, xii, 16–19, 201;
reasons for substitution for persistents, 17–19
Pesticides, persistent: banning in
U.S., xi; characteristics and examples, 15–16; recovery of bird
populations following ban, xi
Pest management advisers, 208
Pest populations: absence of monitoring, 28; control by pesticides
in Culiacán, 33–38
Pest resistance to pesticides, 17,
37–38, 39, 42–43, 65
Petroleum: boom, 225–226, 229;
expropriation, 169–170; Pemex,
71–72; and pesticides, 66

Pheromones, 62, 207, 213
Philippines, 129, 186, 204, 231
Pilots, problems in pesticide application, 23–24
Pioneer Hi-Bred, 172
Plant breeding, 172, 173–174, 176; adaption to nature, 272–279; and genetic engineering, 278; in Green Revolution research, 181–183; as profession, 274
Planters Products Chemical Company, 22
Plant protection, 174, 176
Political stability, in economic development, 233–243
Polyculture, 255–257
Population: decline after Conquest, 127–128, 145; density and agricultural technology, 162; growth, 89, 98–99, 174, 250–251, 294; justification for Green Revolution, 175; and stability, 236; in Taiwan and East Asia, 252
Potatoes, 44
President in Mexican political system, 136–137
Protective measures, absence of, 19–30, 44, 55–56
Protestantism, 115
Puebla Valley, Mexico, 44, 89

Quetzalcoatl, 125–126

Raley's Supermarkets, 198–201
Rancho California, California, 286–288, 298
Rats, 79
Ravicz, Robert S., 108
Reagan, Ronald, 235
Reforma, La (1857 Constitution), 130, 131, 147
Repartimiento, 144
Repetto, Robert, 76
Reproductive failure due to pesticides, 16

Resource management policies, 242–243
Resource pricing, 229–231
Respiratory diseases, among farm workers, 31
Rheumatic fever, related to pesticide exposure, 31
Rhine River, pollution by pesticides, xiii, 244
Rice, 277
Rockefeller Foundation: Carl Sauer, 154; financing agricultural research in Mexico, 6, 66–67, 142–143, 173–187, 281; International Health Division, 179; Survey Commission, 174–177. *See also* Green Revolution
Romero, Felipe, 65
Ronfeldt, David, 269–270
Roosevelt, Franklin Delano, 170; administration, 172, 173
Rostow, Walter W., 231–238

Safeway Supermarkets, 199–200
Safflower, 78
Sahara Desert, 265, 266
Sahel, 265–266
Salinas de Gortari, Carlos, 240, 294, 295
Salubridad y Asistencia, 54
Sánchez, Luis, 33–35
San Cristóbal de las Casas, Chiapas, 48
Sanderson, Steven, 184
San Diego County, California, 286–287
San Diego Union, 289
Sandinista government, 218–219
Sandoz Chemical, pollution of Rhine, xiii, 244
San Francisco, California, 301
Sanidad Vegetal, 59–67, 84–85; control of plant residues, 33–34; cooperation with U.S. agencies, 18, 61; educational efforts, 60, 84–85; field inspection, 28–29;

pesticide monitoring, 44, 59–67, 68–69, 78–81; pesticide promotion, 75, 85; regional committees, 61, and SEDUE, 78–81
San Jerónimo Progreso, Oaxaca, 87–121
San Joaquin Valley, California, 39, 132, 263, 295
San Quintín, Baja California, 40–42, 288
Sarcoma, 205
Sauer, Carl: on Green Revolution, 177–180, 247, 250; on influence of European culture, 285; on triculture, 154–155
Schultes, Richard, 176
Screw-worm fly, 210
Seals, 244
Secondary pest outbreaks, 37
Secretaría de Desarrollo Urbano y Ecología (SEDUE), 76–81
Seed banks, 180–181
Seeds, 156, 178–183; adaptivity, 273–274; chemical dependence, 181–183, 274–275; productivity, 251
Senegal, 219
Sevin (carbaryl), xii
Shapiro, Mark, xi, 18
Shell Chemical, DBCP, xiii
Shilling, Elisabeth, 158–159
Shrimp fishery, and pesticide use, 38
Silacayoápan, Oaxaca, 94, 132
Silent Spring (Carson), xi, xiii, 301
Silk, 129
Simpson-Rodino Act, 294
Sinaloa, 65, 68, 78–81. *See also* Culiacán
Singapore, 228
Sistema Alimentario Mexicano (SAM), 184
Smallpox, 127
Socialist International, 81
Soconusco, Chiapas, 49, 65
Soil erosion: in colonial period,

146, 128, 259; in Mixteca, 93–94, 95, 111, 115, 122–123, 128, 133–135; and political stability, 236; and social stratification, 133–135, 258–259
Soil fertility, 134, 161; and wheat, 181–182
Soil management, 176, 242, 243
Soil types, 162
Sonora, Mexico, 49, 65, 144
Sonoran dynasty, 166–168, 169
Sorghum, 43–44, 228
South Korea, 228, 236, 238
Soviet Union, 265, 267
Soybeans, 78
Sri Lanka, 229
Stakman, E. C., 174–177
State role in the economy, 237
Stavenhagen, Rudolfo, 288
Sterile male release, 209, 213
Sterility, human, 192
Strawberries, 43
Streptococcus, related to pesticide exposure, 31
Subsidies: and concentration of landholding, 248–249; to growth of modernized agriculture, 138–139, 183–185; in Mexico, 73–76, 183–185; to pesticide use in developing countries, xiv
Sudan, 40, 219
Sugarcane, 49, 209, 270
Sustainable agriculture: and capitalism, 283–284; defined, 212–216, 283; and government policy, 276; and research policy, 276; and traditional agriculture, 253–264; in use in U.S., 275–276. *See also* Agricultural technology; Traditional agriculture
Swezey, Sean, 218

Taiwan, 228, 229, 236; as development model, 238, 265, 267–269
Tanzania, 228–229

Technology Assessment, Office of, 282
Tehuacán Valley, 263
Tehuantepec, 144
Tenochtitlán, 158–159
Teotihuacán, 125
Teposcolula, Oaxaca, 133
Terraces, 93, 133–135
Thompson, J. Eric S., 143, 261
Thompson, Richard, 283, 284
Tijuana, Mexico: *El Heraldo*, 2; González family, 286; Mixtecs in, 2, 40, 95, 108, 294–301
Tlatelolco, 225, 229
Tlaxcala, Mexico, 156–157
Tobacco, 42
Toltec, 126
Tonalá, Oaxaca, 92–93
Torres, Quirino, 96–111
Tractor application of pesticides, 29
Traditional agriculture, xii, 46–47, 153–166, 246; and culture, 263–264; and Green Revolution, 184, 253–254; and social systems, 162–166, 258–264; usefulness, 253–264, 269
Trapcropping, 209
Triculture (beans, maize, and squash), 46–47, 153–156; and social systems, 163
Trujillo de Arriaga, Javier: on IPM, 212; on pesticides, 64–67; on triculture, 155–156, 255
Tula, 122, 125, 126
2, 4, 5-T, 141
Typhus, 141
Tzetzales, 47–49
Tzotzil, 289

Union Carbide Corporation, xii, 244
Unión Nacional de Productores de Hortalizas, 73
Union of Soviet Socialist Republics. *See* Soviet Union
United Farm Workers Union (UFW), 193–194, 203, 220, 295, 300–301
UN Food and Agriculture Organization (FAO), 95, 192; pesticide code of conduct, 207–208, 219; and seed banks, 180–181
UN International Labor Organization, 191
UN World Health Organization (WHO), 192; estimates of farm worker poisonings by pesticides, 3–4
United States of America: negative consequences of agricultural technology in, xvi; war with Mexico, 130
U.S. Agency for International Development, 204, 282
U.S. Congress, Office of Technology Assessment, 282
U.S. Department of State, 170, 172, 231, 236, 243; Bureau of Intelligence and Research, 234
U.S. Environmental Protection Agency (EPA): cooperation with Mexico, 18, 61, 63; enforcement of residue standards, 196, 196–199
U.S. Food and Drug Administration (FDA), enforcement of residue and cosmetic standards, 34–35, 196–199
U.S. General Accounting Office (GAO), 196–197, 199, 203
U.S. Immigration and Naturalization Service (INS, *la migra*), 106, 291
U.S. National Environmental Policy Act (NEPA), 282
U.S. Occupational Health and Safety Agency (OSHA), 191
U.S. Supreme Court, 203
Universidad Autónoma de Sinaloa (UAS), 81
University of California, 37, 64, 210, 254–255

Urbanization, 89, 145, 236, 250; in England, 232; and Green Revolution, 185
Ussher, Jorge Ricardo García, 59–62, 64, 78–81

van den Bosch, Robert: cosmetic standards research, 35; pesticide conspiracy theory, xiv
Vargas, Humberto, 62
Vegetable crops, 13–40, 44–45, 59–63, 65, 78, 142; export policy, 230
Velázquez, Luis, 55–57
Ventura County, California, 106
Veracruz, 132
Vietnam, 233, 272
Villa, Francisco (Pancho), 152, 165
Voz del Consumidor, La, 70

Wallace, Henry, 172–174, 178, 281
Wall Street Journal, 188, 189
Water conservation, 157, 161, 242
Waterfowl, and pesticide use, 39
Weir, David, xi, 18
Wheat, 61–62, 142, 184; plant breeding, 181–183

Wildlife, and pesticide use, 39, 68, 79
Wilkes, Garrison, 179–180
Wilkin, Gene, 161–163, 255, 263, 264
Wilson, Pete, 203
Wolf, Eric, 165, 269
World Bank, 171, 203, 222
World Policy Journal, 241

Xochimilco, 144, 159

Yanhuitlán, Oaxaca, 134
Yaquis, 144
Yellow fever, 42, 127
Yosiguchi, Allen, 288
Yucatán, 124, 144
Yugoslavia, 229

Zambia, 228
Zapata, Emiliano, 132, 150–152, 165
Zapotec: early history, 126; farm workers in Culiacán, 1; B. Juárez, 131; resistance, 144; victims of pesticide poisoning, 1, 5